A New Era of Polycyclic Aromatic Hydrocarbons

Kazuichi Hayakawa · Yasunobu Aoki
Editors

A New Era of Polycyclic Aromatic Hydrocarbons

Environmental Pollution and Health Effects

Springer

Editors
Kazuichi Hayakawa
Kanazawa University
Kanazawa, Ishikawa, Japan

Yasunobu Aoki
National Institute for Environmental Studies
Tsukuba, Ibaraki, Japan

ISBN 978-981-96-7694-1 ISBN 978-981-96-7695-8 (eBook)
https://doi.org/10.1007/978-981-96-7695-8

This Springer imprint is published by the registered company Springer Nature Singapore Pte Ltd.
The registered company address is: 152 Beach Road, #21-01/04 Gateway East, Singapore 189721, Singapore

Preface

Air pollution is a global environmental risk threatening human health. A major component of this air pollution is the polycyclic aromatic hydrocarbons (PAHs) and their derivatives, such as nitro-, hydroxy-, and quinoid-PAHs, along with fine particles ($PM_{2.5}$) on which they are adsorbed. These are pollutants emitted from fossil fuel combustion, such as in coal and oil combustion, as well as in biomass burning and agricultural and forest fires.

During the latter half of the 20th century, vehicle emissions were one of the primary contributors of PAHs and NPAHs in urban areas. However, their atmospheric concentrations have been reduced substantially due to advancements in the 2000s in vehicle engine and oil quality. As a result, vehicle emissions are no longer considered the main cause of urban air pollution. In addition, there has been a global shift in energy sources, with for example the adoption of nuclear, solar, and wind energies, as well as biomass fuels, all of which are measures meant to limit carbon dioxide emissions. However, biomass combustion and the presence of forest fires nevertheless still exist and still emit $PM_{2.5}$ and PAHs into the atmosphere. This has meant that emissions have shifted from developed to developing countries, from urban to suburban areas, and from outdoor to indoor environments and that health risks caused by those pollutants will become more of a worldwide issue than before. Emerging evidence suggests that these pollutants are related to not only cancer but also to other diseases, including respiratory diseases (such as asthma), cardiovascular diseases (such as myocardial infarction), and endocrine disruption.

This book covers the current situation of these particular global contaminants and discusses their environmental origins, the health effects, and analytical methods for their detection. It aims to provide an understanding of the latest research on air pollution caused by PAHs and related compounds and foster effective solutions in this new era.

Kanazawa, Japan
Tsukuba, Japan
April 2025

Kazuichi Hayakawa
Yasunobu Aoki

Contents

Typical Pollutions in the World

Characteristic Emissions of Polycyclic Aromatic Hydrocarbons and Nitro-Polycyclic Aromatic Hydrocarbons and Health Risk of Rice Straw Open Burning in Southeast Asia 3
Chau-Thuy Pham and Trung-Dung Nghiem

China's Coal Heating: Research Methods, Emission Factors, and Policy Interventions for PAHs Reduction 31
Yang Xiaoyang and Wang Yicheng

Atmospheric Particulate-Bound Polycyclic Aromatic Hydrocarbons and Nitro-Polycyclic Aromatic Hydrocarbons: Levels, Sources, Health Impacts, and Policy Implications for Africa 59
Augustine Omodieke and Egide Kalisa

Environmental Reaction

Formation of a Quinoid Polycyclic Aromatic Hydrocarbon Promoted by Mineral Dust Aerosols 83
Takayuki Kameda and Daiki Nojo

Polycyclic Aromatic Hydrocarbons from Biomass Burning: Source Contribution Analysis and Risk Assessment 99
Fumikazu Ikemori

Plastics as Sorption Substrates for Polycyclic Aromatic Hydrocarbons 133
Edward Gou Nagato

Toxicity and Adverse Health Effects

Carcinogenic Risk Assessment of Polycyclic Aromatic Hydrocarbons in the Ambient Air Using In Vivo Mutagenicity Data 151
Yasunobu Aoki

Effects and Reactions of Polycyclic Aromatic Hydrocarbons and Their Derivatives in Respiratory System 167
Akiko Honda, Satsuki Takai, Takayuki Kameda, Natsuko Miyasaka, Megumi Nagao, Yasuto Matsui, and Hirohisa Takano

Exposure to PAHs and Asthma/Cough Diseases 181
Akinori Hara and Hiroyuki Nakamura

Analytical Methods

Oxidative Derivatives of Polycyclic Aromatic Hydrocarbons and Their Metabolites 209
Akira Toriba

Instrumental Analysis of Polycyclic Aromatic Hydrocarbons and Nitropolycyclic Aromatic Hydrocarbons 225
Kazuichi Hayakawa

Biomonitoring of PAHs: Animals 247
Masato Honda

The NP Method: A Source Analysis Technique for Combustion-Derived Particulates, Polycyclic Aromatic Hydrocarbons, and Nitro-Derivatives 265
Kazuichi Hayakawa

Typical Pollutions in the World

Characteristic Emissions of Polycyclic Aromatic Hydrocarbons and Nitro-Polycyclic Aromatic Hydrocarbons and Health Risk of Rice Straw Open Burning in Southeast Asia

Chau-Thuy Pham and Trung-Dung Nghiem

Abstract Rice straw open burning (RSOB) is a common agricultural practice in Southeast Asia, significantly contributing to air pollution, health risks, and climate change. This chapter reviews the emission characteristics of polycyclic aromatic hydrocarbons (PAHs) and nitro-polycyclic aromatic hydrocarbons (NPAHs) from RSOB in Southeast Asia, especially in Vietnam, highlighting their emission factors, size distribution, and health risks assessment. Emission factors (EFs) of fine particulate matter ($PM_{2.5}$) emitted from small pile burning in northern Vietnam were significantly higher compared to spread burning in southern Vietnam and other Southeast Asian regions. PAH concentrations in RSOB smoke were significantly higher than background samples, with benzo[*a*]pyrene (B*a*P) and benz[*a*]anthracene (B*a*A) notably accumulating in $PM_{2.5}$. Seasonal and regional variations of PAHs and NPAHs emission were observed, with Northern Vietnam reporting higher levels. Four-ring PAHs shifted to larger particles during the RSOB period (1.34–1.5 μm), while five- and six-ring PAHs exhibited consistent size distribution at 0.8 μm. The highest incremental lifetime cancer risk (ILCR) values were observed in rural Hanoi (3.00×10^{-5} for males and 3.26×10^{-5} for females), with ingestion identified as the most significant health risk than the dermal and inhalation pathways. Despite government efforts to promote alternative straw management practices, RSOB persists due to its low cost and simplicity. This review provides a comprehensive characterization of PAHs and NPAHs in PM of different sizes emitted from traditional local rice straw burning in Vietnam, helping to clarify the emission characteristics and the environmental behavior of PAHs and NPAHs from RSOB in Vietnam and across Southeast Asia.

C.-T. Pham (✉)
University of Engineering and Technology, Vietnam National University, Hanoi, Vietnam
e-mail: thuypc@vnu.edu.vn

T.-D. Nghiem
School of Chemistry and Life Sciences, Hanoi University of Science and Technology, Hanoi, Vietnam
e-mail: dung.nghiemtrung@hust.edu.vn

K. Hayakawa and Y. Aoki (eds.), *A New Era of Polycyclic Aromatic Hydrocarbons*,
https://doi.org/10.1007/978-981-96-7695-8_1

Keywords Rice straw open burning · Polycyclic aromatic hydrocarbons · Nitro-polycyclic aromatic hydrocarbons · Emission characteristics · Health risk · Size distribution · Southeast Asia

1 Introduction

Open burning of agricultural waste is a widespread practice in Asia, which significantly contributes to atmospheric pollution. This method, favored for its low cost and simplicity, has significant environmental consequences, contributing substantially to air pollution in the region. Southeast Asia is one of the world's largest producers of rice, with countries like Thailand, Vietnam, Indonesia, and Philippines generating massive quantities of rice straw annually. The widespread burning of this agricultural residue releases harmful pollutants into the atmosphere, including fine particulate matter (PM), greenhouse gases (GHGs), carbon monoxide (CO), sulfur dioxide (SO_2), nitrogen oxides (NO_x), and toxic organic compounds such as polycyclic aromatic hydrocarbons (PAHs) and nitro-polycyclic aromatic hydrocarbons (NPAHs) (Gadde et al. 2009; Streets et al. 2003). As a result, this activity seriously affects air quality, human health, and climate change (Hoang et al. 2023; Kim Oanh et al. 2011; Lemieux et al. 2004; Pham 2018; Pham et al. 2019; Reid et al. 2005; Sanchis et al. 2014).

Vietnam is among the world's leading rice-producing countries, ranking as the fifth-largest rice exporter and the seventh-largest rice consumer globally, with rice accounting for more than 90% of total cereal production (General Statistics Office of Vietnam 2021). Consequently, substantial amounts of crop residues, such as rice straw and rice stubble, are generated after each harvest. These residues require proper disposal or treatment to prepare the fields for the next planting season (Pham et al. 2021). Rice straw open burning (RSOB) after harvest has long been a common method of crop residue disposal in many parts of Vietnam, similar to other Asian countries. The Vietnamese government has introduced alternatives to RSOB, such as utilizing rice straw for mushroom cultivation, fertilizer production, silage for livestock feed, and biomass fuel. Additionally, strict regulations have been implemented to prohibit RSOB. However, these methods face challenges due to farmers' reluctance and local conditions. As a result, RSOB continues to be a prevalent practice, even in suburban areas of Hanoi, capital of Vietnam. The number of hotspots during rice straw (RS) burning season in Hanoi was shown in Figs. 1 and 2. This activity is a major source of fine particulate matter ($PM_{2.5}$) and many other pollutants such as CO_2, CO, organic compounds, which enhances their ability to penetrate the respiratory system and pose long-term health risks (Andreae 2019; Andreae and Merlet 2001). Among these pollutants, PAHs and NPAHs are pollutants that exist in either the gas phase or particulate phase, depending on their vapor pressure in the ambient air, and are components of PM. Despite being emitted in low concentrations, they possess toxicological properties associated with cancers of the skin, lungs, bladder, liver, and stomach. Health studies have also noted a rise in bronchial asthma among children living near rice fields during burning seasons (Torigoe et al. 2000; Yang et al. 2010).

The environmental impact of RSOB extends beyond local air quality. It contributes significantly to regional haze episodes and transboundary air pollution, particularly during the post-harvest season when burning activities are at their peak. For example, in northern India, RSOB is a key contributor to severe air quality degradation, which experiences elevated $PM_{2.5}$ levels and frequent smog episodes during the winter months (Gadde et al. 2009). Similarly, in countries such as Bangladesh, Pakistan, Thailand, and Philippines, RSOB worsens air pollution in rural and peri-urban areas,

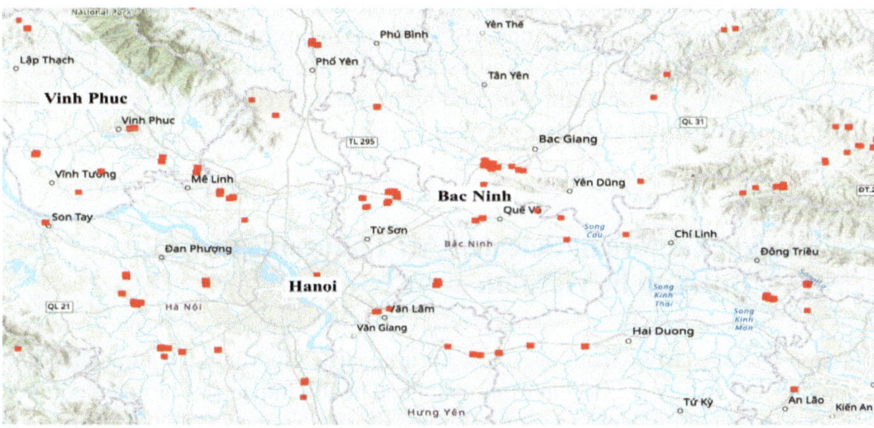

Fig. 1 Fires map over Hanoi during RS burning period (01/06–30/06/2017) (hotspots were produced by MODIS satellite through the US NASA)

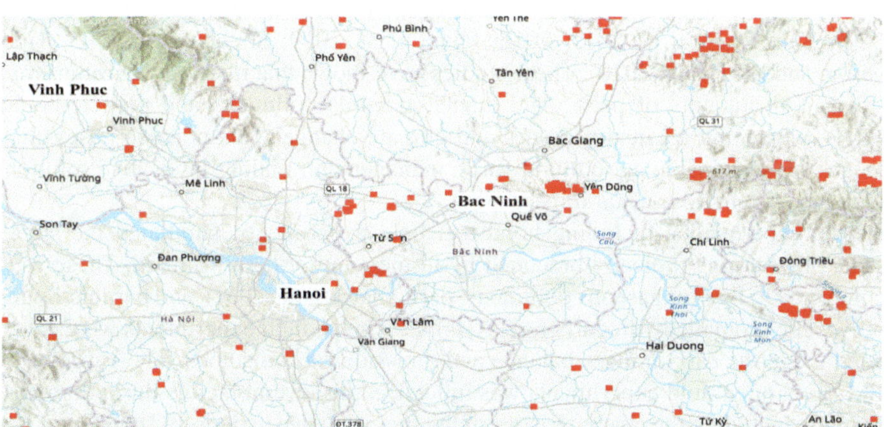

Fig. 2 Fires map over Hanoi during RS burning period (01/01/2020–30/01/2020) (hotspots were produced by MODIS satellite through the US NASA) https://firms.modaps.eosdis.nasa.gov/map/#d:2019-12-03..2020-01-01;l:fires_all,street;v:max;@106.10,21.09,10.32z

affecting both local populations and neighboring regions. Despite the growing awareness of its environmental and health impacts, RSOB remains a prevalent practice in Southeast Asia due to limited access to alternative straw management methods and the absence of stringent regulatory frameworks. Although some countries like Vietnam, Thailand, and India have implemented measures like subsidies for mechanized straw management and penalties for burning, these efforts have seen limited success due to enforcement challenges and economic constraints (Pham et al. 2021; Streets et al. 2003).

This chapter reviews the pollution caused by RSOB in Southeast Asia, with a particular focus on Vietnam, emphasizing the emission characteristics of PM, PAHs, and NPAHs, their health and environmental impacts, as well as the regional challenges in mitigating these emissions. By synthesizing existing studies, this chapter aims to provide a comprehensive understanding of the pollution caused by RSOB in Southeast Asia and highlight potential strategies for reducing its environmental impact and contribution to global warming.

2 Emission Factors of Selected Air Pollutants from Rice Straw Burning in Vietnam

Emission factors (EFs) from biomass burning are important for understanding, managing, and mitigating the environmental and health impacts of biomass combustion. They provide the basis for emission inventories, support policy decisions and strategies aimed at combating climate change and improving air quality globally (Andreae and Merlet 2001). Among many types of biomass burning, rice straw open burning is particularly prevalent and contributes significantly to atmospheric pollution in Asian countries such as Vietnam, Thailand, and the Philippines (Gadde et al. 2009; Kim Oanh et al. 2011, 2015; Pham et al. 2019; Streets et al. 2003; Tipayarom and Kim Oanh 2020). These studies have shown that rice straw burning emits substantial amounts of particulate matter, gaseous pollutants and particle-bound PAHs, worsening air quality issues and climate change. This practice poses challenges for sustainable agricultural waste management, finding innovative solutions to reduce its environmental impact.

The experiments, including hood and field experiments, simulated the small-scale pile burning typically practiced by farmers in northern Vietnam. These were conducted to EFs for gaseous pollutants, particulate matter (PM), and particle-bound polycyclic aromatic hydrocarbons (PAHs). EFs in the field experiments were calculated using the carbon mass balance and emission ratio methods. The variation of PM EF in open-field experiments was influenced by several factors during biomass burning, including the characteristics of the biomass (rice varieties, C content of RS, RS moisture), wind speed and direction, soil surface moisture, prevailing weather conditions and type of burning (pile burning or spread burning) (McMeeking et al. 2009; Reid et al. 2005). Hood experiments were conducted using a double-walled cylindrical cookstove featuring a metallic outer shell and a ceramic inner lining, then eliminating the influence of wind and weather conditions. The results from hood experiments with various rice varieties indicated that the higher emission of PM was associated with increased RS moisture and the higher C content of RS, which also led to greater CO_2 emission (Pham et al. 2021). The average value of emission factor for $PM_{2.5}$ from small piles open burning (34.0 ± 17.6 g kg^{-1} RS) in northern Vietnam was higher compared to values from spread burning in the Mekong Delta, the South of Vietnam (12.1 ± 0.3 for $PM_{2.5}$) (Phuong et al. 2022), Thailand (8.3 ± 2.2 g kg^{-1} RS) (Kim Oanh et al. 2011), 19.2 ± 15.76 g kg^{-1} RS (Kanokkanjana and Garivait 2013), and China (20.3 ± 1.5 g kg^{-1} RS (Zhang et al. 2015). The EF for SO_2 in northern Vietnam, measured at 1.4 ± 1.1 g kg^{-1} RS in field experiments and 1.82 ± 1.77 g kg^{-1} RS in hood experiments, was higher than the values reported in studies from Thailand and China (Pham et al. 2021). The elevated sulfur content in RS could be attributed to the extensive use of high-sulfur fertilizers commonly applied in Vietnam and/or the naturally higher sulfur levels present in the soil (Pham et al. 2021). Data on EF of PM and selected pollutants from RS burning in Vietnam and other countries are presented in Table 1.

In northern Vietnam, the mean emission factors (EFs) for 10 PAHs were 13.5 ± 9.2 mg kg^{-1} RS for $PM_{2.5}$ and 10.5 ± 6.5 mg kg^{-1} RS for TSP. In the Mekong Delta of Vietnam, the EFs for 16 PAHs averaged 12.8 ± 2.5 mg kg^{-1} RS for PM_{10}, 12.5 ± 2.8 mg kg^{-1} RS for TSP (Phuong et al. 2022). The data of PAHs emission factors from RS burning and literature review are displayed in Table 1. The data for field experiments in northern Vietnam found that PAHs were more concentrated on $PM_{2.5}$ and PM_{10} than in total suspended particulates (TSP) during RSOB. Higher molecular weight PAHs, including benzo[*a*]pyrene (B*a*P), dibenz[*a,h*]anthracene (DBA), and indeno[1,2,3-*cd*]pyrene (IDP), exhibited significantly higher EFs on $PM_{2.5}$ compared to TSP, with B*a*P showing a particularly notable increase, being 3.6 times higher. The PAH emission factors observed in hood experiments were lower compared to those measured in field experiments. The EFs of individual PAHs in TSP from hood experiments in northern Vietnam were consistent with previous studies by Keshtkar and Ashbaugh (2007), Sanchis et al. (2014). EFs for PAHs from agricultural residues burning vary significantly across studies and regions due to principally on burning conditions and to a lesser extent on fuel type, efficiency of fuel burned. (Keshtkar and Ashbaugh 2007) found that incomplete combustion conditions significantly increased PAH emissions due to the dominance of pyrolysis over oxidation processes, resulting in an EF of 18.6 mg kg^{-1} for 12 PAHs in TSP during controlled rice straw burning in California. Jenkins et al. (1996) observed a broader range of 5–683 mg kg^{-1} of biomass burning, influenced by burning conditions. In China, Zhang et al. (2011) measured a lower EF of 5.26 mg kg^{-1} for a total of 16 PAHs, including both gas phase and particles phase, under controlled conditions using self-designed combustion stove for rice straw burning. These differences emphasize the significance of combustion efficiency and local practices in determining PAH emissions. The EFs of particle-bound PAHs in Vietnam did not vary significantly between the northern and southern regions. However, northern Vietnam exhibits higher emissions of PM from pile burning compared to spread burning in southern Vietnam, and also

Table 1 Emission factors of PM, gaseous pollutants, and particle-bound PAHs from RS burning at field experiments and literature review

Studies	EF of pollutants[a] (g kg^{-1} dry RS)								MCE ($n = 11$)	%C	%N	%S
	Burning type	CO_2 ($n = 14$)	CO ($n = 14$)	SO_2 ($n = 12$)	NO_2 ($n = 10$)	$PM_{2.5}$ ($n = 10$)	TSP ($n = 4$)	PAHs (mg kg^{-1})				
Pham et al. (2021)	Small piles in field in Hanoi, the North of Vietnam	1160.9 ± 80.9	73.6 ± 46.2	1.4 ± 1.1	1.1 ± 0.9	34.0 ± 17.6	54.6 ± 35.4	10.5 ± 6.5 (in TSP) 13.5 ± 9.2 in PM2.5	0.94 ± 0.01	38.2 ± 5.3		0.24–0.54[b] 0.84[c]
	Hood experiements	1233.8 ± 185.9	53.3 ± 16.6	1.82 ± 1.77	0.10 ± 0.19		10.2 ± 8.5	0.6 ± 0.63				
Phuong et al. (2022)	Mekong Delta, spread burning, the South of Vietnam	1247.5 ± 190		1.4 ± 0.3	1.3 ± 0.3	12.1 ± 0.3	15.5 ± 3.5	12.53 ± 2.83 in PM2.5 12.5 ± 2.8 in TSP	0.92 ± 0.01			
Kim Oanh et al. (2015)	Spread burning, Thailand	1177 ± 140	93 ± 10	0.51 ± 0.32	0.49 ± 0.21	8.3 ± 2.2		0.83 ± 0.14 in TSP	0.93 ± 0.01	49.0 ± 2.7		
Kanokkanjana and Garivait (2013)	Piles in field, Thailand					19.21 ± 15.76				35.4		
Christian et al. (2003)	Small piles in field, Indonesia	1216 ± 97	179.9 ± 39.8			4.2				35.4		
Zhang et al. (2015)	Piles in field, China	1761 ± 30	47 ± 19			20.3 ± 1.5						
Li et al. (2007)	Open burning, wheat straw, China	1470 ± 46	60 ± 23	0.85 ± 0.57	3.3 ± 1.7 (NO_x)					44.8	0.56	0.17

(continued)

Table 1 (continued)

Studies	EF of pollutants[a] ($g\ kg^{-1}$ dry RS)								MCE ($n = 11$)	%C	%N	%S
	Burning type	CO_2 ($n = 14$)	CO ($n = 14$)	SO_2 ($n = 12$)	NO_2 ($n = 10$)	$PM_{2.5}$ ($n = 10$)	TSP ($n = 4$)	PAHs (mg kg^{-1})				
Zhang et al. (2011)	Self-designed combustion stove, Rice straw, China							5.25, 16 PAHs in TSP				
Keshtkar and Ashbaugh (2007)	RS burning in California, Combustion chamber)							18.62 for 12 PAHs 9.9 for 10 PAHs in TSP				

[a]Data show mean ± SD
[b]Sulfur percentage of RS was determined in the present study
[c]Sulfur percentage of RS was recorded by Nguyen et al. (2019)

higher in comparison with those from RSOB in Thailand. Emission factors from RS burning fluctuated with wide ranges. The findings indicate that EFs for PM, SO_2 and particle-bound PAHs are higher in Vietnam compared to those in other countries. This variability is attributed to factors such as local RS burning method (small pile burning), higher RS moisture, higher carbon content, extensive use of high-sulfur fertilizers, and regional differences in fuel and environmental conditions in Vietnam. These findings highlight the need for improved management practices in Vietnam as well as other regions in Southeast Asia to mitigate the environmental impact of rice straw open burning.

3 Emission Characteristics of PAHs and NPAHs in Vietnam

3.1 Emission Characteristics of Polycyclic Aromatic Hydrocarbons from Rice Straw Open Burning in Vietnam

Atmospheric background samples (BG samples before RS burning) and rice straw burning smoke particles (BB samples, during RS burning) were collected in Hanoi, located in northern Vietnam. The sampling was conducted during the harvest season to examine the distribution and contribution of particle-bound PAHs and NPAHs from RSOB in the region. The experiments were carried out in harvested paddy fields in Hanoi during the RS burning seasons (October to June of the following year) from 2016 to 2018. All samplers were placed approximately 1–1.5 m apart, with the sampling inlets positioned 1.5 m above ground level. Samples of both $PM_{2.5}$ and total suspended particulates (TSP) were collected at 14 sampling sites located in suburban areas surrounding the Hanoi capital. Hanoi has two rice straw (RS) burning seasons: the winter-spring burning crop, where rice is harvested and burned in June with frequent rainy weather, and the summer-autumn burning crop, where rice is harvested and burned in October, which is more favorable for burning due to the sunny weather.

There was seasonal variation of PAH concentrations in atmosphere (BG samples) between the two harvest seasons in Hanoi. The total amount of 9 PAHs adsorbed on atmospheric particles in BG samples during RS burning season was averaged in the range of 25.34 ± 9.15 µg g^{-1} and 12.81 ± 7.84 µg g^{-1} for $PM_{2.5}$ and 69.7 ± 44.9 and 32.3 ± 16.8 µg g^{-1} for TSP in summer-autumn and winter-spring crop, respectively (Fig. 3). The concentrations (measured as mass per total aerosol volume) of a total of 9 PAHs on $PM_{2.5}$ prior to RS burning ranged from 0.89 to 18.5 ng m^{-3}, with an average of 14.4 ng m^{-3} in summer-autumn and 2.6 ng m^{-3} in winter-spring, showing a sevenfold difference. For TSP in BG samples, total PAHs varied from 0.97 to 42.5 ng m^{-3}, with an average of 19.92 ng m^{-3} in summer-autumn, which was six times higher than the winter-spring average of 3.41 ng m^{-3} (Fig. 4). The average content of total 9 PAHs adsorbed on particulate matter (PM) in RS burning smokes, was 288.3 and 225.3 µg g^{-1} for $PM_{2.5}$, 246 and 183.3 µg g^{-1} for TSP during the summer-autumn and winter-spring crop seasons, respectively. These values showed no significant seasonal differences for both $PM_{2.5}$ and TSP ($P > 0.05$). Detailed data on the mean contents (in μg g^{-1}) and the mean concentrations (in ng m^{-3}) of individual PAHs in the background environment and burning smoke from the two sampling campaigns conducted during two burning seasons are presented in Tables 2 and 3, respectively. The PAH levels and individual PAH ratios in $PM_{2.5}$ from rice straw (RS) burning in northern Vietnam were about ten times lower than those reported by Kim Oanh et al. (2011) in Thailand, likely due to differences in RS composition, moisture content, and burning methods (pile vs. spread burning). Notably, the RS used in this study had a lower carbon content (39.4 ± 7%) than that reported in Thailand (49%) (Ly 2006). Additionally, variations in moisture content and burning method, which influence combustion temperature, further contributed to reduction in PAH emissions.

The average concentrations of the 9 PAHs in RS burning smokes examined in Hanoi were 4487 and 3064 ng m^{-3} for $PM_{2.5}$, and 2700 and 2340 ng m^{-3} for TSP during the summer-autumn and winter-spring crop seasons, respectively (Fig. 4). The average concentrations of PAHs in RS burning smoke were significantly higher than those in background (BG) samples, frequently exceeding 100 times for the majority of compounds in both $PM_{2.5}$ and TSP. Benzo[*a*]pyrene (B*a*P) and benz[*a*]anthracene (B*a*A) accumulation was more prominent in $PM_{2.5}$ than in TSP. The average concentrations of B*a*P and B*a*A in $PM_{2.5}$ from RS smokes were 726 and 1200 times higher, respectively, compared to BG samples. In TSP, these concentrations were only 116

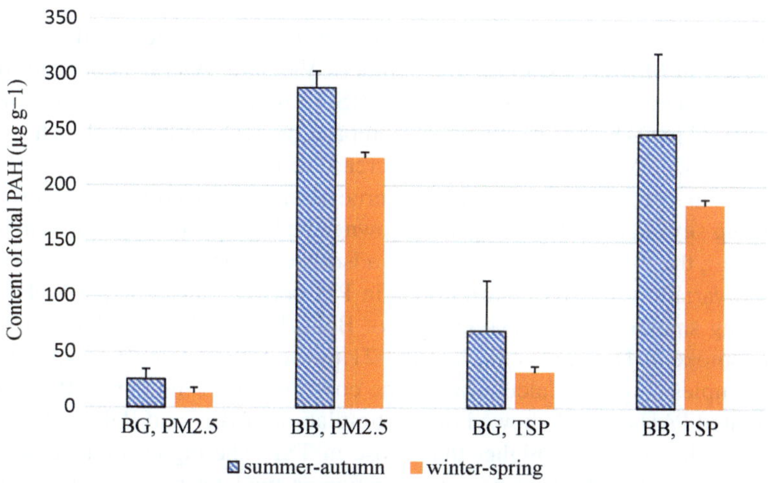

Fig. 3 Content of total PAHs in PM in Hanoi. *Note* BG: Background samples, BB: RS Burning samples. Samples were collected from 2016 to 2018. Data show mean ± SD of total 9 PAHs in PM in summer-autumn and winter-spring. 9 PAHs including: Flu: Fluoranthene; Pyr: Pyrene; B*a*A: Benz[*a*]anthracene; Chr: Chrysene; B*b*F: benzo[*b*]fluoranthene; B*k*F: Benzo[*k*]fluoranthene; B*a*P: Benzo[*a*]pyrene; DBA: Dibenz[*a,h*]anthracene; IDP: Indeno[1,2,3-*cd*]pyrene; $PM_{2.5}$: Particulate matter with diameter < 2.5 μm; TSP: Total suspended particulate

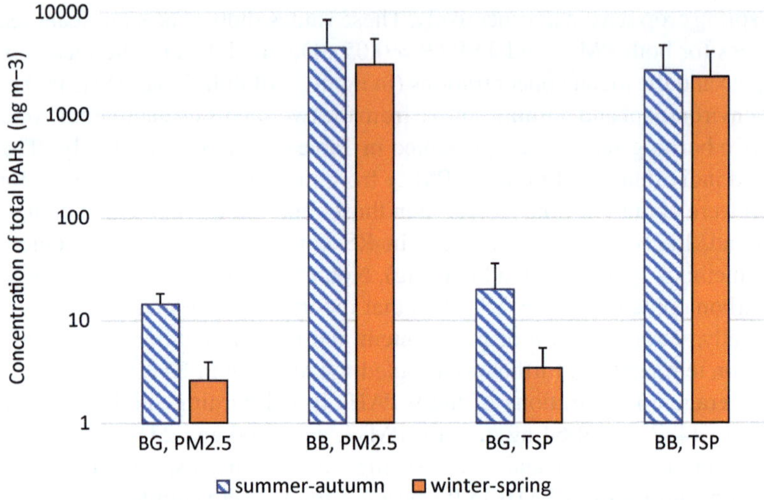

Fig. 4 Concentrations of total PAHs in Hanoi. *Note* BG: Background samples, BB: RS Burning samples. Samples were collected from 2016 to 2018. Data show mean ± SD of total 9 PAHs in PM in summer-autumn and winter-spring. 9 PAHs including: Flu: Fluoranthene; Pyr: Pyrene; B*a*A: Benz[*a*]anthracene; Chr: Chrysene; B*b*F: benzo[*b*]fluoranthene; B*k*F: Benzo[*k*]fluoranthene; B*a*P: Benzo[*a*]pyrene; DBA: Dibenz[*a,h*]anthracene; IDP: Indeno[1,2,3-*cd*]pyrene; $PM_{2.5}$: Particulate matter with diameter < 2.5 μm; TSP: Total suspended particulate

and 192 times higher. These results fall within the same range as other measurements taken in Mekong Delta region, southern Vietnam. Average concentration of total PAHs bound in TSP and PM_{10} during RSOB in Mekong Delta region was 809 times and 974 times higher, respectively, compared to background levels. The smaller the size of PM (PM_{10}), the greater the increase in the concentration of total PAHs (Phuong et al. 2022). These findings are crucial in understanding health effects of RS burning smoke, as $PM_{2.5}$, PM_{10} are known to be highly harmful to human health. Considering the difference in PAH ratios adsorbed on particles (μg g^{-1}) to understand the variation of PAH contributions in PM between airborne particulates and RS burning smoke. The average content of B*a*A, B*a*P, and Chr in $PM_{2.5}$ from RS burning smoke in Hanoi was 28, 27, and 21 times higher, respectively, than those in BG samples (Fig. 5). Calculation of the contribution of individual PAH in total accumulated PAHs in PM showed the accumulation of B*a*P and IDP in $PM_{2.5}$ from burning smoke was much higher than those in TSP. The higher molecular weight PAHs tend to preferentially concentrate on fine particulates rather than coarse ones emitted from sources of RSOB in Hanoi. This pattern was consistent with the findings observed in the previous reports by Duan et al. (2005), Keshtkar and Ashbaugh (2007), López Cancio et al. (2004).

Airborne particles of various sizes (six fractions) at three sampling sites were collected in additional sampling campaigns in RSOB areas in Hanoi in October 2018, November 2019, and January 2020. Each sampling device was placed 8–11 m

Table 2 PAHs content in PM (μg/g PM)

PAHs (μg g^{-1})	BG[a], $PM_{2.5}$		BB[b], $PM_{2.5}$		BG, TSP		BB, TSP	
	Summer-Autumn	Winter-Spring	Summer-Autumn	Winter-Spring	Summer-Autumn	Winter-Spring	Summer-Autumn	Winter-Spring
Flu	7.91 ± 3.71	1.57 ± 1.94	71.28 ± 41.16	46.36 ± 31.87	5.29 ± 3.90	2.59 ± 1.96	54.18 ± 32.98	72.37 ± 36.80
Pyr	5.35 ± 2.03	2.21 ± 1.95	68.15 ± 40.91	44.97 ± 19.82	6.68 ± 3.59	4.10 ± 2.06	39.01 ± 16.28	42.28 ± 22.83
B*a*A	1.33 ± 1.13	1.76 ± 1.88	55.15 ± 13.19	37.48 ± 11.77	10.60 ± 8.93	2.15 ± 2.28	65.51 ± 40.82	15.20 ± 12.28
Chr	1.26 ± 0.74	0.76 ± 0.47	30.07 ± 9.27	14.88 ± 4.89	11.57 ± 8.50	3.45 ± 1.78	38.32 ± 37.21	16.25 ± 10.44
B*b*F	2.04 ± 1.13	1.98 ± 0.91	12.31 ± 4.29	22.89 ± 8.31	15.26 ± 10.08	5.99 ± 4.56	19.21 ± 12.32	15.22 ± 11.74
B*k*F	0.99 ± 0.57	0.80 ± 0.38	6.52 ± 2.32	2.74 ± 1.02	7.35 ± 4.67	2.97 ± 1.66	9.19 ± 7.30	1.82 ± 1.08
B*a*P	1.72 ± 1.68	0.68 ± 0.52	27.67 ± 8.91	35.26 ± 8.97	5.32 ± 3.13	3.39 ± 3.03	12.94 ± 5.78	10.15 ± 7.47
DBA	0.67 ± 0.50	0.19 ± 0.15	2.64 ± 1.62	6.66 ± 6.29	0.88 ± 0.65	0.50 ± 0.41	1.37 ± 0.66	3.46 ± 3.31
IDP	4.06 ± 2.26	2.87 ± 1.47	14.53 ± 5.77	14.04 ± 4.58	6.73 ± 3.01	7.18 ± 3.75	7.01 ± 3.58	6.59 ± 6.75
Total	25.34 ± 9.15	12.81 ± 7.84	288.30 ± 114.76	225.28 ± 48.05	69.68 ± 44.97	32.33 ± 16.81	246.74 ± 72.94	183.34 ± 95.79

[a]Background samples, [b]RS Burning samples

Abbreviations: Flu: Fluoranthene; Pyr: Pyrene; B*a*A: Benz[*a*]anthracene; Chr: Chrysene; B*b*F: benzo[*b*]fluoranthene; B*k*F: Benzo[*k*]fluoranthene; B*a*P: Benzo[*a*]pyrene; DBA: Dibenz[*a,h*]anthracene; IDP: Indeno[1,2,3-*cd*]pyrene; $PM_{2.5}$: Particulate matter with diameter < 2.5 μm; TSP: Total suspended particulate (Pham et al. 2019, with several modifications)

Table 3 Concentrations of PAHs in ng m^{-3}

PAHs (ng m^{-3})	BG[a], $PM_{2.5}$		BB[b], $PM_{2.5}$		BG, TSP		BB, TSP	
	Summer-Autumn	Winter-Spring	Summer-Autumn	Winter-Spring	Summer-Autumn	Winter-Spring	Summer-Autumn	Winter-Spring
Flu	4.46 ± 1.71	0.28 ± 0.34	1,219.6 ± 1,208.1	923.1 ± 911.2	1.52 ± 1.28	0.27 ± 0.20	735.5 ± 678.4	996.4 ± 788.9
Pyr	3.09 ± 0.84	0.44 ± 0.32	1,166.0 ± 1,150.5	695.8 ± 666.0	1.87 ± 1.22	0.43 ± 0.24	571.7 ± 499.2	605.0 ± 507.8
B*a*A	0.71 ± 0.57	0.34 ± 0.31	793.4 ± 566.3	455.1 ± 250.3	3.08 ± 2.99	0.22 ± 0.22	668.1 ± 572.3	164.9 ± 125.0
Chr	0.69 ± 0.37	0.16 ± 0.09	409.1 ± 292.2	185.7 ± 134.6	3.37 ± 2.95	0.37 ± 0.21	301.9 ± 219.7	177.8 ± 112.2
B*b*F	1.14 ± 0.55	0.42 ± 0.23	172.5 ± 132.4	261.0 ± 139.5	4.36 ± 3.31	0.64 ± 0.51	160.1 ± 91.0	158.8 ± 96.0
B*k*F	0.55 ± 0.29	0.17 ± 0.08	93.9 ± 75.2	32.9 ± 21.3	2.09 ± 1.53	0.32 ± 0.21	70.0 ± 42.8	21.4 ± 14.9
B*a*P	1.04 ± 0.80	0.15 ± 0.09	390.2 ± 289.7	424.2 ± 233.7	1.52 ± 1.11	0.37 ± 0.35	109.9 ± 63.6	110.9 ± 65.5
DBA	0.46 ± 0.18	0.05 ± 0.02	42.0 ± 42.7	74.1 ± 78.6	0.25 ± 0.22	0.05 ± 0.04	13.1 ± 10.5	28.3 ± 10.2
IDP	2.35 ± 1.23	0.61 ± 0.33	201.0 ± 151.6	144.0 ± 104.9	1.87 ± 1.08	0.75 ± 0.40	69.8 ± 68.1	77.3 ± 55.1
Total	14.41 ± 3.69	2.60 ± 1.31	4487.6 ± 3850.1	3064.1 ± 2370.3	19.92 ± 15.37	3.41 ± 1.93	2700.1 ± 1965.3	2340.8 ± 1732.8

[a]Background samples, [b]RS Burning samples

Abbreviations: Flu: Fluoranthene; Pyr: Pyrene; B*a*A: Benz[*a*]anthracene; Chr: Chrysene; B*b*F: benzo[*b*]fluoranthene; B*k*F: Benzo[*k*]fluoranthene; B*a*P: Benzo[*a*]pyrene; DBA: Dibenz[*a,h*]anthracene; IDP: Indeno[1,2,3-*cd*]pyrene; $PM_{2.5}$: Particulate matter with diameter < 2.5 μm; TSP: Total suspended particulate (Pham et al. 2019, with several modifications)

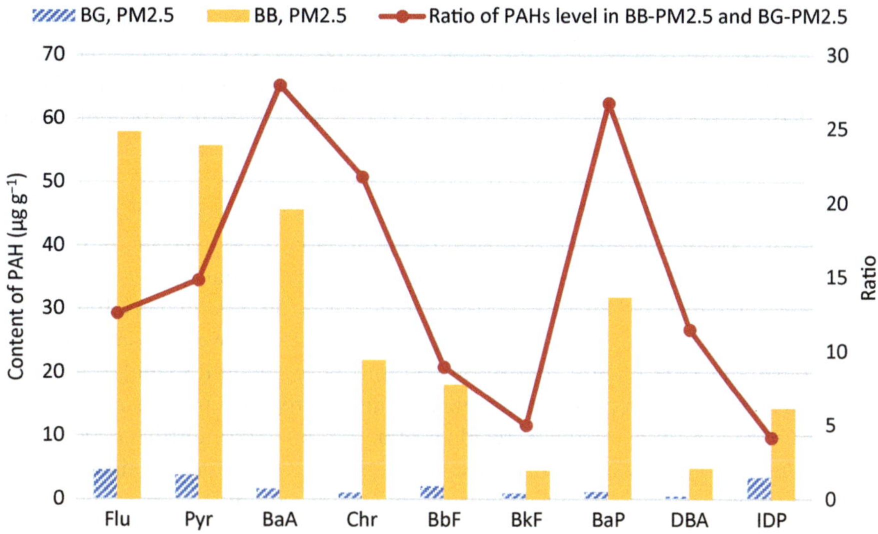

Fig. 5 Comparison of PAH levels in $PM_{2.5}$ between BG and BB samples in Hanoi. *Note* The values were the mean values of individual PAH content ($\mu g\ g^{-1}$) from BG (Background samples) and BB samples (RS Burning samples). Samples were collected from 2016 to 2018

above the ground on the rooftops of residential buildings. Average content of total 10 PAHs bound to all particles size fractions at the three sampling sites ranged from 420.7 to 1732.6 $\mu g\ g^{-1}$, with the lowest levels found at the site farthest from RS burning source and the highest at the site closest to the RS burning source (Pham et al. 2024). The higher PAH content in this sampling campaign in Hanoi compared to above sampling for BG samples may be attributed to factors such as the timing of sampling and the sampling height. These samples were collected at a higher elevation (8–11 m), whereas BG samples were taken closer to ground level (1.5 m). The difference in sampling height can influence the dispersion and accumulation of pollutants. Additionally, during atmospheric transport, PAHs on airborne particles may undergo processes such as diffusion, condensation, and coagulation, as well as mix with emissions from other sources, which can further impact their concentration at various altitudes.

The relative concentration ratios of each PAHs to a relatively stable PAH (benzo[*k*]fluoranthene-B*k*F) were examined in order to observe the contribution of PAHs from RS burning smoke to atmospheric pollution. The concentration ratio of B*a*P/B*k*F in Hanoi atmosphere during RS burning season was observed to fluctuate at higher levels compared to the values reported in Thailand (Kim Oanh et al. 2011) and Taiwan (Chen et al. 2008) (Table 4). The concentration ratios of B*a*A, Chr, and B*a*P to B*k*F in $PM_{2.5}$ from RS burning smoke in Hanoi were significantly elevated ($P < 0.001$) compared to background samples. Specifically, the B*a*A/B*k*F, Chr/B*k*F, and B*a*P/B*k*F ratios were 4.8, 4.44, and 7.43 times higher in burning smoke, respectively ($P < 0.001$). On the other hand, the ratios of Flu, Pyr, DBA, and IDP to B*k*F showed no significant increase between burning smoke and background samples (*P* > 0.02). These results emphasize the notable contributions of B*a*P, B*a*A, and Chr in $PM_{2.5}$ to the emissions from RS burning smoke. The concentration ratios of B*a*P/B*k*F and B*a*A/B*k*F in RS burning smoke in Hanoi were significantly higher than those reported in previous research Kim Oanh et al. (2011), being 3.5 and 4.8 times greater, respectively. These ratios were also considerably greater than those observed in the atmosphere by Kim Oanh (Kim Oanh et al. 2015). In addition, the average concentration ratio of B*a*P/B*k*F observed in the hood burning experiment in Hanoi (3.4 ± 1.2) was several times higher than that recorded in similar RS burning experiments conducted in an enclosure (1.65) or in a tubular furnace system in China (0.37) (Hays et al. 2005; Lu et al. 2009) (Table 4). Despite the lower overall contribution of PAHs in PM, the PAH-to-B*k*F concentration ratios observed in RS smokes in Hanoi were higher than those reported in Thailand. This discrepancy may be attributed to differences in the burning methods, residues of pesticides and/or fertilizers in the rice straw from cultivation, variations in rice varieties, and/or weather conditions influencing the presence of PAHs. Further experiments are required to better understand these variations.

Of the detected PAHs, the contribution of B*a*P in RS burning smoke in Hanoi was measured at 18.1 ± 2.7%, significantly higher than the 2.2 ± 1.6% observed in PM from motorcycle exhaust, which was collected on motorcycle test bed of internal combustion engine laboratory (Pham et al. 2013). The average B*a*P/B*k*F ratio in RS burning smoke (3.4 ± 1.2) was greater than the ratio observed in motorcycle exhaust (2.6 ± 1.1). Although B*a*P has a shorter lifespan than B*k*F, its concentration in RS burning smoke was significantly higher compared to other PAHs. This notable contribution of B*a*P to the total PAHs in RS burning emissions aligns with findings by Orakij et al. (2017), which studied PAHs from indoor biomass burning during cooking activities.

In summary, RSOB contributes significantly PAHs to the atmosphere, with the elevated levels of B*a*A, Chr, and B*a*P in $PM_{2.5}$ from burning smoke compared to background samples. B*a*P notably contributed to the total PAHs in RSOB smoke in Hanoi. The B*a*P/B*k*F ratio in RSOB smoke in Hanoi was notably higher than those in studies from Thailand and China, indicating regional variation in PAH emissions. These findings highlight potential health risks for those exposed to RSOB smoke in Hanoi due to high toxicity of high molecular weight, such as B*a*P, which is classified as a Group 1 (carcinogenic to human) (IARC 2012).

3.2 Emission Characteristics of Nitro-Polycyclic Aromatic Hydrocarbons from Rice Straw Open Burning in Northern Vietnam

Nitro-polycyclic aromatic hydrocarbons (NPAHs) are derivatives of PAHs formed during combustion processes, including biomass burning. Biomass burning, such

Table 4 Summary of PAHs profiles relative to B*k*F in $PM_{2.5}$ and TSP

The ratios of each PAHs relative to B*k*F for $PM_{2.5}$							
Studies/Ratios[a]	Flu/B*k*F	Pyr/B*k*F	B*a*A/B*k*F	Chr/B*k*F	B*a*P/B*k*F	DBA/B*k*F	IDP/B*k*F
Atmospheric particles (BG samples) before RS burning (Pham et al. 2019)	5.55 ± 6.88	4.97 ± 4.98	2.43 ± 2.70	1.15 ± 0.42	1.23 ± 0.76	0.48 ± 0.30	3.82 ± 1.13
Atmospheric particles during RS burning period in Taiwan (Chen et al. 2008) (ng/m^3)			0.25	0.5	3.25		
Atmospheric particles during non-RS burning period in Taiwan (Chen et al. 2008)			0.5	0.83	1.5		
Atmospheric particles in AIT, Thailand (Kim Oanh et al. 2015) (ng/m^3)			0.38	0.47	0.89		
RSOB smoke, pile burning in Hanoi (Pham et al. 2019)	14.84 ± 11.0	13.08 ± 9.54	11.69 ± 9.77	5.11 ± 3.36	9.14 ± 6.60	2.11 ± 4.06	3.83 ± 1.74
BG samples in Mekong Delta (Can Tho) (Phuong et al. 2022)			0.44	1.65	5.56		
RSOB smokes, spread burning in Mekong Delta (Can Tho), (Phuong et al. 2022)			3.24	4.60	3.87		
RSOB in Thailand, spread burning (Kim Oanh et al. 2011)	10	5.8	2.4	3.4	2.6		
RS open burn simulations in an enclosure in China (Hays et al. 2005)			2.1	1.85	1.65		
Tubular furnace system, RS burning in 500 °C (Lu et al. 2009)	5.27	4.50	5.23	3.72	0.37	0.13	0.14
RS hood burning in Hanoi (Pham et al. 2021)			1.0 ± 0.6	1.4 ± 0.6	3.4 ± 1.2		

Note All ratios were calculated from the PAHs content in PM (mg/g) or the PAHs concentration (ng/m^3)

as agricultural residue combustion, wildfires, and household use of wood or crop residues, coal, is a significant source of NPAHs in the atmosphere (Keyte et al. 2013; Yang et al. 2010). During these processes, incomplete combustion leads to the formation of NPAHs directly or through atmospheric reactions involving emitted PAHs with radical species such as OH and NO_3 radicals and nitrogen oxides (NO_x) (Atkinson and Arey 2003; Reisen and Arey 2005). Among biomass burning emissions, NPAHs like 1-nitropyrene (1-NP) are of particular concern due to their high mutagenic and carcinogenic properties (IARC 2012). Studies have shown that biomass burning contributes significantly to NPAH pollution in rural and agricultural regions, especially during seasonal open-field burning of crop residues (Reisen and Arey 2005). These emissions are further influenced by factors such as combustion temperature, biomass type/composition, and the presence of nitrogen-containing fertilizers or pesticides on the biomass (Bamford and Baker 2003). The understanding of NPAHs from biomass burning is crucial for assessing their environmental and health impacts, as these emissions can travel long distances and contribute to air quality degradation in both local and distant regions (Durant et al. 1996; Keyte et al. 2013). Many studies have reported nitro-PAHs in particulate matter originating from vehicular emissions (Albinet et al. 2008; Hattori et al. 2007; Pham et al. 2013; Yang et al. 2010), but few have considered the emissions from biomass burning (Orakij et al. 2017; Souza et al. 2014), especially NPAHs from RSOB (Pham et al. 2019).

The first observation of serval NPAHs in RS burning smokes was conducted in Hanoi with low concentrations (Table 5). The average concentrations of 1-nitropyrene (1-NP) in $PM_{2.5}$ in Hanoi atmosphere during RS burning season (BG samples) were 0.34 ± 0.31 μg g^{-1}, which was significantly higher compared to the concentration in RS burning smokes (0.05 ± 0.03 μg g^{-1}). 6-nitrochrysene (6-NC) was not detected in $PM_{2.5}$ from BG samples but was identified in rice straw burning smoke, with an average concentration of 0.09 ± 0.06 μg g^{-1} for $PM_{2.5}$ and 0.04 ± 0.03 μg g^{-1} for TSP. These NPAHs come both from original source such as diesel engine, coal burning, and wood burning and from reaction of the parent PAHs in the atmosphere (Bamford and Baker 2003; Ciccioli et al. 1989; Yang et al. 2010). The level of NPAHs in RS burning smokes in Hanoi was comparable to those from wood burning and significantly lower than those from motorcycle exhaust emissions (Yang et al. 2010). Secondary NPAHs, including 2-nitrofluorene (2-NF) and 2-nitropyrene (2-NP), could be formed through secondary reactions between parent PAHs and radical agents over time (Hayakawa et al. 2002; Murahashi et al. 1999). Their production was not significantly different between atmospheric particles and RS burning smoke in Hanoi. Of the detected NPAHs, a clear seasonal variation in 2-NF concentrations was observed in both BG and BB samples for $PM_{2.5}$ and TSP. The concentration of 2-NF in $PM_{2.5}$ during the summer-autumn harvest season (6.2 ± 2.1 μg g^{-1} and 7.1 ± 4.9 μg g^{-1} in BG and BB samples, respectively) was significantly higher than that in the winter-spring season (0.5 ± 0.7 μg g^{-1} and 0.09 ± 0.01 μg g^{-1} in BG and BB samples, respectively). This suggests that the higher formation of 2-NF during the summer-autumn harvest season is due to the increased sunlight, which is favorable for photochemical reactions. This finding aligns with the fact that the winter-spring crop (harvested in June) is often followed by rain after harvest. In contrast, the sunny weather during the summer-autumn crop (harvested in October) is more favorable for the formation of NPAHs through photochemical reactions with OH radicals during the day.

The concentration ratio of 2-NF to 1-NP has been utilized to distinguish between the relative contributions of primary sources and atmospheric gas-phase formation of NPAHs (Ciccioli et al. 1989, 1996; Feilberg et al. 2001; Marino et al. 2000). The 2-NF/1-NP concentration ratio exceeded 5, indicating the notable contribution of gas-phase formation processes to the production of 2-NF (Ciccioli et al. 1989). The [2-NF]/[1-NP] ratio was above 5 at nearly all sampling sites in Hanoi, emphasizing the predominance of secondary formation in NPAH production. The [2-NF]/[1-NP] ratio during the summer-autumn crop was significantly higher than during the winter-spring crop in both BG and BB samples for $PM_{2.5}$ and TSP. This finding suggests that the formation of 2-NF is more favorable in the summer-autumn crop compared to the winter-spring crop. This could be attributed to the higher levels of OH and/or NO_3 radicals present in the summer-autumn crop compared to the winter-spring crop. The [1NP]/[Pyr] ratio was used to identify the primary sources of NPAHs from parent PAHs in the atmosphere. The [1NP]/[Pyr] ratio in Hanoi showed a significant decrease, dropping from 0.095 ± 0.086 in atmospheric particles before burning to 0.0019 ± 0.0024 in $PM_{2.5}$ from RS burning smoke, suggesting 1-NP from primary source of RSOB (Tang et al. 2005). This ratio for RSOB in Hanoi closely matched values reported for coal burning, attributed to similar combustion temperatures, with RS burning ranging from 300 to 700 °C and coal burning around 900 °C (Nassar et al. 2011). These findings suggest that [NPAH]/[PAH] ratios in RS burning particulates are comparable to those from coal burning. The results from RS burning in Hanoi aligned with previous studies, lower combustion temperatures result in reduced NPAH yields and lower [NPAH]/[PAH] ratios (Hayakawa et al. 2020; Tang et al. 2005; Yang et al. 2010).

4 Size Distribution of PAHs in Southeast Asia

Understanding the size distribution of airborne particulate matter (PM) is essential for assessing their environmental behavior and their effects on the human respiratory system. The factors influencing changes in size distribution of particulate PAHs, including emission sources characteristics, combustion processes, ambient temperature, meteorological conditions, the physical and chemical properties of PAHs (Zhang et al. 2020). Previous research has shown that a substantial fraction of higher molecular weight PAHs, which are more toxic, tends to associate with smaller particles such as $PM_{0.1}$ and $PM_{2.5}$. The variations in the size distribution of particle-bound PAHs during RSOB and non-RSOB periods, as well as at varying distances from the RS burning source (both near and far), were examined in Hanoi. It was observed that the size distribution for four-ring PAHs exhibited noticeable variations depending on the proximity to the burning source and the burning period. In contrast, the five- and six-ring PAHs consistently showed the same size distribution of 0.8 μm across all

Table 5 NPAH concentrations and the NPAHs/PAHs ratios in $PM_{2.5}$ and TSP

NPAHs ($\mu g\ g^{-1}$)	BG[a]-$PM_{2.5}$			BB[b]-$PM_{2.5}$		
	Summer-Autumn	Winter-Spring	Average[a]	Summer-Autumn	Winter-Spring	Average[c]
1-NP	0.51 ± 0.25	0.04 ± 0.03	0.34 ± 0.31	0.07 ± 0.04	0.03 ± 0.01	0.05 ± 0.03
2-NF	6.2 ± 2.1	0.5 ± 0.7	5.33 ± 2.84	7.1 ± 4.9	0.09 ± 0.01	5.07 ± 5.22
2-NP	2.06 ± 3.19	2.34 ± 3.3	2.71 ± 2.92	0.24 ± 0.28	1.23 ± 0.77	0.76 ± 0.74
6-NC	ND[d]	ND[d]	ND[d]	0.04 ± 0.02	0.15 ± 0.04	0.09 ± 0.06
NPAHs ($\mu g\ g^{-1}$)	*BG-TSP*			*BB-TSP*		
1-NP	0.035 ± 0.027	0.018 ± 0.008	0.03 ± 0.02	0.18 ± 0.14	0.18 ± 0.15	0.18 ± 0.14
2-NF	3.73 ± 3.27	1.13 ± 1.12	1.99 ± 2.17	15.8 ± 12.82	0.02 ± 0.02	6.33 ± 10.76
2-NP	0.13 ± 0.195	0.74 ± 0.35	0.46 ± 0.42	0.27 ± 0.37	0.77 ± 0.42	0.54 ± 0.46
6-NC	0.047 ± 0.042	0.06 ± 0.049	0.06 ± 0.05	0.03 ± 0.02	0.04 ± 0.02	0.04 ± 0.03

[a]Background samples; [b]RS Burning samples; [c]The average values were calculated from all samples in sampling campaign; [d]Not detected. Abbreviations: 1-NP: 1-nitropyrene; 2-NF: 2-nitrofluorene; 2-NP: 2-nitropyrene; 6-NC: 6-nitrochrysene; Pyr: Pyrene (Pham et al. 2019, with several modifications)

cases, regardless of their proximity to the RS burning source or whether it was during the RSOB or non-RSOB periods. The normalized concentration of four-ring PAHs peaked at a particle size of 0.87 μm during the non-RS burning period. However, during the RS burning period, the peak concentration shifted to larger particles, reaching its maximum at 1.5 μm (Pham et al. 2024) (Fig. 6). Similarly, near the RS burning source, the peak concentration of total four-ring PAHs was observed in larger particles (1.34 μm), while at locations farther from the source, the peak occurred at a smaller particle size of 1 μm (Fig. 7). The evaporation capacity of each PAH influences its distribution across particles of varying sizes. Five- and six-ring PAHs with low volatility tend to attach to smaller particles from combustion sources, allowing them to remain in the atmosphere for longer periods. In contrast, four-ring PAHs, which are more volatile, tend to evaporate more easily, leading to their accumulation on larger particles within the RS burning area. This experiment highlights the influence of RS burning on the particle size distribution of four-ring PAHs and the impact of proximity to the burning source.

Of the six size fractions measured in Hanoi, the $PM_{0.1–0.5}$ fraction accounted for the largest portion of the total PAH mass. During the RS burning period and near the RS burning source, it contributed 38.0% and 38.9%, respectively. In contrast, during the non-RS burning period and farther from the RS burning source, it accounted for 32.5% and 51.8%, respectively (Fig. 8). The prevalence of total PAHs in the $PM_{0.1–0.5}$ fraction during RSOB is consistent with the PAH distribution observed in smoke emissions from RS burning in Lampang Province, Northern Thailand (Phoothiwut and Junyapoon 2013). A comparison of size distribution of PAHs in Hanoi and provinces in other countries is shown in Table 6. The unimodal size distribution of four- to six-ring PAHs in particulate, with the highest concentration in the accumulation mode observed in Hanoi, aligns with findings from previous studies conducted

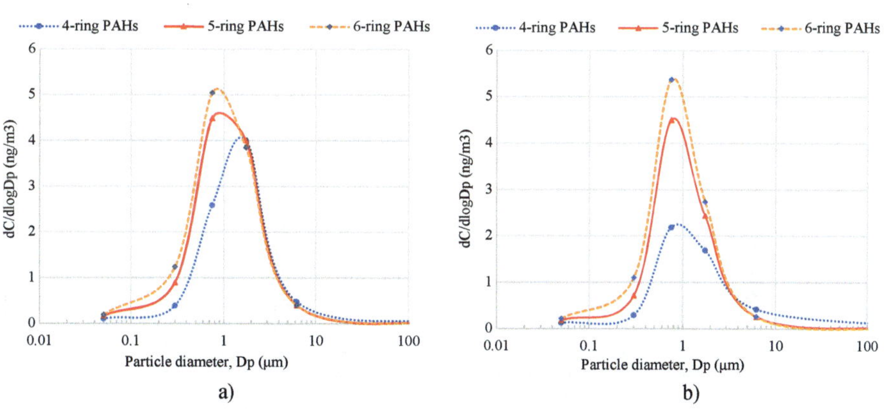

Fig. 6 Size distribution of total of four-ring, five-ring, six-ring PAHs normalized concentrations during RS burning period (**a**) and non-burning period (**b**) at DA site in Hanoi. Note: d*C* is the concentration on each filter (ng/m^3), *C* is the sum concentration on all filters, and dlogDp is the logarithmic size interval for each impactor stage in aerodynamic diameter (Dp) (Pham et al. 2024, with several modifications)

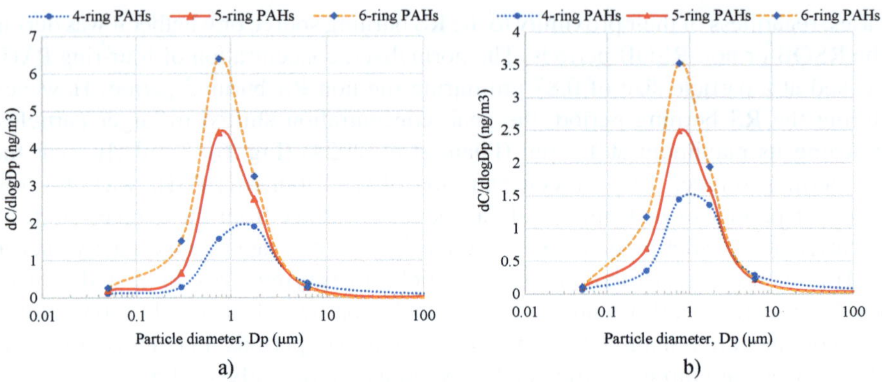

Fig. 7 Size distribution of total of four-ring, five-ring, six-ring PAHs normalized concentrations at QO_near-RS burning site (**a**) and TM_far-RS burning site (**b**) in Hanoi. *Note* dC is the concentration on each filter (ng/m^3), C is the sum concentration on all filters, and dlogDp is the logarithmic size interval for each impactor stage in aerodynamic diameter (Dp) (Pham et al. 2024, with several modifications)

at rural locations in China (Li et al. 2007), agricultural crop residue burning in China (Zhang et al. 2011), wood burning in India (Venkataraman et al. 2002), and fresh combustion smoke and ambient air (Zhang et al. 2020). Zhang et al. (2011) reported that PAHs emitted from the burning of agricultural crop residues were associated with fine particles ($PM_{2.5}$), with particle size distributions peaking at 0.1, 0.15, and 0.15 μm for rice, wheat, and corn straws, respectively.

On the other hand, several studies in East Asia have reported a distinct bimodal size distribution of PAHs from biomass burning (Chen et al. 2008; Keshtkar and Ashbaugh 2007; Shen et al. 2011, 2013; Yang et al. 2006) or even a trimodal distribution (Wu et al. 2006). Two peaks of 0.32–0.56 μm and 3.2–5.6 μm were observed during RS

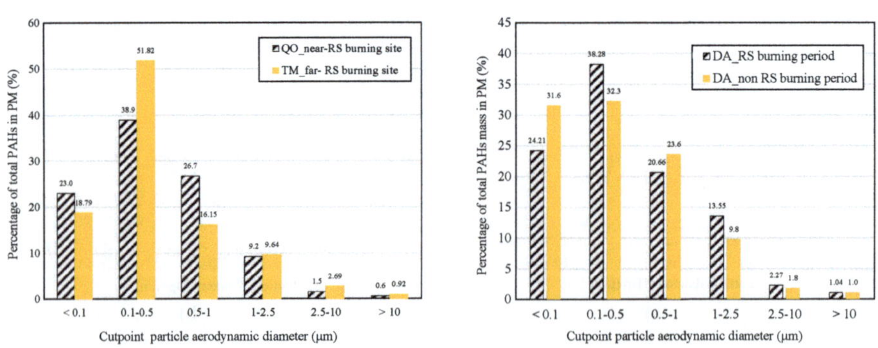

a) The percentage of total PAHs in size-segregated particles at the QO_near-RS burning site and the TM_far-RS burning site.

b) The percentage of total PAHs in size-segregated particles during the RS burning period and the non-RS burning period

Fig. 8 Percentage of total PAHs in size-segregated particles in Hanoi (Pham et al. 2024, with several modifications)

Table 6 Comparison of PAHs size distribution in Southeast Asia and other countries

Studies	Areas	Fuel type, type of experiments	PAHs emission factors (mg/kg)	Number of PAHs	Size distribution of PAHs (μm)
Pham et al. (2024)	Suburban area of Hanoi, Vietnam	RS open burning at field	10.5 ± 6.5 (TSP) 13.5 ± 9.2 for PM2.5	9 PAHs	Unimodal, peak at 1.34 or 1.5 μm during the RSOB period And 0.87 μm or 1 μm during non-burning period and far RS burning source
Li et al. (2007)	Rural area, Shandong province, China	Open burning of wheat straw and corn stover			Single mode, peaks at 0.26–0.38 μm
Zhang et al. (2011)	China	Chamber experiment, rice, wheat, and corn straws	5.26 mg/kg for rice straw, 1.37 mg/kg for wheat	16 PAHs	Unimodal mode, peaks at 0.10, 0.15, and 0.15 μm
Shen et al. (2011)	China	Crop residues	35 ± 23 mg/kg	16 PAHs	Bimodal 2.1–10 μm and < 2.1 μm
Venkataraman et al. (2002)	India	Wood	0.40–1.01 mg/kg		2.0–3.2 μm
Shen et al. (2013)		Fuel wood	1.79 ± 1.55 mg/kg	28 PAHs	< 0.4 (26.7%) 0.4–1.1 (39.4%)
		Brush wood	12.1 ± 9.1		0.4–1.1 (28.0%) 2.1–10 (35.0%)
		Bamboo	5.36 ± 4.46		< 0.4 (21.8%) 0.4–1.1 (39.4%)
Yang et al. (2006)	Taiwan	Straw burning		21 PAHs	Bimodal size distribution accumulation mode and coarse mode 0.1–3.2 μm
Chen et al. (2008)	Taiwan	RS burning and non-burning			Bimodal 0.32–0.56 μm and 3.2–5.6 μm

(continued)

Table 6 (continued)

Studies	Areas	Fuel type, type of experiments	PAHs emission factors (mg/kg)	Number of PAHs	Size distribution of PAHs (μm)
Keshtkar and Ashbaugh (2007)	California	Two common agricultural waste residues (biofuels), chamber experiments			Bimodal size distribution in accumulation mode coarse mode
Wu et al. (2006)	Rural and urban sites in Tianjin, China				Trimodal distribution with the major peaks occurring at 0.43–2.1 and 9.0–10.0 μm

burning and non-burning conditions at both the rural and urban sites located in the top five rice-producing counties in central Taiwan (Chen et al. 2008). This finding aligns with results from a study conducted in a small town, a key rice production area in central Taiwan, which revealed that during rice straw burning periods, the PAH size distribution displayed both accumulation and coarse modes (Yang et al. 2006). These findings were consistent with those reported by Keshtkar and Ashbaugh (2007), which was conducted for two common agricultural waste residues (biofuels) in California. Shen et al. (2013) investigated three types of residential wood and found that PAHs predominantly associate with fine particulate matter in the size ranges of 0.4–1.1 μm and below 0.4 μm during fuelwood and bamboo burning, consistent with the findings of Venkataraman et al. (2002) (Table 6). The size distribution of atmospheric PAHs is controlled by three mechanisms: adsorption onto the available aerosol surface area for 5-, 6-, and 7-ring PAHs; absorption into the available aerosol organic mass for all PAHs; and multilayer adsorptive accumulation in coarse-mode aerosols for 3- and 4-ring PAHs (Duan et al. 2005). These mechanisms can explain for the accumulation of high molecular PAHs in fine mode and low molecular PAHs in course mode observed in RS burning season in Hanoi, and in other biomass burning areas in other countries. These fine particles are known to enhance respiratory penetration and facilitate long-range transport. The variation in size distribution of particle-bound PAHs across different countries may be related to multiple factors such as meteorological condition, fuel combustion process, sampling methods. Further research is needed to better understand these findings and the underlying factors contributing to such variations.

5 Health Risk Assessments of PAHs and NPAHs in Hanoi

RSOB in Hanoi contributes significantly to air pollution, which may poses serious health risks due to the emission of fine particulate matter ($PM_{2.5}$), PAHs, and NPAHs. These pollutants, especially high-molecular-weight PAHs like B*a*P, which are carcinogenic and mutagenic, are released during RSOB. $PM_{2.5}$ serves as the primary carrier for these harmful compounds, allowing them to penetrate deeper into the respiratory system and increasing their toxicity (Pham et al. 2024; Hao et al. 2023). Long-term exposure to PAHs and NPAHs has been linked to bronchial asthma, cardiovascular diseases, and lung cancer (Durant et al. 1996; Torigoe et al. 2000). In Hanoi, elevated levels of PAHs and NPAHs are largely attributed to biomass burning, vehicular emissions, and industrial activities (Pham et al. 2019, 2021; Hao et al. 2023). These studies have shown that high-molecular-weight PAHs, such as B*a*P, B*a*A, strongly attach to fine particulate matter ($PM_{2.5}$), resulting in elevated concentrations in the atmosphere and subsequently increasing health risks for those exposed. B*a*P and its derivatives are known to cause long-term cancer risks when concentrations exceed international safety thresholds (IARC 2017; Kelly et al. 2021; Keyte et al. 2013). Efforts to mitigate these risks should include stricter controls on biomass burning, vehicular emissions, and industrial activities (Gadde et al. 2009; Kim Oanh et al. 2011; Pham et al. 2024).

Hao Zhang et al. (2023) performed a comprehensive analysis of both PAHs and NPAHs across different regions of Hanoi (downtown, suburban, and rural areas) to assess health risk. Three exposure routes in Hanoi including ingestion, inhalation, and dermal contact were assessed incremental lifetime cancer risk (ILCR) using the total toxic equivalent (TEQ). The mean ILCR values for both males and females in Hanoi exceeded the acceptable threshold set by the U.S. EPA (1×10^{-6}). The highest ILCR values were found in the rural site (DA), with values of 3.00×10^{-5} for males and 3.26×10^{-5} for females, followed by suburban area (GL site), with values of 1.37×10^{-5} for males and 1.49×10^{-5} for females, indicating a higher cancer risk in rural areas, particularly for females. This suggests that approximately 30 cancer cases per one million males and 33 cancer cases per one million females may occur in rural Hanoi. These number in Hanoi were lower as compared in Chiang Mai, Thailand, and Delhi, India, where 100 and 102–788 cancer cases, respectively, might occur among one million residents (Shen et al. 2019; Yadav et al. 2020). The risk rankings for the three exposure routes examined in Hanoi showed that ingestion posed the highest health risk, followed by dermal exposure and inhalation. Females exhibited slightly higher risks than males, likely due to their longer life expectancy (80 years compared to 75 years for males). The ILCR values in this study were higher than those found in previous research conducted in Hanoi, where the ILCR values ranged from 7.5×10^{-6} to 2.3×10^{-5}, which calculated by using TEF-BaP-EQs (Trung et al. 2021), indicating that the inclusion of ingestion and dermal exposure routes provides a more accurate assessment of health risks. However, the ILCR values in this study were lower than those in a previous study using PAH-CALUX assays, 1.0×10^{-4} to 2.8×10^{-4} for adults and from 6.4×10^{-5} to 1.8×10^{-4} for children which could be more specific in identifying the risks associated with PAHs (Pieterse et al. 2013; Trung et al. 2021). This finding emphasizes the necessity for Vietnam to implement stricter pollution controls, adopt alternative RSOB management practices, and fully assess health risks.

References

Albinet A, Leoz-Garziandia E, Budzinski H, Villenave E, Jaffrezo JL (2008) Nitrated and oxygenated derivatives of polycyclic aromatic hydrocarbons in the ambient air of two French alpine valleys. Part 1: Concentrations, sources and gas/particle partitioning. Atmos. Environ. 42 (1): 43–54. https://doi.org/10.1016/j.atmosenv.2007.10.009

Andreae MO (2019) Emission of trace gases and aerosols from biomass burning – an updated assessment. 19: 8523–8546. https://doi.org/10.5194/acp-19-8523-2019

Andreae MO, Merlet P (2001) Emission of trace gases and aerosols from biomass burning. Global Biogeochem. Cycles 15 (4): 955–966

Atkinson R, Arey J (2003) Atmospheric Degradation of Volatile Organic Compounds. Chem. Rev. 103 (12): 4605–4638. https://doi.org/10.1021/cr0206420

Bamford HA, Baker JE (2003) Nitro-polycyclic aromatic hydrocarbon concentrations and sources in urban and suburban atmospheres of the Mid-Atlantic region. Atmos. Environ. 37 (15): 2077–2091. https://doi.org/10.1016/S1352-2310(03)00102-X

Chen KS, Wang HK, Peng YP, Wang WC, Chen CH, Lai CH (2008) Effects of open burning of rice straw on concentrations of atmospheric polycyclic aromatic hydrocarbons in Central Taiwan. J. Air Waste Manag. Assoc. 58 (10): 1318–1327. https://doi.org/10.3155/1047-3289.58.10.1318

Christian TJ, Kleiss B, Yokelson RJ, Holzinger R, Crutzen PJ, Hao WM, Saharjo BH, Ward DE (2003) Comprehensive laboratory measurements of biomass-burning emissions: 1. Emissions from Indonesian, African, and other fuels. J Geophys Res 108(D23):4719. https://doi.org/10.1029/2003JD003704

Ciccioli P, Cecinato A, Brancaleoni E, Draisci R, Liberti A (1989) Evaluation of nitrated polycydic aromatic hydrocarbons in anthropogenic emission and air samples: A possible means of detecting reactions of carbonaceous particles in the atmosphere. Aerosol Sci. Technol. 10 (2): 296–310. https://doi.org/10.1080/02786828908959266

Ciccioli P, Cecinato A, Brancaleoni E, Frattoni M, Zacchei P, Miguel AH, de Castro Vasconcellos P (1996) Formation and transport of 2-nitrofluoranthene and 2-nitropyrene of photochemical origin in the troposphere. J. Geophys. Res. Atmos. 101 (D14): 19567–19581. https://doi.org/10.1029/95jd02118

Duan J, Bi X, Tan J, Sheng G, Fu J (2005) The differences of the size distribution of polycyclic aromatic hydrocarbons (PAHs) between urban and rural sites of Guangzhou, China. Atmos. Res. 78 (3–4): 190–203. https://doi.org/10.1016/j.atmosres.2005.04.001

Durant JL, Busby WF, Lafleur AL, Penman BW, Crespi CL (1996) Human cell mutagenicity of oxygenated, nitrated and unsubstituted polycyclic aromatic hydrocarbons associated with urban aerosols. Mutat. Res. Toxicol. 371 (3): 123–157. https://doi.org/10.1016/S0165-1218(96)90103-2

Feilberg A, Poulsen MWB, Nielsen T, Skov H (2001) Occurrence and sources of particulate nitro-polycyclic aromatic hydrocarbons in ambient air in Denmark. Atmos. Environ. 35 (2): 353–366. https://doi.org/10.1016/S1352-2310(00)00142-4

Gadde B, Bonnet S, Menke C, Garivait S (2009) Air pollutant emissions from rice straw open field burning in India, Thailand and the Philippines. Environ. Pollut. 157 (5): 1554–1558. https://doi.org/10.1016/j.envpol.2009.01.004

General Statistics Office of Vietnam (2021) *The Statistical Yearbook of Vietnam 2021* Statistical Publishing House: Hanoi, Vietnam

Hattori T, Tang N, Tamura K, Hokoda A, Yang X, Igarashi K, Ohno M, Okada Y, Kameda T, Toriba A, Hayakawa K (2007) Particulate polycyclic aromatic hydrocarbons and their nitrated derivatives in three cities in Liaoning Province, China. Environ. Forensics 8 (1–2): 165–172. https://doi.org/10.1080/15275920601180701

Hayakawa K, Tang N, Akutsu K, Murahashi T, Kakimoto H, Kizu R, Toriba A (2002) Comparison of polycyclic aromatic hydrocarbons and nitropolycyclic aromatic hydrocarbons in airborne particulates collected in downtown and suburban Kanazawa, Japan. Atmos. Environ. 36 (35): 5535–5541. https://doi.org/10.1016/S1352-2310(02)00252-2

Hayakawa K, Tang N, Toriba A, Nagato EG (2020) Calculating sources of combustion-derived particulates using 1-nitropyrene and pyrene as markers. Environ. Pollut. 265: 114730. https://doi.org/10.1016/j.envpol.2020.114730

Hays MD, Fine PM, Geron CD, Kleeman MJ, Gullett BK (2005) Open burning of agricultural biomass : Physical and chemical properties of particle-phase emissions. Atmos. Environ. 39: 6747–6764. https://doi.org/10.1016/j.atmosenv.2005.07.072

Hoang AQ, Tran TL, Tuyen LH, Nguyen TAH, Pham DM, Nguyen TC, Nguyen TN, Phan DQ, Nguyen MK, Tran VQ, Pham CT, Do Bui Q, Nguyen TQH (2023) Perfluoroalkyl substances in food contact materials: preliminary investigation in Vietnam and global comparison. Environ. Sci. Pollut. Res. 30 (47): 104181–104193. https://doi.org/10.1007/s11356-023-29746-5

IARC (2012) IARC Monographs on the Evaluation of Carcinogenic Risks to Humans. WHO Press

IARC (2017) *IARC monographs on the evaluation of the carcinogenic risks to humans. WHO, Geneva*

Jenkins BM, Jones AD, Turn SQ, Williams RB (1996) Particle concentrations, gas-particle partitioning, and species intercorrelations for polycyclic aromatic hydrocarbons (PAH) emitted during biomass burning. Atmos. Environ. 30 (22): 3825–3835. https://doi.org/10.1016/1352-2310(96)00084-2

Kanokkanjana K, Garivait S (2013) Alternative Rice Straw Management Practices to Reduce Field Open Burning in Thailand. Int. J. Environ. Sci. Dev. 4 (2): 119–123. https://doi.org/10.7763/IJESD.2013.V4.318

Kelly JM, Ivatt PD, Evans MJ, Kroll JH, Hrdina AIH, Kohale IN, White FM, Engelward BP, Selin NE (2021) Global Cancer Risk From Unregulated Polycyclic Aromatic Hydrocarbons. 5 (9): e2021GH000401. https://doi.org/10.1029/2021GH000401

Keshtkar H, Ashbaugh LL (2007) Size distribution of polycyclic aromatic hydrocarbon particulate emission factors from agricultural burning. Atmos. Environ. 41 (13): 2729–2739. https://doi.org/10.1016/j.atmosenv.2006.11.043

Keyte IJ, Harrison RM, Lammel G (2013) Chemical reactivity and long-range transport potential of polycyclic aromatic hydrocarbons – a review. Chem. Soc. Rev. 42 (24): 9333–9391. https://doi.org/10.1039/C3CS60147A

Kim Oanh NT, Ly BT, Tipayarom D, Manandhar BR, Prapat P, Simpson CD, Sally Liu L-J (2011) Characterization of particulate matter emission from open burning of rice straw. Atmos. Environ. 45 (2): 493–502. https://doi.org/10.1016/j.atmosenv.2010.09.023

Kim Oanh NT, Tipayarom A, Ly B-T, Tipayarom D, Simpson CD, Hardie D, Liu L-JS (2015) Characterization of gaseous and semi-volatile organic compounds emitted from field burning of rice straw. Atmos. Environ. 119: 182–191. https://doi.org/10.1016/j.atmosenv.2015.08.005

Lemieux PM, Lutes CC, Santoianni DA (2004) Emissions of organic air toxics from open burning: a comprehensive review. Prog. Energy Combust. Sci. 30 (1): 1–32. https://doi.org/10.1016/j.pecs.2003.08.001

Li X, Wang S, Duan L, Hao J, Li C, Chen Y, Yang L (2007) Particulate and trace gas emissions from open burning of wheat straw and corn stover in China. Environ. Sci. Technol. 41 (17): 6052–6058. https://doi.org/10.1021/es0705137

López Cancio JA, Vera Castellano A, Santana Martín S, Santana Rodrîguez JF (2004) Size distributions of PAHs in ambient air particles of two areas of Las Palmas De Gran Canaria. Water. Air. Soil Pollut. 154 (1–4): 127–138. https://doi.org/10.1023/B:WATE.0000022935.85890.e0

Lu H, Zhu L, Zhu N (2009) Polycyclic aromatic hydrocarbon emission from straw burning and the influence of combustion parameters. Atmos. Environ. 43 (4): 978–983. https://doi.org/10.1016/j.atmosenv.2008.10.022

Ly BT (2006) *Characterization of Aerosol Emission from Rice Straw Open Burning for Assessment of Impact on Air Quality and Climate* Asian Institute of Technology: Bangkok, Thailand

Marino F, Cecinato A, Siskos PA (2000) Nitro-PAH in ambient particulate matter in the atmosphere of Athens. Chemosphere 40 (5): 533–537. https://doi.org/10.1016/S0045-6535(99)00308-2

McMeeking GR, Kreidenweis SM, Baker S, Carrico CM, Chow JC, Collett JL, Hao WM, Holden AS, Kirchstetter TW, Malm WC, Moosmüller H, Sullivan AP, Cyle E. W (2009) Emissions of trace gases and aerosols during the open combustion of biomass in the laboratory. J. Geophys. Res. Atmos. 114 (19): 1–20. https://doi.org/10.1029/2009JD011836

Murahashi T, Kizu R, Kakimoto H, Toriba A, Hayakawa K (1999) 2-Nitrofluoranthene, 1-, 2- and 4-Nitropyrenes and 6-Nitrochrysene in Diesel-Engine Exhaust and Airborne Particulates. J. Heal. Sci. 45 (5): 244–250. https://doi.org/10.1248/jhs.45.244

Nassar HF, Tang N, Kameda T, Toriba A, Khoder MI, Hayakawa K (2011) Atmospheric concentrations of polycyclic aromatic hydrocarbons and selected nitrated derivatives in Greater Cairo, Egypt. Atmos. Environ. 45 (39): 7352–7359. https://doi.org/10.1016/j.atmosenv.2011.07.043

Nguyen TN, Nguyen MN, Mcnamara M, Dultz S, Meharg A, Nguyen VT (2019) Encapsulation of lead in rice phytoliths as a possible pollutant source in paddy soils. Environ Exp Bot 162:58–66. https://doi.org/10.1016/j.envexpbot.2019.02.009

Orakij W, Chetiyanukornkul T, Kasahara C, Boongla Y, Chuesaard T, Furuuchi M, Hata M, Tang N, Hayakawa K, Toriba A (2017) Polycyclic aromatic hydrocarbons and their nitro derivatives from indoor biomass-fueled cooking in two rural areas of Thailand: a case study. Air Qual. Atmos. Heal. 10 (6): 747–761. https://doi.org/10.1007/s11869-017-0467-y

Pham CT (2018) *Polycyclic Aromatic Hydrocarbons and Nitropolycyclic Aromatic Hydrocarbons in Motorcycle Exhaust BT - Polycyclic Aromatic Hydrocarbons: Environmental Behavior and Toxicity in East Asia* (K. Hayakawa (ed.); pp. 137–153) Springer Singapore. https://doi.org/10.1007/978-981-10-6775-4_11

Pham CT, Boongla Y, Nghiem TD, Le HT, Tang N, Toriba A, Hayakawa K (2019) Emission characteristics of polycyclic aromatic hydrocarbons and nitro-polycyclic aromatic hydrocarbons from open burning of rice straw in the north of vietnam. Int. J. Environ. Res. Public Health 16 (13): 2343. https://doi.org/10.3390/ijerph16132343

Pham CT, Kameda T, Toriba A, Hayakawa K (2013) Polycyclic aromatic hydrocarbons and nitropolycyclic aromatic hydrocarbons in particulates emitted by motorcycles. Environ. Pollut. 183: 175–183. https://doi.org/10.1016/j.envpol.2013.01.003

Pham CT, Ly B-T, Nghiem T-D, Pham TH-P, Minh N-T, Tang N, Hayakawa K, Toriba A (2021) Emission factors of selected air pollutants from rice straw burning in Hanoi, Vietnam. Air Qual. Atmos. Heal. https://doi.org/10.1007/s11869-021-01050-6

Pham CT, Nghiem TD, Le HT, Chu HD, Tran Viet T, Sekiguchi K, Tang N, Hayakawa K, Toriba A (2024) Size distribution of airborne particle-bound polycyclic aromatic hydrocarbons during rice straw open burning in Hanoi, Vietnam. Atmos. Pollut. Res. 15 (6): 102115. https://doi.org/10.1016/j.apr.2024.102115

Phoothiwut S, Junyapoon S (2013) Size distribution of atmospheric particulates and particulate-bound polycyclic aromatic hydrocarbons and characteristics of PAHs during haze period in Lampang Province, Northern Thailand. Air Qual. Atmos. Heal. 6 (2): 397–405. https://doi.org/10.1007/s11869-012-0194-3

Phuong PTH, Nghiem TD, Pham TMT, Nguyen TD (2022) Emission factors of selected air pollutants from rice straw open burning in the Mekong Delta of Vietnam. Atmos. Pollut. Res. 13 (3). https://doi.org/10.1016/j.apr.2022.101353

Pieterse B, Felzel E, Winter R, van der Burg B, Brouwer A (2013) PAH-CALUX, an Optimized Bioassay for AhR-Mediated Hazard Identification of Polycyclic Aromatic Hydrocarbons (PAHs) as Individual Compounds and in Complex Mixtures. Environ. Sci. Technol. 47 (20): 11651–11659. https://doi.org/10.1021/es403810w

Reid JS, Koppmann R, Eck TF, Eleuterio DP (2005) A review of biomass burning emissions part II: intensive physical properties of biomass burning particles. Atmos. Chem. Phys. 5 (3): 799–825. https://doi.org/10.5194/acp-5-799-2005

Reisen F, Arey J (2005) Atmospheric Reactions Influence Seasonal PAH and Nitro-PAH Concentrations in the Los Angeles Basin. Environ. Sci. Technol. 39 (1): 64–73. https://doi.org/10.1021/es0354541

Sanchis E, Ferrer M, Calvet S, Coscollà C, Yusà V, Cambra-López M (2014) Gaseous and particulate emission profiles during controlled rice straw burning. Atmos. Environ. 98: 25–31. https://doi.org/10.1016/j.atmosenv.2014.07.062

Shen G, Tao S, Chen Y, Zhang Y, Wei S, Xue M, Wang B, Wang R, Lu Y, Li W, Shen H, Huang Y, Chen H (2013) Emission Characteristics for Polycyclic Aromatic Hydrocarbons from Solid Fuels Burned in Domestic Stoves in Rural China. Environ. Sci. Technol. 47 (24): 14485–14494. https://doi.org/10.1021/es403110b

Shen G, Wang W, Yang Y, Ding J, Xue M, Min Y, Shen H, Li W, Wang B, Wang R, Wang X, Tao S, Russell AG (2011) partitioning. 45 (4): 1206–1212. https://doi.org/10.1021/es102151w.Emissions

Shen R, Wang Y, Gao W, Cong X, Cheng L, Li X (2019) Size-segregated particulate matter bound polycyclic aromatic hydrocarbons (PAHs) over China: Size distribution, characteristics and health risk assessment. Sci. Total Environ. 685: 116–123. https://doi.org/10.1016/j.scitotenv.2019.05.436

Souza KF, Carvalho LRF, Allen AG, Cardoso AA (2014) Diurnal and nocturnal measurements of PAH, nitro-PAH, and oxy-PAH compounds in atmospheric particulate matter of a sugar cane burning region. Atmos. Environ. 83: 193–201. https://doi.org/10.1016/j.atmosenv.2013.11.007

Streets DG, Yarber KF, Woo JH, Carmichael GR (2003) Biomass burning in Asia: Annual and seasonal estimates and atmospheric emissions. Global Biogeochem. Cycles 17 (4). https://doi.org/10.1029/2003gb002040

Tang N, Hattori T, Taga R, Igarashi K, Yang X, Tamura K, Kakimoto H, Mishukov VF, Toriba A, Kizu R, Hayakawa K (2005) Polycyclic aromatic hydrocarbons and nitropolycyclic aromatic hydrocarbons in urban air particulates and their relationship to emission sources in the Pan-Japan Sea countries. Atmos. Environ. 39 (32): 5817–5826. https://doi.org/10.1016/j.atmosenv.2005.06.018

Tipayarom A, Kim Oanh NT (2020) Influence of rice straw open burning on levels and profiles of semi-volatile organic compounds in ambient air. Chemosphere 243: 125379. https://doi.org/10.1016/j.chemosphere.2019.125379

Torigoe K, Hasegawa S, Numata O, Yazaki S, Matsunaga M, Boku N, Hiura M, Ino H (2000) Influence of emission from rice straw burning on bronchial asthma in children. Pediatr. Int. 42: 143–150. https://doi.org/10.1046/j.1442-200X.2000.01196.x

Trung NT, Anh HQ, Tue NM, Suzuki G, Takahashi S, Tanabe S, Khai NM, Hong TT, Dau PT, Thuy PC, Tuyen LH (2021) Polycyclic aromatic hydrocarbons in airborne particulate matter samples from Hanoi, Vietnam: Particle size distribution, aryl hydrocarbon ligand receptor activity, and implication for cancer risk assessment. Chemosphere 280 (April): https://doi.org/10.1016/j.chemosphere.2021.130720

Venkataraman C, Negi G, Brata Sardar S, Rastogi R (2002) Size distributions of polycyclic aromatic hydrocarbons in aerosol emissions from biofuel combustion. J. Aerosol Sci. 33 (3): 503–518. https://doi.org/10.1016/S0021-8502(01)00185-9

Wu SP, Tao S, Liu WX (2006) Particle size distributions of polycyclic aromatic hydrocarbons in rural and urban atmosphere of Tianjin, China. Chemosphere 62 (3): 357–367. https://doi.org/10.1016/j.chemosphere.2005.04.101

Yadav A, Behera SN, Nagar PK, Sharma M (2020) Spatio-seasonal concentrations, source apportionment and assessment of associated human health risks of pm2.5-bound polycyclic aromatic hydrocarbons in delhi, india. Aerosol Air Qual. Res. 20 (12): 2805–2825. https://doi.org/10.4209/aaqr.2020.04.0182

Yang HH, Tsai CH, Chao MR, Su YL, Chien SM (2006) Source identification and size distribution of atmospheric polycyclic aromatic hydrocarbons during rice straw burning period. Atmos. Environ. 40 (7): 1266–1274. https://doi.org/10.1016/j.atmosenv.2005.10.032

Yang X-Y, Igarashi K, Tang N, Lin J-M, Wang W, Kameda T, Toriba A, Hayakawa K (2010) Indirect- and direct-acting mutagenicity of diesel, coal and wood burning-derived particulates and contribution of polycyclic aromatic hydrocarbons and nitropolycyclic aromatic hydrocarbons. Mutat. Res. Toxicol. Environ. Mutagen. 695 (1–2): 29–34. https://doi.org/10.1016/j.mrgentox.2009.10.010

Zhang Hao, Pham C-T, Chen B, Zhang X, Wang Y, Bai P, Zhang L, Nagao S, Toriba A, Nghiem T-D, Tang N (2023) Main Emission Sources and Health Risks of Polycyclic Aromatic Hydrocarbons and Nitro-Polycyclic Aromatic Hydrocarbons at Three Typical Sites in Hanoi. Atmosphere (Basel). 14 (5):. https://doi.org/10.3390/atmos14050782

Zhang Hefeng, Hu D, Chen J, Ye X, Wang SX, Hao JM, Wang L, Zhang R, An Z (2011) Particle size distribution and polycyclic aromatic hydrocarbons emissions from agricultural crop residue burning. Environ. Sci. Technol. 45 (13): 5477–5482. https://doi.org/10.1021/es1037904

Zhang L, Yang L, Zhou Q, Zhang X, Xing W, Wei Y, Hu M, Zhao L, Toriba A, Hayakawa K, Tang N (2020) Size distribution of particulate polycyclic aromatic hydrocarbons in fresh combustion smoke and ambient air: A review. J. Environ. Sci. (China) 88: 370–384. https://doi.org/10.1016/j.jes.2019.09.007

Zhang T, Wooster MJ, Green DC, Main B (2015) New field-based agricultural biomass burning trace gas, PM2.5, and black carbon emission ratios and factors measured in situ at crop residue fires in Eastern China. Atmos. Environ. 121: 22–34. https://doi.org/10.1016/j.atmosenv.2015.05.010

China's Coal Heating: Research Methods, Emission Factors, and Policy Interventions for PAHs Reduction

Yang Xiaoyang and Wang Yicheng

Abstract With China's rapid economic development, coal has long dominated its energy structure, serving as the primary energy source for industrial production and residential heating. However, coal combustion releases significant amounts of pollutants, particularly polycyclic aromatic hydrocarbons (PAHs), posing serious threats to the environment and human health. This study reviews research methodologies on PAHs emissions from coal combustion in China, including field sampling, laboratory combustion experiments, and modeling simulations, establishing a comprehensive emission factors (EFs) framework. It also highlights measures adopted in recent years to mitigate coal-related pollution, such as the application of flue gas purification technologies in industry and the implementation of clean energy policies like "coal-to-electricity" and "coal-to-gas" for residential heating. These efforts have significantly reduced PAHs emissions.

Keywords Coal consumption · Polycyclic aromatic hydrocarbons · Emission factors · Flue gas purification

1 Introduction

1.1 The Role of Coal in China's Energy Structure

Coal has always played a dominant role in China's energy mix. According to data from the China Energy Statistical Yearbook, as shown in Table 1 (National Bureau of Statistics of China 2025), although China's energy consumption structure has been adjusted in recent years and the proportion of coal consumption has decreased, its share of energy consumption in 2022 is still high at about 56% (Table 2). An

Y. Xiaoyang (✉) · W. Yicheng
Institute of Atmospheric Environment, Chinese Research Academy of Environmental Sciences, Chaoyang, Beijing, China
e-mail: yangxy@craes.org.cn

K. Hayakawa and Y. Aoki (eds.), *A New Era of Polycyclic Aromatic Hydrocarbons*,
https://doi.org/10.1007/978-981-96-7695-8_2

extrapolation of the data yields the projected share of each energy source in the total energy consumption.

This proportion is much higher than the global average, highlighting China's high dependence on coal. The widespread use of coal is mainly due to China's vast industrial system, electricity demand, and heating demand in northern China (Miller 2004).

The core position of coal in China is not only reflected in the total consumption, but also in its supporting role in China's economic development (Bloch et al. 2012). As a cheap energy source, coal has provided a strong impetus for China's industrialization process (Wang and Li 2016). However, this dependence has also brought significant environmental problems, such as air pollution, water pollution, and soil pollution (Munawer 2018).

Compared with Japan and South Korea, China's reliance on coal is much higher in its energy consumption mix (Zhu et al. 2024). Although the proportion of coal consumption in Japan and South Korea has declined in recent years, coal remains an

Table 1 China's total energy consumption and composition (National Bureau of Statistics of China 2025)

Years	Total energy consumption (10,000 tons of standard coal)	Composition			
		Coal	Oil	Natural gas	Primary electricity and other energy sources
2005	261,369	189,231	46,524	6273	19,341
2006	286,467	207,402	50,132	7735	21,199
2007	311,442	225,795	52,945	9343	23,358
2008	320,611	229,237	53,542	10,901	26,931
2009	336,126	240,666	55,125	11,764	28,571
2010	360,648	249,568	62,753	14,426	33,901
2011	387,043	271,704	65,023	17,804	32,512
2012	402,138	275,465	68,363	19,303	39,007
2013	416,913	280,999	71,292	22,096	42,525
2014	428,334	281,844	74,102	23,987	48,402
2015	434,113	276,964	79,877	25,179	52,094
2016	441,492	274,608	82,559	26,931	57,394
2017	455,827	276,231	86,151	31,452	61,992
2018	471,925	278,436	89,194	35,866	68,429
2019	487,488	281,281	92,623	38,999	74,586
2020	498,314	283,541	93,683	41,858	79,232
2021	525,896	293,976	97,817	46,279	87,825
2022	540,956	302,935	97,372	45,440	95,208
2023	572,000	316,316	104,676	48,620	102,388

Table 2 China's energy share

Year	Share of coal in total energy consumption (%)	Share of oil in total energy consumption (%)	Share of natural gas in total energy consumption (%)	Share of primary electricity and other energy sources in total energy consumption (%)
2005	72	18	2	7
2006	72	18	3	7
2007	72	17	3	7
2008	72	17	3	8
2009	72	16	3	9
2010	69	17	4	9
2011	70	17	5	8
2012	69	17	5	10
2013	67	17	5	10
2014	66	17	6	11
2015	64	18	6	12
2016	62	19	6	13
2017	61	19	7	14
2018	59	19	8	14
2019	58	19	8	15
2020	57	19	8	16
2021	56	19	9	17
2022	56	18	8	18
2023	55	18	9	18

important part of their energy structure, especially in power generation and industry (Yun 2024; Sabūnas 2024; Lee et al. 2024; Yasuda 2024). According to the International Energy Agency (IEA), Japan and South Korea have gradually reduced the share of coal in their energy and promoted energy transformation by increasing the use of natural gas and renewable energy (Eweade et al. 2024). For example, South Korea's share of coal in power generation is higher than the IEA average, but it is gradually reducing its reliance on coal by accelerating the development of renewable energy, improving energy efficiency, and promoting carbon neutrality goals (Cho et al. 2024). In Japan, with the reduction of nuclear energy use, coal consumption has increased, but it has also significantly increased the share of natural gas power generation and renewable energy in recent years (Eweade et al. 2024). Nevertheless, compared with China, Japan and South Korea have lower total coal consumption and their energy structures also show a faster trend of clean transformation (Kang 2024).

China's coal consumption is several times higher than Japan's (Wang et al. 2024b) and even dozens of times higher than South Korea's. This is mainly because China's population size and economic scale are much larger than those of Japan and South Korea. In addition, China's energy resource endowment also determines the dominant position of coal in energy consumption (Song et al. 2018; Wang et al. 2022; Qian et al. 2021). China is a country with abundant coal and little oil. It has abundant coal reserves but relatively scarce oil and natural gas resources, which further aggravates China's dependence on coal.

1.2 Changes in Coal Consumption Year by Year

China's coal consumption has undergone significant changes since the 1990s (Shen et al. 2012). During the period of accelerated industrialization in the 2000s, coal consumption showed a rapid growth trend. However, in recent years, with the Chinese government's emphasis on environmental protection and energy structure optimization, as well as the development and promotion of clean energy technologies, the growth rate of coal consumption has slowed down and even shown a downward trend (Wang et al. 2024b; Sinton and Fridley 2000). In the future, with the further adjustment of China's economic structure and the continuous development of clean energy technology, the proportion of coal in China's energy consumption is expected to further decrease (Chai et al. 2019).

During the 1990s and 2000s, China's economy experienced rapid growth, the industrialization process accelerated, and the demand for energy rose sharply. As China's main energy source, coal consumption showed a rapid growth trend during this period. According to data from the China Energy Statistical Yearbook, China's coal consumption increased from hundreds of millions of tons to billions of tons during this period, with an average annual growth rate of double digits. This rapid growth is mainly attributed to China's vast industrial system, electricity demand, and heating demand in northern China (Xiong et al. 2015; Ma and Stern 2008).

In recent years, as the Chinese government has attached great importance to environmental protection and optimization of energy structure, the growth rate of coal consumption has slowed down and even shown a downward trend. This change is mainly affected by many factors:

Firstly, the Chinese government has increased its control over coal use (Zhang 2010), through the implementation of a series of policy measures to limit coal consumption. For example, the promotion of policies such as "coal-to-electricity" and "coal-to-gas" (Chen and Chen 2019) effectively reduces the proportion of coal used in heating and industrial production. These policies not only help reduce coal consumption, but also help improve air quality and reduce environmental pollution (Wang 2023).

Secondly, with the continuous development of clean energy technology and the reduction of costs (Adelekan et al. 2024), clean energy such as renewable energy and natural gas has gradually increased its share in China's energy consumption (Wang et al. 2023). The substitution effect of these clean energy sources has further reduced the growth rate of coal consumption.

In addition, the adjustment and transformation and upgrading of China's economic structure have also had an impact on coal consumption. With the rapid development of the service industry and high-tech industries, the proportion of traditional heavy industry in the national economy has gradually declined (Xu and Xu 2023), which has reduced the demand for coal to a certain extent.

2 Research Methods for PAHs Emissions from Coal Combustion

2.1 *Characteristics of PAHs*

The coal combustion process produces a wide range of emissions, including polycyclic aromatic hydrocarbons (PAHs), a class of organic compounds with two or more benzene ring structures. PAHs are highly stable, difficult to degrade, persist in the environment for long periods, and accumulate and amplify through the food chain (Gbeddy et al. 2020). In addition, PAHs are also carcinogenic, teratogenic, and mutagenic (Kumar et al. 2016; Ifegwu and Anyakora 2015), posing a serious threat to human health. Therefore, the study of coal-fired PAHs emissions is of great significance to environmental protection and human health.

The research methods for coal-fired PAHs emissions primarily consist of three approaches (Hsu et al. 2016): field sampling, laboratory analysis, and model simulation. These methods complement and verify each other and together constitute a complete system for the study of coal-fired PAHs emissions. In the future, with the continuous development and innovation of science and technology, the research methods of coal-fired PAHs emissions will continue to be improved and optimized, providing stronger support for environmental protection and human health.

2.2 *Field Sampling Methods*

On-site sampling captures the actual emissions generated during coal combustion, providing data support for subsequent laboratory analysis and model simulation (Zhang et al. 2018a). On-site sampling usually includes two methods: flue gas sampling (Zhao et al. 2019) and particulate matter sampling (Zhao et al. 2019; Yang et al. 2021). Flue gas sampling is mainly used to obtain gaseous pollutants generated during coal combustion, while particulate matter sampling is used to obtain solid pollutants generated during coal combustion. During the sampling process, the sampling conditions need to be strictly controlled to ensure the accuracy and reliability of the sampling results.

On-site sampling usually includes the following key steps:

1. **Sampling point selection**: Select appropriate sampling points according to the research purpose and actual situation (Zhao et al. 2019). The sampling points

should be able to represent the main characteristics of coal combustion emissions and avoid interference from other pollution sources.

2. **Sampling equipment preparation**: Prepare the sampling equipment, including sampler, filter cartridge, absorption solution, etc. The sampling equipment should be calibrated to ensure its accuracy and reliability.
3. **Sampling Operation**: When the combustion equipment is in normal operation, use a sampler to introduce the flue gas into the sampling equipment. During the sampling process, the sampling flow rate and time should be strictly controlled to ensure that a sufficient number of samples are collected.
4. **Sample storage and transportation**: The collected samples should be sealed and stored immediately to avoid contamination and volatilization (Yang et al. 2021). At the same time, the samples should be sent to the laboratory for analysis as soon as possible to reduce changes in the samples during transportation and storage.

2.3 Laboratory Analysis Tests

Laboratory analysis is performed on samples collected on site (Mentes et al. 2022), the concentration and type of PAHs can be quantitatively determined, and the pollution degree and environmental impact of coal-fired PAHs emissions can be evaluated. Laboratory analysis usually uses modern analytical techniques (Hong and Choi 2015; Vistnes et al. 2022) such as high-performance liquid chromatography (HPLC) and gas chromatography-mass spectrometry (GC–MS). These technologies have the advantages of high sensitivity, good accuracy, and high separation efficiency and can meet the needs of coal-fired PAHs emission research.

Laboratory combustion tests usually use equipment such as small fixed-bed furnaces or tubular furnaces to simulate coal combustion under controlled combustion conditions. During the test, researchers record the flue gas composition, with particular focus on PAH concentrations and their changes over time. At the same time, by adjusting parameters such as coal type, combustion temperature, and combustion method, the effects of these factors on PAHs emissions can be examined, such as:

1. **Coal type** (Zhao et al. 2020): There are significant differences in the types and concentrations of PAHs produced by different types of coal during combustion. In general, anthracite and bituminous coal produce higher PAHs concentrations during combustion, while lignite produces relatively lower PAHs concentrations. This may be related to the type and content of organic matter in coal.
2. **Combustion temperature** (Han et al. 2020): Combustion temperature is one of the key factors affecting the generation and emission of PAHs. As the combustion temperature increases, the amount of PAHs generated generally increases. However, when the combustion temperature is too high, some PAHs may undergo pyrolysis or oxidation reactions, resulting in a decrease in emission concentration.
3. **Combustion method** (Fakinle et al. 2022): The combustion method also has an important impact on PAHs emissions. For example, fluidized bed combustion

technology can more effectively reduce PAHs emissions than fixed bed combustion technology. This may be related to the more complete oxygen mixing and more uniform temperature distribution during fluidized bed combustion.

2.4 *Model Simulation*

Model simulation can simulate the generation, migration and transformation of PAHs during coal combustion by constructing mathematical models or computer simulation systems and then predict the pollution trend and environmental impact of coal-fired PAHs emissions. Model simulation usually includes physical models, chemical models and comprehensive models. Different types of models have different scopes of application and advantages and disadvantages.

For example, the Greenhouse Gas-Air Pollution Interactions and Synergies (GAINS) model is a widely used tool to assess the effectiveness of reducing air pollutants and greenhouse gas emissions (Apsimon et al. 2009). The model quantifies the changes in PAHs emissions under different policy scenarios by integrating coal-fired emission factors (EFs), regional emission activity data, and meteorological conditions. In a study, the GAINS model was used to simulate the impact of the "coal-to-electricity" policy in northern China on PAHs emissions (Xue et al. 2016). The results showed that after residents switched to clean energy for heating, PAHs emissions in Beijing dropped significantly, with an average reduction of 35–45%.

In addition, the Weather Research and Forecasting model with Chemistry (WRF-Chem) model combines meteorological forecasts and chemical transport processes to more accurately simulate the migration and deposition behavior of pollutants in the atmosphere (Clima and Te 2016). In a study focusing on Northeast Asia, the WRF-Chem model was used to analyze the cross-border transport of PAHs emitted from coal combustion in winter (Mao et al. 2024). The results show that PAHs emissions from Northeast China contribute more than 15% to pollution on the Korean Peninsula during the winter when wind speeds are high. This precise simulation provides an important reference for formulating regional pollution control strategies.

The data required for model simulation includes coal emission factors, meteorological conditions, and statistical data. Coal emission factors are obtained through laboratory combustion tests and literature research, including the emission characteristics of PAHs under different coal types, combustion temperatures, and methods. Meteorological data come from meteorological observation stations or reanalysis data, including parameters such as temperature, humidity, and wind speed. Statistical data are used to verify the accuracy and reliability of model simulation results, including PAHs concentration monitoring data from environmental monitoring stations.

The simulation analysis process is shown below:

1. **Model configuration**: According to the characteristics and needs of the study area, the GAINS model and the WRF-Chem model are configured. The model's grid resolution, time step, boundary conditions, and other parameters are set to

ensure that the model can accurately simulate the meteorological conditions and pollutant transport processes in the study area.

2. **Input data preparation**: Collate the input data such as coal emission factors, meteorological conditions, and statistical data into the format required by the model and import them into the model. Ensure the accuracy and completeness of the data to avoid adverse effects on the simulation results.
3. **Model run**: Start the model for simulation analysis. The GAINS model will calculate the emissions of PAHs based on the input data and simulate their diffusion behavior in the atmosphere in combination with meteorological conditions. The WRF-Chem model will further consider the chemical transformation and deposition process of pollutants and provide more detailed simulation results.
4. **Result output and analysis**: After running the model, the simulation results are generated and analyzed. The accuracy and reliability of the model are evaluated by comparing the simulation results with environmental monitoring data. At the same time, the simulation results are analyzed in depth to explore the impact of different factors on PAHs emissions and diffusion.

Based on the simulation results, targeted coal-burning pollution control measures can be formulated. For example, the emission concentration of PAHs can be reduced by optimizing the combustion method and improving the combustion efficiency; the impact of PAHs on the environment and human health can be reduced by strengthening the construction of meteorological monitoring and early warning systems. In addition, the model simulation results can also provide a scientific basis for the formulation and evaluation of environmental policies and promote the in-depth development of coal-burning pollution control and environmental protection work.

3 Toxicity and EFs of PAHs

3.1 Toxicity of PAHs

PAHs constitute a class of organic compounds characterized by the presence of two or more benzene rings fused together in a condensed ring system. These compounds exhibit chemical stability, high melting and boiling points, and solubility in various organic solvents such as acetone, benzene, and dichloromethane, while being insoluble in water (Barbosa et al. 2023). Due to their lipophilic nature, PAHs have the ability to accumulate within biological organisms and are resistant to natural degradation in the environment (Sahoo et al. 2020), thus widely recognized as persistent organic pollutants (POPs).

From a toxicological perspective, PAHs exhibit potent carcinogenic, teratogenic, and mutagenic effects on organisms, collectively referred to as the "three-effect toxicity." PAHs can enter the body through various routes, including skin contact, inhalation, and ingestion, causing damage to multiple systems (Barbosa et al. 2023). For instance, PAHs can induce tissue hyperplasia, leading to impairments in the nervous system, immune system, liver, kidneys, and adrenal glands. Extensive epidemiological data and animal studies have confirmed the carcinogenic potential of PAHs (Sun et al. 2021a), particularly those with four to six fused rings, which demonstrate significant carcinogenicity in experimental animals.

It is noteworthy that the toxicity of PAHs is closely linked to their chemical structure. Generally, PAHs with more fused rings have stronger toxicity (Sun et al. 2021a). For example, benzo[*a*]pyrene and benzo[*a,h*]anthracene, known for their potent carcinogenicity, represent some of the most toxic members within the PAHs family.

3.2 EFs of PAHs

PAHs are among the most stable forms of hydrocarbons, distinguished by their low hydrogen-to-carbon ratio. They are typically present as complex mixtures rather than isolated compounds. The formation of PAHs primarily arises from the incomplete combustion and pyrolysis of fossil fuels, particularly coal, which serves as a major contributor. Additionally, PAHs are released through the burning of wood and other fuels, as well as from petroleum products (Ruwei et al. 2017). Secondary sources include petroleum spills, oil seepage, and the diagenesis of organic matter in anoxic sediments.

Table 3 (Ravindra et al. 2008) presents comprehensive PAH emission rates from coal combustion, alongside rates from other fuel sources for comparison. For the purpose of facilitating comparison, emission rates from a selection of other fuel combustion sources have also been included in the table. Subsequently, an in-depth discussion of these rates, categorized according to the principal sources—domestic, mobile, industrial, agricultural, and natural—will be undertaken, as outlined below.

To ascertain the relative contributions of diverse emission sources and formulate efficacious control strategies, precise inventories are frequently compiled, taking into account the intensities of emission activities and EFs. Typically, EFs exhibit variability contingent upon factors such as fuel type, fuel origin, stove configuration, combustion parameters, dilution techniques, among others. In the study conducted by Yang et al. (2016), the emission factors of PAHs from an industrial coal-fired boiler were investigated.

In the study (Yang et al. 2016), gases were extracted and two-stage diluted from the flue using a dilution system (FPS 4000, Dekati Ltd., Finland). The flue gas temperature was then cooled to room temperature (about 20 °C), and the dilution ratio was controlled between 14 and 20. PM_{10} samples were collected by a double-channel cyclone PM sampler on quartz fiber filters (47 mm diameter, Advantec, Japan) during normal operation conditions of the boilers. Each sample had a sampling time duration of about 150 min, and a total of four samples were collected. To evaluate particulate matter (PM) removal rate, PM_{10} samples were collected before and after precipitation. Additionally, PAHs in the samples collected after precipitation were analyzed.

Table 3 Estimated PAHs emission rate from various sources

PAHs sources	Emission rate	Remarks
Domestic		
Wood/coal (Kim Oahnh et al. 1999) briquettes/charcoal	25–100 mg kg^{-1}	18 PAHs highest from wood
Coal (Oros and Simoneit 2000)	0.95 mg kg^{-1}	Average of 4 coal ranks
Anthracite coal (Lee et al. 2005)	2 mg kg^{-1}	12 different cook stoves were used
Natural gas home appliances (Rogge et al. 1993b)	1–2000 pg kg^{-1}	pg/kg of natural gas burned
Pine (Rogge et al. 1998)	6.9 mg kg^{-1}	Wood combustion in residential fireplaces
Wood/root-fuel (Gupta et al. 1998)	5.3–13.2mg kg^{-1}	B[*a*]P
Pine (Schauer et al. 2001)	13.7 mg kg^{-1}	Wood combustion in residential fireplaces
Domestic burning (Lee et al. 2005)	~ 100 mg kg^{-1}	Controlled burning experiments for space heating
Mobile		
Gasoline engine (Rogge et al. 1993a)	350 or 26 μg km^{-1}	Catalytic converter equipped light vehicles (average of 7)
Aircrafts (EPAQS 1998)	1.24 mg	Per landing-take off cycle for B[*a*]P
Helicopter (Chen et al. 2006)	63.4 mg l^{-1} Fuel	22 PAHs
Ships (Westerholm and Li 1994)	500 mg kg^{-1}	Marine diesel engine, sum of 25 PAHs
Industrial		
Industrial stacks (Yang et al. 1998)	77–3970 μg kg^{-1} 2–16 μg kg^{-1}	9 different stacks For B[a]P
Industrial boiler (Li et al. 1999)	13,300 μg kg^{-1}	Lower molecular weight PAHs
Heavy oil + natural gas	2900 μg kg^{-1}	Dominate in the stack flue gas
Coke oven gas + blast furnace gas boiler (Pisupati et al. 2000)		
Agricultural		
Open burning (Kakareka and Kukharchyk 2003)	240–571 mg kg^{-1}	Simulation of agricultural debris

The EF for PAHs based on fuel mass is expressed in terms of milligrams (mg) of PAHs emitted per kilogram (kg) of fuel combusted. The EF for species *X*, on a fuel-mass basis, can be determined using the subsequent equation:

$$EF_X = \frac{[X] \times D \times Q \times T}{M} \tag{1}$$

Table 4 EFs of PAHs in coal-fired boiler bituminous (units: mg/kg)

Name	Compound	EFs
Fluoranthene	Flu	0.0044
Pyrene	Pyr	0.0014
Benzo[*a*]anthracene	BaA	0.0006
Chrysene	Chr	0.0011
Benzo[*b*]fluoranthene	BbF	0.0022
Benzo[*k*]fluoranthene	BkF	0.0007
Benzo[*a*]pyrene	BaP	0.0006
Indeno[*1,2,3-cd*] pyrene	IDP	0.0003
Dibenz[*a,h*]anthracene	DBA	0.0003
Benzo[*ghi*]perylene	BghiPe	0.0013
Sum(10-PAHs)		0.0128

where $[X]$ (mg/m^3) is the concentration of species X on the sample filter, Q (m^3/min) is the flow rate of the flue gas in stack, T (min) is the sampling time, D is the dilution factor, and M (kg) is the weight of fuel burned during the sampling time.

The EFs for the ten identified PAHs present within the particulate samples were meticulously calculated on a fuel-weight basis, and the resultant values are displayed below (Yang et al. 2016) (Table 4).

The calculated average EFs for the total concentration of PAHs emanating from coal-fired boilers were 0.0125 mg/kg. In parallel, as per the regulatory standards established by Beijing, the prescribed maximum concentration threshold for BaP is 0.3 mg/m^3 (DB11/501–2007). Nonetheless, the emissions from the tested bituminous coal were found to contain a BaP concentration of 0.56 mg/m^3, thereby underscoring the substantial role of coal-fired boiler combustion in exacerbating atmospheric pollution. In light of these findings, a research proposal is advanced, advocating for the implementation of more advanced precipitation systems as an effective measure to mitigate PAHs emissions.

In a separate investigation (Yang et al. 2017), an examination was conducted to ascertain the emission factors of PAHs that are emitted during the combustion process of household coal stoves prevalent in rural regions of China. The scope of this study encompassed a diverse array of four distinct raw coal types, specifically comprising three bituminous coals and one anthracite coal, in addition to a single type of honeycomb briquette. In this study, four distinct coal samples were employed, sourced from Ningxia Province, Inner Mongolia Autonomous Region, Shanxi Province, and Guizhou Province in China. The samples were specifically designated as follows: YC (Yinchuan) for the Ningxia Province sample, DS (Dongsheng) for the Inner Mongolia sample, DT (Datong) for the Shanxi Province sample, and ZJ (Zhijin) for the Guizhou Province sample. Furthermore, an additional coal sample was processed into a honeycomb briquette form for comparative analysis in the experimental setting, and this sample was denoted as DT-HB (Datong Honeycomb Briquette). Through the application of rigorous experimental methodologies, the emission factors for different PAHs species were meticulously determined when these fuels were ignited within typical rural Chinese cooking stoves.

The EFs pertaining to the PAHs present in the comprehensive samples, which encompass both the gas and particulate phases and are subsequently referred to as "total EFs," as well as those found in the gas phase samples and particulate phase samples, have been meticulously documented in Table 5 (Yang et al. 2017). These data were derived from the testing of four distinct coal samples and one honeycomb briquette sample, all of which were evaluated on a fuel-weight basis.

[a]emission factor for the gas phase; [b]emission factor for the particulate phase.

The PAHs emission factors obtained for the five materials, which encompass four distinct types of raw coal and one honeycomb briquette, reveal a notable trend: the PAHs emission factors associated with the three bituminous coal samples are considerably higher than those observed for the anthracite coal sample. Within the context of the four raw coal samples, the DS sample stands out as having the highest PAHs emission factor. Conversely, the honeycomb briquette sample, designated as DT-HB, exhibits the lowest PAHs emission factor among the materials tested. These findings underscore the existence of substantial disparities in PAHs emission factors across different fuel types, a phenomenon that may be closely linked to the variations in the volatile component content inherent to each fuel.

4 Toxicity and EFs of Nitrogen-Containing Polycyclic Aromatic Hydrocarbons (NPAHs)

4.1 Toxicity of NPAHs

NPAHs represent a class of organic compounds characterized by the incorporation of nitrogen atoms within their multiple cyclic structures. The distinctive feature of their molecular architecture lies in the embedding of one or more nitrogen atoms within the aromatic ring system, which confers upon NPAHs unique chemical properties and biological activities. Some of the toxic NPAHs species are listed in Table 6 (Yang et al. 2017):

In terms of chemical properties, the stability of NPAHs is influenced by their molecular structure and environmental conditions (Wang et al. 2017). The incorporation of nitrogen atoms imparts increased molecular polarity to NPAHs compared to conventional PAHs, affecting their solubility and mobility in the environment. Furthermore, the stability of NPAHs is susceptible to various factors, including light exposure, temperature, pH, and interactions with other chemicals (Nowakowski et al. 2022). For instance, under certain conditions, NPAHs may undergo photochemical reactions, leading to structural transformations or decomposition.

The toxic effects of NPAHs cannot be overlooked (Wang et al. 2017). They exert significant toxicity on biological organisms, primarily attributed to the nitrogen atoms and aromatic ring systems in their molecular structures. NPAHs can enter

Table 5 EFs for the PAHs in the particulate phase and gas phase for the different fuels (mg/kg)

Anthracite coal				Bituminous coal											
	ZJ			YC			DS			DT			DT-HB		
	EF_g [a]	EF_p [b]	Total	EF_g	EF_p	Total	EF_g	EF_p	Total	EF_g	EF_p	Total	EF_g	EF_p	Total
Ace	0.092	N.D	0.092	0.050	N.D	0.050	0.12	N.D	0.12	0.046	N.D	0.046	0.028	N.D	0.028
Fle	0.14	N.D	0.14	0.079	N.D	0.079	0.20	N.D	0.20	0.23	N.D	0.23	0.042	N.D	0.042
Phe	0.052	N.D	0.052	0.048	N.D	0.048	1.54	N.D	1.54	0.022	N.D	0.022	0.069	N.D	0.069
Flu	0.42	2.8	3.22	0.33	2.9	3.23	3.4	10	13.4	0.49	2.5	2.99	0.13	0.87	1.00
Pyr	0.36	1.7	2.06	0.30	2.5	2.80	3.3	7.7	11.0	0.13	2.1	2.23	0.11	0.66	0.77
BaA	N.D	0.53	0.53	N.D	0.61	0.61	N.D	2.8	2.8	N.D	0.48	0.48	N.D	0.30	0.30
Chr	N.D	0.52	0.52	N.D	0.62	0.62	N.D	2.3	2.3	N.D	0.50	0.50	N.D	0.26	0.26
BbF	N.D	0.37	0.37	N.D	0.40	0.40	N.D	4.6	4.6	N.D	0.44	0.44	N.D	0.19	0.19
BkF	N.D	0.19	0.19	N.D	0.19	0.19	N.D	2.1	2.1	N.D	0.77	0.77	N.D	0.11	0.11
BaP	N.D	0.40	0.40	N.D	0.36	0.36	N.D	4.0	4.0	N.D	0.78	0.78	N.D	0.23	0.23
DBA	N.D	N.D	N.D	N.D	N.D	N.D	N.D	N.D	N.D	N.D	N.D	N.D	N.D	N.D	N.D
BghiPe	N.D	0.13	0.13	N.D	0.16	0.16	N.D	4.0	4.0	N.D	0.60	0.60	N.D	0.039	0.039
IDP	N.D	0.24	0.24	N.D	0.21	0.21	N.D	2.2	2.2	N.D	0.18	0.18	N.D	0.11	0.11
$\sum PAH_{13}$	1.064	6.88	7.944	0.807	7.95	8.757	8.56	39.7	48.26	0.918	8.35	9.268	0.379	2.769	3.148

Table 6 Toxic NPAHs

No.	Name	Abbreviation
1	1,3-dinitropyrenes	1,3-DNPs
2	1,6-dinitropyrenes	1,6-DNPs
3	1,8-dinitropyrenes	1,8-DNPs
4	1-nitropyrene	1-NP
5	1-nitrofluoranthene	1-NFR
6	3-nitrofluoranthene	3-NFR
7	4-nitrophenanthrene	4-NPh
8	9-nitrophenanthrene	9-NPh
9	2-nitropyrene	2-NP
10	4-nitropyrene	4-NP
11	2-nitrofluoranthene	2-NFR
12	6-nitrochrysene	6-NC
13	3-nitrobenz[*a*]anthracene	3-NBaA
14	5-nitroacenaphthene	5-NAc
15	2-nitrofluorene	2-NF
16	7-nitrobenz[*a*]anthracene	7-NBaA
17	10-nitrobenz[*a*]anthracene	10-NBaA
18	1-nitroperylene	1-NPer
19	3-nitroperylene	3-NPer
20	6-nitrobenzo[*a*]pyrene	6-NBaP
21	2-nitroanthracene	2-NA
22	9-nitroanthracene	9-NA
23	2-nitrotriphenylene	2-NTP

organisms through various routes, including inhalation, ingestion, and skin contact, subsequently posing health hazards. Studies have demonstrated that NPAHs possess carcinogenic, mutagenic, and teratogenic properties, which are primarily mediated through their interactions with biomacromolecules within organisms. For example, NPAHs can form adducts with DNA, interfering with normal DNA replication and transcription processes, thereby inducing gene mutations and cellular oncogenesis.

4.2 EFs of NPAHs

NPAHs exhibit distinct emission characteristics that are influenced by a multitude of factors (Yang et al. 2017). Their formation is particularly sensitive to temperature, as nitration reactions are more favorable at higher temperatures. Furthermore, the emission levels of NPAHs vary significantly across different fuel types and combustion conditions; for instance, coal-fired stoves generally emit lower quantities of NPAHs compared to diesel vehicles. Additionally, some NPAHs are not only present in the emissions from their sources but can also be generated from parent PAHs through reactions with atmospheric OH or NO_3 radicals. The generation pathways of NPAHs are more complex than those of PAHs, involving a greater number of chemical reactions and influencing factors such as temperature and NO_x concentration. During the combustion of certain fuels, specific NPAHs like 2-NP and 2-NFR are produced not directly in the emissions but rather form in the atmosphere through reactions with gaseous OH radicals followed by subsequent NO_2 addition.

In the course of prior experimental investigations focused on determining the emission factors of PAHs emanating from the combustion processes of household coal stoves prevalent in rural China, an exploration of NPAHs was also undertaken. Utilizing rigorous experimental methodologies, the emission factors for a total of 23 distinct NPAHs species, produced through the combustion of these fuels in typical rural Chinese cooking stoves, were meticulously ascertained. The EFs for the NPAHs are listed in Table 7 (Yang et al. 2017).

A detailed examination of the NPAHs emission factors (Yang et al. 2017) for the four distinct types of raw coal and one honeycomb briquette sample reveals marked disparities in both the types and quantities of NPAHs generated during the combustion of various fuel types. While the scope of NPAHs species detected within the experimental framework was somewhat constrained, the results nevertheless disclosed that a mere four NPAHs were identified in the honeycomb briquette sample, contrasting sharply with the significantly higher number of NPAHs detected in the other fuel samples. Additionally, the presence of secondary NPAHs in the emissions of certain fuels was observed, hinting at the possibility of chemical reactions occurring within the dilution tunnel situated between the stove and the sampler. Collectively, these findings offer invaluable insights that may inform and guide future research endeavors aimed at enhancing combustion technologies and mitigating the emissions of harmful substances.

5 Emissions of PAHs

5.1 Estimation of PAHs Emissions from Coal Combustion in China

As the world's largest coal consumer, China's coal-related PAHs emissions have always attracted considerable attention. In order to estimate China's annual average coal-fired PAHs emissions, a combination of coal combustion and PAHs emission factors is usually used (Wang et al. 2024a).

Based on the existing research data, the average annual emissions of PAHs from coal combustion in China are quite considerable, which can be estimated by measuring the PAHs emission factors generated from the combustion of different

Table 7 EFs for the NPAHs in the particulate phase for the different fuels (μg/kg)

	Anthracite coal	Bituminous coal			
	ZJ	YC	DS	DT	DT-HB
1,6-DNP	N.D	N.D	N.D	N.D	N.D
1,8-DNP	0.052	0.039	0.27	0.028	0.021
1,3-DNP	0.015	0.0046	0.054	0.0085	0.011
10-NBA	N.D	N.D	N.D	N.D	N.D
3-NBA	N.D	N.D	N.D	N.D	N.D
5-NAc	N.D	N.D	N.D	N.D	N.D
2-NF	N.D	N.D	N.D	N.D	N.D
9-Nph	N.D	N.D	N.D	N.D	N.D
2-NA	0.18	0.12	0.52	0.65	0.23
9-NA	N.D	N.D	N.D	N.D	N.D
4-Nph	N.D	N.D	N.D	N.D	N.D
2-NP + 2-NFR	N.D	2.9	N.D	2.1	N.D
4-NP	N.D	0.25	N.D	N.D	N.D
3-NFR	N.D	N.D	N.D	N.D	N.D
1-NP	0.17	0.12	N.D	0.35	N.D
1-NFR	2.13	0.55	N.D	0.94	N.D
2-NTP	N.D	N.D	N.D	N.D	N.D
6-NC	N.D	N.D	N.D	N.D	N.D
7-NBaA	N.D	N.D	N.D	N.D	N.D
3-NPer	0.054	0.027	0.16	0.067	0.056
6-NBaP	0.47	0.32	4.1	N.D	N.D
1-NPer	N.D	0.032	0.22	N.D	N.D
$\sum NPAH_{23}$	3.071	4.3626	5.324	4.1435	0.318

coal types in China and combining them with the data on the amount of coal burned in China. Research shows that although PAHs emission factors vary by types of coal, China's overall PAHs emissions from coal combustion remain high (Li et al. 2021).

The distribution of PAHs emissions in China shows distinct regional patterns. Due to the uneven distribution of coal resources and differences in industrial layout, there are significant differences in the PAHs emissions from coal combustion in different regions of China (Liang et al. 2022; Du et al. 2020). In general, areas with abundant coal resources and intensive industries have higher emissions of coal-burning PAHs, while areas with scarce coal resources and relatively backward industries have lower emissions.

In addition, seasonal factors (Shin et al. 2022) and meteorological conditions (Hao et al. 2024) can significantly affect PAH emissions. For example, in winter, due to the increase in heating demand, the amount of coal burned increases accordingly, resulting in an increase in PAHs emissions; while in summer, due to the decrease in coal burning, PAHs emissions decrease accordingly. At the same time, meteorological conditions such as wind speed and humidity will also affect the transmission and diffusion of PAHs in the atmosphere, thereby affecting its emission distribution.

In recent years, the Chinese government has adopted a series of policy interventions to reduce emissions of coal-fired PAHs (Liu et al. 2018). These measures mainly include flue gas purification measures in industrial coal-fired facilities and "coal-to-electricity" and "coal-to-gas" residential heating (Chen and Chen 2019). Flue gas purification measures for industrial coal-fired facilities have effectively reduced the emission of pollutants (Sun et al. 2021b) such as PAHs generated during coal combustion by installing dust collectors, desulfurization, denitrification devices, and other equipment. The implementation of these measures has not only improved the quality of the atmospheric environment, but also promoted the sustainable development of industrial enterprises (Omer 2008). The promotion of policies such as "coal-to-electricity" and "coal-to-gas" for residential heating has further reduced the emission of PAHs from coal combustion. By replacing traditional coal-fired heating with clean energy, not only has the emission of pollutants been reduced, but the quality of life and health of residents has also been improved. The implementation of these policies is of great significance to promoting China's energy structure adjustment and environmental protection work. As can be seen from Fig. 1, the PAHs emissions have decreased from 112.92 Gg in 2013 to 100.09 Gg in 2017; it has decreased by 11.36% from 2013 to 2017 (Wang et al. 2021), which is enough to prove the effectiveness and determination of the Chinese government in combating PAHs.

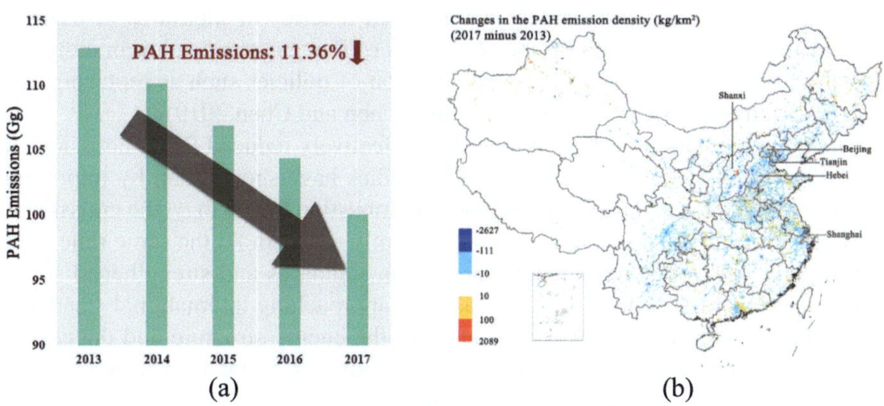

Fig. 1 Changes in PAH concentration in China from 2013 to 2017 (Wang et al. 2021)

5.2 Comparison with East Asian Peers

Why China's PAHs Emissions Are Higher than Japan's and South Korea's

In comparison with Japan and South Korea, China's emissions of PAHs are significantly greater, a circumstance largely attributable to the fact that China's consumption of coal vastly surpasses that of these two nations (Zhao et al. 2019). Coal stands as the primary source of energy consumption within China, and the process of its combustion results in the generation of substantial quantities of pollutants, including PAHs.

According to the IEA, although global coal demand will decline in the next few years, China and India will still account for more than 70% of global coal consumption. This shows that China's scale of coal consumption far exceeds that of other East Asian countries, resulting in a corresponding increase in PAHs emissions. In addition, factors such as China's industrial structure and energy efficiency have also exacerbated PAHs emissions to a certain extent.

China's PAHs Emissions Reduction Is Close to International Standards

Despite China's historically high emissions of PAHs, significant progress has been made in reducing such emissions in recent years through the stringent implementation of policies. The government has adopted a series of measures to mitigate coal-related pollution, including the promotion of flue gas purification in industrial coal-fired facilities and the widespread adoption of policies such as replacing coal with electricity or gas for residential heating (Chen and Chen 2019).

The implementation of these policies has effectively reduced PAHs emissions in China. For example, Zhang et al. (2021)'s studies have shown that, by improving coal-fired efficiency and adopting advanced combustion technology, the emission of PAHs during coal combustion can be significantly reduced. At the same time, with increasingly stringent environmental protection standards and strengthened supervision, China's PAHs emissions are gradually approaching international standards. In the future, with the further optimization of the energy structure and the continuous strengthening of environmental protection policies, China is expected to make greater progress in reducing PAHs emissions.

6 China's Efforts to Reduce Coal-Burning Pollution and Its Results

6.1 Strengthening Industrial Flue Gas Purification

In the field of industrial coal burning, the Chinese government and enterprises have taken a series of effective measures to reduce coal-burning pollution, among which strengthening the application of flue gas desulfurization, denitrification, and dust removal technology is one of the key measures (Li et al. 2022). The widespread application of these technologies has significantly reduced the emission of harmful substances such as PAHs, playing a positive role in improving air quality and protecting the ecological environment.

Dust removal technology is to remove particulate matter from coal-fired flue gas through physical methods (Wang et al. 2014). Particulate matter is an important component of air pollution, among which particulate matter such as $PM_{2.5}$ and PM_{10} are particularly harmful to human health. China's industrial coal-fired facilities generally use high-efficiency dust removal technologies such as electrostatic precipitators and bag filters, which effectively reduce the emission concentration of particulate matter and also reduce the emission of harmful substances such as PAHs attached to particulate matter.

The widespread application of these technologies has not only significantly reduced the concentration of pollutants emitted during industrial coal combustion, but also promoted technological progress and upgrading of related industries (Beer 2000). According to relevant research, by strengthening the application of flue gas desulfurization, denitrification, and dust removal technologies, the concentration of pollutants emitted by China's industrial coal-fired facilities has been greatly reduced, gradually approaching or reaching international advanced levels.

To summarize, China has achieved remarkable results in reducing coal pollution, of which strengthening industrial flue gas purification is one of the key initiatives. In the future, with the continuous progress of technology and the continuous promotion of policies, China is expected to make more progress in reducing coal-fired pollution and to make greater contributions to the improvement of air quality and the protection of the ecological environment.

6.2 Policy Intervention: Implementation of "Coal-to-Electricity" and "Coal-to-Gas" Initiatives and Their Environmental Benefits

Since the early twenty-first century, China has progressively integrated atmospheric pollution control into its national development strategy. With the acceleration of industrialization and urbanization, the pollution caused by PAHs from household coal combustion has become increasingly prominent. To address this challenge, the Chinese government systematically advanced the "coal-to-electricity" and "coal-to-gas" initiatives (collectively termed the "Dual Replacement" policy) starting in 2013, significantly reducing emissions of PAHs and other pollutants through the substitution of clean energy for traditional household coal combustion. This section systematically examines the phased characteristics of policy implementation and their environmental benefits based on a policy timeline (Fig. 2) and empirical data.

In 2013, the Chinese government introduced the Air Pollution Prevention and Control Action Plan (commonly known as the "Ten Measures"), marking a transition to a more scientific and systematic phase of air pollution governance. As the first national-level policy framework, the "Ten Measures" prioritized the management of household coal combustion, mandating pilot programs for "coal-to-electricity" and "coal-to-gas" in key regions such as the Beijing-Tianjin-Hebei (BTH) area. The policy required the elimination of decentralized coal-fired boilers and promoted electric heating and natural gas. During the initial phase, technical validation and regional pilot projects were emphasized. Data indicated that EFs in pilot areas decreased by 30–40% compared to traditional coal combustion (Yang et al. 2016), laying a technical foundation for subsequent large-scale implementation.

In 2017, the launch of the "Blue Sky Defense Campaign" accelerated the nationwide implementation of the "Dual Replacement" policy. Following the completion of the "Ten Measures" phased targets, the "2 + 26" city cluster in northern China (Beijing, Tianjin, and 26 surrounding cities) became the core implementation zone (Zhang et al. 2018b). By the end of 2017, the BTH region had completed clean energy retrofits for coal-fired boilers totaling 15,000 steam tons in capacity, achieved coal-free status in Beijing's plain areas, it also reduced the number of coal-fired boilers nationwide from 520,000 to 100,000. Studies revealed that PAHs emission factors from rural household coal combustion reached 48.26 mg/kg (Yang et al. 2017), while the "Dual Replacement" policy, covering nearly 4 million households, led to a significant reduction in winter PAHs emissions. GAINS model simulations demonstrated that $PM_{2.5}$ concentrations in Beijing decreased by over 35% compared to 2013 levels, with BaP concentrations declining by 40% (Wang et al. 2021).

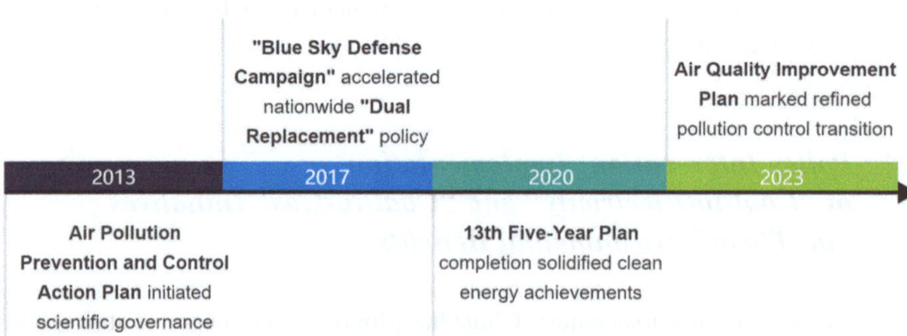

Fig. 2 Timeline of China's atmospheric governance policies

By 2020, the conclusion of the 13th Five-Year Plan solidified achievements in clean energy substitution. The national coal share in primary energy consumption dropped to 56.2%, while renewable energy installed capacity surpassed coal-fired power for the first time. The "Dual Replacement" policy expanded to emerging key regions such as the Fenwei Plain. Rural clean heating adoption rates rose to 70%, with electricity and natural gas demonstrating significant substitution effects over household coal. Environmental monitoring data indicated that winter PAHs emissions in Beijing decreased by 35%–45% compared to 2013, with BaP concentrations falling below 0.3 ng/m^3 (Xue et al. 2016), complying with the Beijing Municipal Integrated Emission Standard for Air Pollutants (DB11/501-007). Concurrently, the adoption rate of new energy vehicles reached 25.6% of new vehicle sales, reducing mobile-source PAHs emissions and highlighting policy synergies.

In 2023, the introduction of the Air Quality Continuous Improvement Action Plan signified a transition to refined pollution governance. The plan established coordinated control targets for $PM_{2.5}$ and ozone, requiring a 10% reduction in nitrogen oxides (NO_x) and volatile organic compounds (VOCs) emissions by 2025, alongside enhanced post-implementation oversight of "coal-to-gas" initiatives. Policy focus shifted to industrial facility upgrades and rural energy structure optimization. Ultra-low emission retrofits for coal-fired power units achieved 90% coverage, reducing PAHs emission factors to 0.0125 mg/kg (Table 4). The promotion of DT-HB in rural areas decreased PAHs emission factors by 80% compared to household coal, with technological advancements and regulatory enforcement jointly driving emission reduction efficiency.

Comprehensive environmental benefit analyses confirm the significant contributions of the "Dual Replacement" policy to PAHs reduction and air quality improvement. From 2013 to 2017, China's total PAHs emissions decreased from 112.92 to 100.09 Gg, an 11.36% reduction, with emissions of highly carcinogenic compounds such as BbF declining by 30%–45% (Wang et al. 2021). The annual average $PM_{2.5}$ concentration in the BTH region dropped from 89.5 μg/m^3 in 2013 to below 30 μg/m^3 by 2022, accompanied by a 90% reduction in heavy pollution days (Wang et al. 2023). Health risk assessments indicate that the policy implementation prevented over 100,000 annual cases of respiratory diseases linked to PAHs exposure (Zhang et al. 2022), underscoring its public health value.

Looking forward, China must further optimize its clean energy supply system, strengthen policy enforcement in rural areas, and promote synergistic control of PAHs and greenhouse gases. Through technological innovation and policy refinement, China is poised to achieve dual objectives of energy transition and air pollution governance, offering scientific paradigms and practical experience for global PAHs mitigation.

6.3 China's Achievements

In recent years, China has made remarkable progress in reducing coal-related pollution (Chai et al. 2019), including significant efforts in controlling emissions of PAHs, a class of hazardous organic pollutants produced during coal combustion. PAHs, known for their toxicity, persistence, and carcinogenic properties, have been a major focus in China's environmental protection initiatives. This shift not only marks China's substantial progress in environmental protection but also demonstrates its determination and effectiveness as an important participant in global environmental protection.

To tackle PAHs emissions, China has prioritized technological advancements in flue gas purification for industrial coal-burning facilities. The widespread adoption of high-efficiency dust removal, desulfurization, and denitrification technologies has significantly reduced emissions of particulate matter, which often serves as a carrier for PAHs. For instance, coal-fired industrial boilers equipped with advanced emission control systems have shown substantial reductions in PAHs concentrations, with studies reporting up to a 50–60% decline in BaP, one of the most toxic PAHs.

China has long been the world's largest coal consumer, and the pollutant emissions from its coal combustion have posed a severe challenge to the global environment. However, as the government's awareness of environmental protection deepens and policies are constantly adjusted, China has begun to actively seek and practice feasible paths to reduce and replace coal consumption (Zhao and Alexandroff 2019). China's policies of "coal-to-electricity" and "coal-to-gas" have played a vital role in curbing PAHs emissions from residential coal combustion, which historically contributed to elevated PAHs levels, especially in northern China during winter. The shift to cleaner heating systems has reduced residential emissions of PAHs such as Flu, Pyr, and Chr. Studies suggest that these measures have contributed to a 30–45% reduction in total PAHs emissions in key areas like Beijing-Tianjin-Hebei.

As coal consumption decreases, China's air pollutant emissions also show a clear downward trend (Song et al. 2017). The emissions of major pollutants such as SO_2, nitrogen oxides, and particulate matter have all decreased, and air quality has improved significantly. In particular, in key pollution areas such as the Beijing-Tianjin-Hebei region, the Yangtze River Delta, and the Pearl River Delta, the environmental improvement effect is particularly prominent due to more centralized policy implementation. The $PM_{2.5}$ concentration in these areas has dropped significantly, the number of days with heavy pollution has been greatly reduced, and the quality of life of residents has been significantly improved.

The continuous improvement of China's air quality and environmental conditions is not only due to the reduction in coal consumption, but also inseparable from the government's continued efforts in pollution control. China has increased its investment in environmental protection infrastructure (Zhang and Wen 2008), strengthened environmental supervision, and promoted the innovation and application of environmental protection technologies. These measures have brought about significant changes in China's environmental conditions and shown a positive momentum of improvement.

China's achievements in reducing PAHs from coal combustion serve as a valuable model for other coal-dependent regions globally. The combination of policy enforcement, technological innovation, and energy structure optimization demonstrates a path toward minimizing the health and environmental impacts of PAHs.

In conclusion, through targeted measures and policies, China has made significant strides in reducing PAHs emissions from both industrial and residential sources. These achievements reflect the country's commitment to addressing environmental challenges and improving air quality while contributing to global efforts to reduce persistent organic pollutants.

References

Adelekan, O. A., Ilugbusi, B. S., Adisa, O., Obi, O. C., Awonuga, K. F., Asuzu, O. F., & Ndubuisi, N. L. (2024). Energy transition policies: a global review of shifts towards renewable sources. *Engineering Science & Technology Journal*, *5*(2), 272–287.

Apsimon, H., Amann, M., Åstrom, S., Oxley., T., (2009). Synergies in addressing air quality and climate change. *Climate Policy*, (6), 669–680.

Barbosa Jr, F., Rocha, B. A., Souza, M. C., Bocato, M. Z., Azevedo, L. F., Adeyemi, J. A., ... & Campiglia, A. D. (2023). Polycyclic aromatic hydrocarbons (PAHs): updated aspects of their determination, kinetics in the human body, and toxicity. *Journal of Toxicology and Environmental Health, Part B*, *26*(1), 28–65.

Beer, J. M. (2000). Combustion technology developments in power generation in response to environmental challenges. *Progress in Energy and Combustion Science*, *26*(4–6), 301–327.

Bloch, H., Rafiq, S., & Salim, R. (2012). Coal consumption, CO2 emission and economic growth in China: Empirical evidence and policy responses. *Energy Economics*, *34*(2), 518–528.

Chai, J., Du, M., Liang, T., Sun, X. C., Yu, J., & Zhang, Z. G. (2019). Coal consumption in China: how to bend down the curve?. *Energy Economics*, *80*, 38–47.

Chen, H., & Chen, W. (2019). Potential impact of shifting coal to gas and electricity for building sectors in 28 major northern cities of China. *Applied Energy*, *236*, 1049–1061.

Chen, Y. C., Lee, W. J., Uang, S. N., Lee, S. H., & Tsai, P. J. (2006). Characteristics of polycyclic aromatic hydrocarbon (PAH) emissions from a UH-1H helicopter engine and its impact on the ambient environment. *Atmospheric Environment*, *40*(39), 7589–7597.

Cho, S., Jeong, Y. S., & Huh, J. H. (2024). Is South Korea's 2050 Carbon-Neutral scenario sufficient for meeting greenhouse gas emissions reduction goal? *Energy for Sustainable Development*, *80*, 101447.

CLIMA, T., & TE, W. (2016). Coupled Chemistry-Meteorology/Climate Modelling (CCMM): Status and relevance for numerical weather prediction, atmospheric pollution and climate research (Symposium materials). *WMO GAW Report, Geneva, Switzerland. [Available online at* www.wmo.int/pages/prog/arep/gaw/documents/Final_GAW_226_10_May.pdf.*]*.

Du, W., Yun, X., Chen, Y., Zhong, Q., Wang, W., Wang, L., ... & Tao, S. (2020). PAHs emissions from residential biomass burning in real-world cooking stoves in rural China. *Environmental Pollution*, *267*, 115592.

EPAQS (Expert Panel on Air Quality Standards), 1998. Polycyclic Aromatic Hydrocarbons. Final Draft Report for the Department of the Environment, Transport and the Regions, October 1998.

Eweade, B. S., Uzuner, G., Akadiri, A. C., & Lasisi, T. T. (2024). Japan energy mix and economic growth nexus: Focus on natural gas consumption. *Energy & Environment*, *35*(2), 692–724.

Fakinle, B. S., Odekanle, E. L., Ike-Ojukwu, C., Sonibare, O. O., Falowo, O. A., Olubiyo, F. W., … & Aremu, C. O. (2022). Quantification and health impact assessment of polycyclic aromatic hydrocarbons (PAHs) emissions from crop residue combustion. *Heliyon, 8*(3).

Gbeddy, G., Goonetilleke, A., Ayoko, G. A., & Egodawatta, P. (2020). Transformation and degradation of polycyclic aromatic hydrocarbons (PAHs) in urban road surfaces: Influential factors, implications and recommendations. *Environmental Pollution, 257*, 113510.

Gupta, S., Saksena, S., Shankar, V. R., & Joshi, V. (1998). Emission factors and thermal efficiencies of cooking biofuels from five countries. *Biomass and Bioenergy, 14*(5–6), 547–559.

Han, Y., Chen, Y., Feng, Y., Song, W., Cao, F., Zhang, Y., … & Chen, J. (2020). Different formation mechanisms of PAH during wood and coal combustion under different temperatures. *Atmospheric Environment, 222*, 117084.

Hao, W., Liang, B., Chen, J., Chen, Y., Wang, Z., Zhao, X., … & Yang, F. (2024). Secondary formation of oxygenated and nitrated polycyclic aromatic compounds under stagnant weather conditions: Drivers and seasonal variation. *Science of the Total Environment, 929*, 172487.

Hong, J. R., & Choi, K. M. (2015). Analytical characteristics of GC/MS and HPLC according to the concentration distribution of PAHs. *Journal of Korean Society of Occupational and Environmental Hygiene, 25*(3), 312-321.

Hsu, W. T., Liu, M. C., Hung, P. C., Chang, S. H., & Chang, M. B. (2016). PAH emissions from coal combustion and waste incineration. *Journal of Hazardous Materials, 318*, 32–40.

Ifegwu, O. C., & Anyakora, C. (2015). Polycyclic aromatic hydrocarbons: part I. Exposure. *Advances in Clinical Chemistry, 72*, 277–304.

Kakareka, S. V., & Kukharchyk, T. I. (2003). PAH emission from the open burning of agricultural debris. *Science of the total environment, 308*(1–3), 257–261.

Kang, D. (2024). The establishment of evaluation systems and an index for energy superpower. *Applied Energy, 356*, 122344.

Kim Oanh, N. T., Bætz Reutergårdh, L., & Dung, N. T. (1999). Emission of polycyclic aromatic hydrocarbons and particulate matter from domestic combustion of selected fuels. *Environmental Science & Technology, 33*(16), 2703–2709.

Kumar, V., Kothiyal, N. C., Saruchi, Vikas, P., & Sharma, R. (2016). Sources, distribution, and health effect of carcinogenic polycyclic aromatic hydrocarbons (PAHs)–current knowledge and future directions. *Journal of the Chinese Advanced Materials Society, 4*(4), 302–321.

Lee, J. H., Hyun, M. K., & Yoo, S. H. (2024). Economic impacts of the coal extraction sector on the South Korean national economy: An input-output analysis. *The Extractive Industries and Society, 17*, 101436.

Lee, R. G., Coleman, P., Jones, J. L., Jones, K. C., & Lohmann, R. (2005). Emission factors and importance of PCDD/Fs, PCBs, PCNs, PAHs and PM_{10} from the domestic burning of coal and wood in the UK. *Environmental Science & Technology, 39*(6), 1436–1447.

Li, C. T., Mi, H. H., Lee, W. J., You, W. C., & Wang, Y. F. (1999). PAH emission from the industrial boilers. *Journal of Hazardous Materials, 69*(1), 1–11.

Li, X., Han, J., Liu, Y., Dou, Z., & Zhang, T. A. (2022). Summary of research progress on industrial flue gas desulfurization technology. *Separation and Purification Technology, 281*, 119849.

Li, X., Wang, Z., & Guo, T. (2021). Emission of PM2. 5-bound polycyclic aromatic hydrocarbons from biomass and coal combustion in China. *Atmosphere, 12*(9), 1129.

Liang, M., Liang, H., Gao, P., Rao, Z., & Liang, Y. (2022). Characterization and risk assessment of polycyclic aromatic hydrocarbon emissions by coal fire in northern China. *Environmental Geochemistry and Health, 44*(3), 933–942.

Liu, Y., Zhang, Y., Li, C., Bai, Y., Zhang, D., Xue, C., & Liu, G. (2018). Air pollutant emissions and mitigation potential through the adoption of semi-coke coals and improved heating stoves: Field evaluation of a pilot intervention program in rural China. *Environmental Pollution, 240*, 661–669.

Ma, C., & Stern, D. I. (2008). China's changing energy intensity trend: a decomposition analysis. *Energy Economics, 30*(3), 1037–1053.

Mao, K., Yao, Y., Wang, K., Liu, C., Tang, G., Feng, S., … & Li, Z. (2024). Analysis of PM2.5 Pollution Transport Characteristics and Potential Sources in Four Chinese Megacities During 2022: Seasonal Variations. *Atmosphere, 15*(12), 1482.

Mentes, D., Kováts, N., Muránszky, G., Hornyák-Mester, E., & Póliska, C. (2022). Evaluation of flue gas emission factor and toxicity of the PM-bounded PAH from lab-scale waste combustion. *Journal of Environmental Management, 324*, 116371.

Miller, B. G. (2004). *Coal Energy Systems*. Elsevier.

Munawer, M. E. (2018). Human health and environmental impacts of coal combustion and post-combustion wastes. *Journal of Sustainable Mining, 17*(2), 87–96.

National Bureau of Statistics of China. (2025). *Energy consumption total and composition* . Retrieved Month Day, Year, from https://data.stats.gov.cn/tablequery.htm?code=AA0701.

Nowakowski, M., Rykowska, I., Wolski, R., & Andrzejewski, P. (2022). Polycyclic aromatic hydrocarbons (PAHs) and their derivatives (O-PAHs, N-PAHs, OH-PAHs): Determination in suspended particulate matter (SPM)–A review. *Environmental Processes, 9*(1), 2.

Omer, A. M. (2008). Energy, environment and sustainable development. *Renewable and sustainable energy reviews, 12*(9), 2265–2300.

Oros, D. R., & Simoneit, B. R. T. (2000). Identification and emission rates of molecular tracers in coal smoke particulate matter. *Fuel, 79*(5), 515–536.

Pisupati, S. V., Wasco, R. S., & Scaroni, A. W. (2000). An investigation on polycyclic aromatic hydrocarbon emissions from pulverized coal combustion systems. *Journal of Hazardous Materials, 74*(1-2), 91–107.

Qian, X., Wang, D., Wang, J., & Chen, S. (2021). Resource curse, environmental regulation and transformation of coal-mining cities in China. *Resources Policy, 74*, 101447.

Ravindra, K., Sokhi, R., & Van Grieken, R. (2008). Atmospheric polycyclic aromatic hydrocarbons: source attribution, emission factors and regulation. *Atmospheric Environment, 42*(13), 2895–2921.

Rogge, W. F., Hildemann, L. M., Mazurek, M. A., Cass, G. R., & Simoneit, B. R. (1998). Sources of fine organic aerosol. 9. Pine, oak, and synthetic log combustion in residential fireplaces. *Environmental Science & Technology, 32*(1), 13–22.

Rogge, W. F., Hildemann, L. M., Mazurek, M. A., Cass, G. R., & Simoneit, B. R. (1993a). Sources of fine organic aerosol. 2. Noncatalyst and catalyst-equipped automobiles and heavy-duty diesel trucks. *Environmental science & technology, 27*(4), 636–651.

Rogge, W. F., Mazurek, M. A., Hildemann, L. M., Cass, G. R., & Simoneit, B. R. (1993b). Quantification of urban organic aerosols at a molecular level: identification, abundance and seasonal variation. *Atmospheric Environment. Part A. General Topics, 27*(8), 1309–1330.

Ruwei Wang, Ruoyu Sun, Guijian Liu, Balal Yousaf, Dun Wu, Jian Chen & Hong Zhang. (2017). A review of the biogeochemical controls on the occurrence and distribution of polycyclic aromatic compounds (PACs) in coals. *Earth-Science Reviews*, 400–418.

Sabūnas, A. (2024). The Developments of Japan's Climate Policy Since the Paris Agreement: The Progress in a Quest for a Net Zero Target.

Sahoo, B. M., Ravi Kumar, B. V., Banik, B. K., & Borah, P. (2020). Polyaromatic hydrocarbons (PAHs): structures, synthesis and their biological profile. *Current Organic Synthesis, 17*(8), 625–640.

Schauer, J. J., Kleeman, M. J., Cass, G. R., & Simoneit, B. R. (2001). Measurement of emissions from air pollution sources. 3. C1 – C29 organic compounds from fireplace combustion of wood. *Environmental science & technology, 35*(9), 1716–1728.

Shen, L., Gao, T. M., & Cheng, X. (2012). China's coal policy since 1979: A brief overview. *Energy Policy, 40*, 274–281.

Shin, S. M., Lee, J. Y., Shin, H. J., & Kim, Y. P. (2022). Seasonal variation and source apportionment of oxygenated polycyclic aromatic hydrocarbons (Opahs) and polycyclic aromatic hydrocarbons (Pahs) in Pm2. 5 in Seoul, Korea. *Atmospheric Environment, 272*, 118937.

Sinton, J. E., & Fridley, D. G. (2000). What goes up: recent trends in China's energy consumption. *Energy Policy, 28*(10), 671–687.

Song, C., Wu, L., Xie, Y., He, J., Chen, X., Wang, T., Mao., H., (2017). Air pollution in China: Status and spatiotemporal variations. *Environmental Pollution*, 334–347.

Song, M., Wang, J., & Zhao, J. (2018). Coal endowment, resource curse, and high coal-consuming industries location: Analysis based on large-scale data. *Resources, Conservation and Recycling, 129*, 333–344.

Sun, K., Song, Y., He, F., Jing, M., Tang, J., & Liu, R. (2021a). A review of human and animals exposure to polycyclic aromatic hydrocarbons: Health risk and adverse effects, photo-induced toxicity and regulating effect of microplastics. *Science of The Total Environment, 773*, 145403.

Sun, S., Liu, W., Guan, W., Zhu, S., Jia, J., Wu, X., … & He, Y. (2021b). Effects of air pollution control devices on volatile organic compounds reduction in coal-fired power plants. *Science of the Total Environment, 782*, 146828.

Vistnes, H., Sossalla, N. A., Røsvik, A., Gonzalez, S. V., Zhang, J., Meyn, T., & Asimakopoulos, A. G. (2022). The determination of polycyclic aromatic hydrocarbons (PAHs) with HPLC-DAD-FLD and GC-MS techniques in the dissolved and particulate phase of road-tunnel wash water: A case study for cross-array comparisons and applications. *Toxics, 10*(7), 399.

Wang, A., Song, Q., Tu, G., Wang, H., Yue, Y., & Yao, Q. (2014). Influence of flue gas cleaning system on characteristics of PM 2.5 emission from coal-fired power plants. *International Journal of Coal Science & Technology, 1*, 4–12.

Wang, C., Yang, J., Zhu, L., Yan, L., Lu, D., Zhang, Q., … & Li, Z. (2017). Never deem lightly the "less harmful" low-molecular-weight PAH, NPAH, and OPAH—disturbance of the immune response at real environmental levels. *Chemosphere, 168*, 568–577.

Wang, J., Xia, L., & Wang, R. (2024a). Provincial emission inventory of atmospheric polycyclic aromatic hydrocarbons (PAHs) in China. *Atmospheric Environment, 334*, 120729.

Wang, Q. (2023). Research on the Impact of the "Coal to Gas" and "Coal to Electricity" Policies on Urban Air Quality. In *E3S Web of Conferences* (Vol. 438, p. 01004). EDP Sciences.

Wang, Q., & Li, R. (2016). Journey to burning half of global coal: Trajectory and drivers of China's coal use. *Renewable and Sustainable Energy Reviews, 58*, 341–346.

Wang, Q., Guo, J., Li, R., & Jiang, X. T. (2023). Exploring the role of nuclear energy in the energy transition: A comparative perspective of the effects of coal, oil, natural gas, renewable energy, and nuclear power on economic growth and carbon emissions. *Environmental Research, 221*, 115290.

Wang, Q., Zhang, X., Guo, C., Zhou, H., Zhao, Y., & Lin, B. (2024b). Analysis of carbon emission differences between Chinese and Japanese buildings based on qualitative and quantitative comparisons. *Journal of Building Engineering, 89*, 109260.

Wang, T., Li, B., Liao, H., & Li, Y. (2021). Spatiotemporal distribution of atmospheric polycyclic aromatic hydrocarbon emissions during 2013–2017 in mainland China. *Science of the Total Environment, 789*, 148003.

Wang, Y., Deng, X., Zhang, H., Liu, Y., Yue, T., & Liu, G. (2022). Energy endowment, environmental regulation, and energy efficiency: Evidence from China. *Technological Forecasting and Social Change, 177*, 121528.

Westerholm, R., & Li, H. (1994). A multivariate statistical analysis of fuel-related polycyclic aromatic hydrocarbon emissions from heavy-duty diesel vehicles. *Environmental science & technology, 28*(5), 965–972.

Xiong, W., Wang, Y., Mathiesen, B. V., Lund, H., & Zhang, X. (2015). Heat roadmap China: New heat strategy to reduce energy consumption towards 2030. *Energy, 81*, 274–285.

Xu, B., & Xu, R. (2023). Assessing the carbon intensity of the heavy industry in China: Using a nonparametric econometric model. *Environmental Impact Assessment Review, 98*, 106925.

Xue, Y., Zhou, Z., Nie, T., Wang, K., Nie, L., Pan, T., … & Shao, P. (2016). Trends of multiple air pollutants emissions from residential coal combustion in Beijing and its implication on improving air quality for control measures. *Atmospheric Environment, 142*, 303–312.

Yang, H. H., Lee, W. J., Chen, S. J., & Lai, S. O. (1998). PAH emission from various industrial stacks. *Journal of Hazardous materials, 60*(2), 159–174.

Yang, W., Pudasainee, D., Gupta, R., Li, W., Wang, B., & Sun, L. (2021). An overview of inorganic particulate matter emission from coal/biomass/MSW combustion: Sampling and measurement, formation, distribution, inorganic composition and influencing factors. *Fuel Processing Technology, 213*, 106657.

Yang, X., Geng, C., Sun, X., Yang, W., Wang, X., & Chen, J. (2016). Characteristics of particulate-bound polycyclic aromatic hydrocarbons emitted from industrial grade biomass boilers. *Journal of Environmental Sciences, 40*, 28–34.

Yang, X., Liu, S., Xu, Y., Liu, Y., Chen, L., Tang, N., & Hayakawa, K. (2017). Emission factors of polycyclic and nitro-polycyclic aromatic hydrocarbons from residential combustion of coal and crop residue pellets. *Environmental Pollution, 231*, 1265–1273.

Yasuda, Y. (2024). Quantifying the reduction in coal and increase in renewables in OECD (Organization for Economic Co-operation and Development) countries: Proposal for a coal-renewable energy index and map. *Renewable and Sustainable Energy Reviews, 198*, 114424.

Yun, Y. (Ed.). (2024). Recent Advances for Coal Energy in the 21st Century.

Zhang, C., Liu, C., Li, X., Liu, P., Zhang, Y., Du, Q., ... & Mu, Y. (2021). A novel clean combustion technology for solid fuels to efficiently reduce gaseous and particulate emissions. *Journal of Cleaner Production, 320*, 128864.

Zhang, J., Wang, W., Gao, Lei., Deng, Z., Tian, Yu. (2022). Can the Coal-to-Gas/Electricity Policy Improve Air Quality in the Beijing–Tianjin–Hebei Region?—Empirical Analysis Based on the PSM-DID. *Atmosphere*, (6), 879–879.

Zhang, K. M., & Wen, Z. G. (2008). Review and challenges of policies of environmental protection and sustainable development in China. *Journal of Environmental Management, 88*(4), 1249–1261.

Zhang, X., Wang, P., Han, Q., Li, H., Wang, T., & Ding, M. (2018a). Metal–organic framework based in-syringe solid-phase extraction for the on-site sampling of polycyclic aromatic hydrocarbons from environmental water samples. *Journal of separation science, 41*(8), 1856–1863.

Zhang, Y., Liu, C., Li, K., & Zhou, Y. (2018b). Strategy on China's regional coal consumption control: A case study of Shandong province. *Energy Policy, 112*, 316–327.

Zhang, Z. (2010). China in the transition to a low-carbon economy. *Energy Policy, 38*(11), 6638–6653.

Zhao, N., Li, B., Chen, D., Bahargul, T., Wang, R., Zhou, Y., ... & Ju, X. (2020). The effect of coal size on $PM_{2.5}$ and PM-bound polycyclic aromatic hydrocarbon (PAH) emissions from a domestic natural cross-draft stove. *Journal of the Energy Institute, 93*(2), 542–551.

Zhao, S., & Alexandroff, A. (2019). Current and future struggles to eliminate coal. *Energy Policy, 129*, 511–520.

Zhao, S., Pudasainee, D., Duan, Y., Gupta, R., Liu, M., & Lu, J. (2019). A review on mercury in coal combustion process: Content and occurrence forms in coal, transformation, sampling methods, emission and control technologies. *Progress in Energy and Combustion Science, 73*, 26–64.

Zhu, K., Ali, A., Zhang, T., & Zada, M. (2024). An empirical investigation of the impact of energy consumption, globalization and natural resources on ecological footprint and economic growth, evidence from China, Japan, South Korea and China Taiwan. *Energy & Environment*, 0958305X241251421.

Atmospheric Particulate-Bound Polycyclic Aromatic Hydrocarbons and Nitro-Polycyclic Aromatic Hydrocarbons: Levels, Sources, Health Impacts, and Policy Implications for Africa

Augustine Omodieke and Egide Kalisa

Abstract Polycyclic aromatic hydrocarbons (PAHs) and their nitro derivatives (NPAHs) are part of the hazardous organic component of particulate matter air pollutants, with significant environmental and health impacts. Most of the literature on the characterization and sources of PAHs and NPAHs has focused on the Global North, and the literature from the Global South shows that research from South Asian contexts is more common than analyses of Africa. Monitoring atmospheric PAHs and NPAHs in Africa is critical but has remained a significant challenge because of several factors, including resource constraints, methodological inconsistencies, a dearth of data, and limited technical capacity. However, current analyses focusing on African contexts are proliferating and provide key insights for this analysis. Vegetation fires and agricultural biomass burning dominate rural landscapes. At the same time, emissions from old vehicles and dusty roads are the main sources of PAH/NPAH in urban centers as a result of rapid urbanization, population and economic growth. This chapter explores and details the characterization and key sources of PAHs and NPAHs in air particulate matter in Africa and provides a basis for future research and the need for comprehensive monitoring while highlighting policy interventions to address the growing public health and environmental challenges.

Keywords Polycyclic aromatic hydrocarbon · Nitro polycyclic aromatic hydrocarbon · Particulate matter · Vegetation · Vehicle emission · Africa

---

A. Omodieke · E. Kalisa (✉)
Department of Epidemiology and Biostatistics, Schulich School of Medicine and Dentistry, Western University, London, ON, Canada
e-mail: ekalisa2@uwo.ca

K. Hayakawa and Y. Aoki (eds.), *A New Era of Polycyclic Aromatic Hydrocarbons*,
https://doi.org/10.1007/978-981-96-7695-8_3

1 Introduction

In low-and middle-income countries, less than 1% of those living in urban areas breathe air that meets the World Health Organization (WHO) air quality guidelines, resulting in millions of premature deaths each year. The growth of industry, urbanization, home heating and cooking, commercial activities, deforestation and transportation are all contributors to air pollution. Among these, wildfire, domestic biomass burning, and transport emissions are often the most intractable sources and continue to dominate air pollution in Africa. Africa is experiencing rapid economic development and population growth. It is thus faced with the challenge of decoupling social and economic development from the negative externalities of economic growth, including air pollution. Particulate matter air pollution is approximately ten times higher than the annual WHO recommended limit of 5 $\mu g/m^3$ (World Health Organization 2021) in most African countries, emphasizing the need for action.

Polycyclic aromatic hydrocarbons (PAHs) are part of the organic component of fine particulate matter ($PM_{2.5}$). They are toxic organic pollutants produced by incomplete combustion of organic materials. In Africa, PAHs and their nitro derivatives (NPAHs) are increasingly becoming a concern due to rapid urbanization, deforestation, increasing vehicular traffic, and sustained traditional practices such as open-fire cooking—posing significant risks to more than 1.2 billion people (Health Effects Institute (HEI) 2022). PAHs are linked to severe health effects, including cancer, respiratory issues, and reproductive harm, making them a critical environmental and public health concern for African countries (Bortey-Sam et al. 2017; Morakinyo et al. 2020). Vegetation fires play an essential role in many African ecosystems and farming methods, including those in savannas, forests, and the combustion of agricultural waste. These fires, which are typically seasonal and fueled by both natural and human-induced events, substantially impact atmospheric PAH and NPAH levels. A key contributor to air pollution in cities is vehicular traffic, which has worsened as a result of the importation of used cars, a lack of regulations on fuel quality, and the rapid expansion of cities themselves.

There is still a lot we do not know about the sources, distribution, and effects of PAHs and NPAHs in Africa, even though they are becoming more and more acknowledged as important pollutants. To fill this void, this chapter examines sources, levels and health impacts caused by atmospheric PAHs and NPAHs in the African context. The chapter highlights the research gaps and future research directions and emphasizes the need for policies and strategies that are adaptable and specific to the contexts in Africa to address these emissions, as well as the opportunities and challenges that come with these interventions.

2 Particulate Matter ($PM_{2.5}$) Air Pollution in Africa

The evaluation of ambient air pollution in Africa typically presents measurements of particulate matter ($PM_{2.5}$ and PM_{10}) as metrics. Scientific publications have informed government control measures to address concerns about ambient air pollution in developed countries. However, little to no attention is given to this in African countries despite the significant problems it poses to human health from a public health point of view. Atmospheric PM is a complex mixture of solid particles and liquid droplets found in the air. It is composed of chemical components, including sulfates, nitrates, PAHs, NPAHs, heavy metals, dust, minerals, and soot, and biological components, including viruses, bacteria, and fungi (Morakinyo et al. 2016) (Fig. 1), which are not commonly measured by government monitoring programs or included in air quality standards in Africa.

Available air data in Africa are sparse but indicate that the levels of ambient air pollutants such as particulate matter ($PM_{2.5}$) are substantially higher than WHO air quality guidelines (Bililign et al. 2024). In 2021, based on new evidence linking air pollution with health effects at lower concentrations, the World Health Organization (WHO) revised its guidelines, lowering the recommended limit for the annual mean concentration of $PM_{2.5}$ to 5 $\mu g/m^3$ (WHO 2021). According to the HEI Africa has one of the highest average annual particulate matter (PM) exposures (HEI 2022, 2024). The continent is home to five of the top ten countries in the world (Niger, Nigeria, Egypt, Mauritania, and Cameroon), with the highest population-weighted average annual exposures to $PM_{2.5}$ (HEI 2020), which are six to eight times higher than the annual exposure limits recommended by the WHO (5 $\mu g/m^3$). In addition, a growing body of research consistently indicates that most African cities have poor ambient $PM_{2.5}$ (Agbo et al. 2021).

Available satellite estimate data analyzed from 54 countries from 1990 to 2019 showed that the mean annual concentration of $PM_{2.5}$ was highest in West Africa,

Fig. 1 **a** A schematic representation of the primary source of particulate matter-bound polycyclic aromatic hydrocarbons (PAHs) and their nitro-derivatives (NPAHs), and **b** associated inhalation health risks in Africa

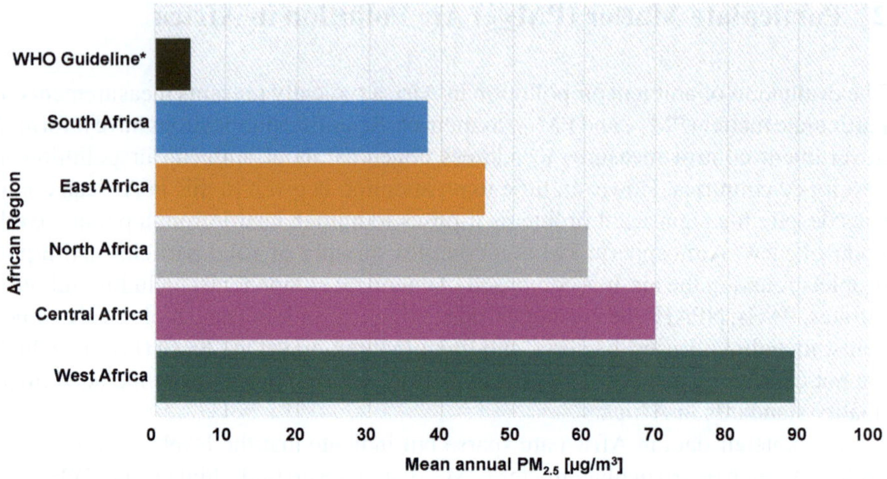

Fig. 2 Mean annual $PM_{2.5}$ concentrations in African regions (1990–2019) based on satellite data benchmark compared to the revised WHO. State of Global Air 2020. The data source is the Institute for Health Metrics and Evaluation (IHME 2019). *WHO guideline revised in 2021

followed by Central, North, East and South Africa, and significantly exceeded the annual WHO guidelines (Fig. 2).

To compare, we have also analyzed $PM_{2.5}$ ground-monitored data obtained from a mini-reference station operated by the US Embassy in different African countries from 2016 to 2024 (Fig. 3). The results indicate that the mean annual averages of the ground-monitored $PM_{2.5}$ data in most of the major African cities exceed the new WHO guidelines for $PM_{2.5}$ levels by up to 15 times, with Egypt, Chad, Mali, Democratic Republic of Congo and Uganda accounting for the highest levels of these pollutants in the Northern, Western and Eastern regions of Africa respectively. While all of this suggests how alarming the problem of air pollution in Africa must be, knowledge about air pollution and its impact on health is not well advanced in most African countries, partly because of the lack of reliable monitoring data relating to air pollution at ground level in African regions.

3 Atmospheric PAHs and NPAHs in Africa

PAHs and NPAHs are linked to cancer and chronic lung disorders (asthma) even at very low concentrations, but little is known about their occurrence in ambient and indoor environments in Africa. While the chemical composition of PM and associated health effects has been widely studied in developed countries, there have been limited studies in the African continent. The limited number of studies on PAHs and NPAHs in Africa has made it difficult to determine how much of a problem the continent is dealing with and, therefore, limits evidence-based recommendations to policy

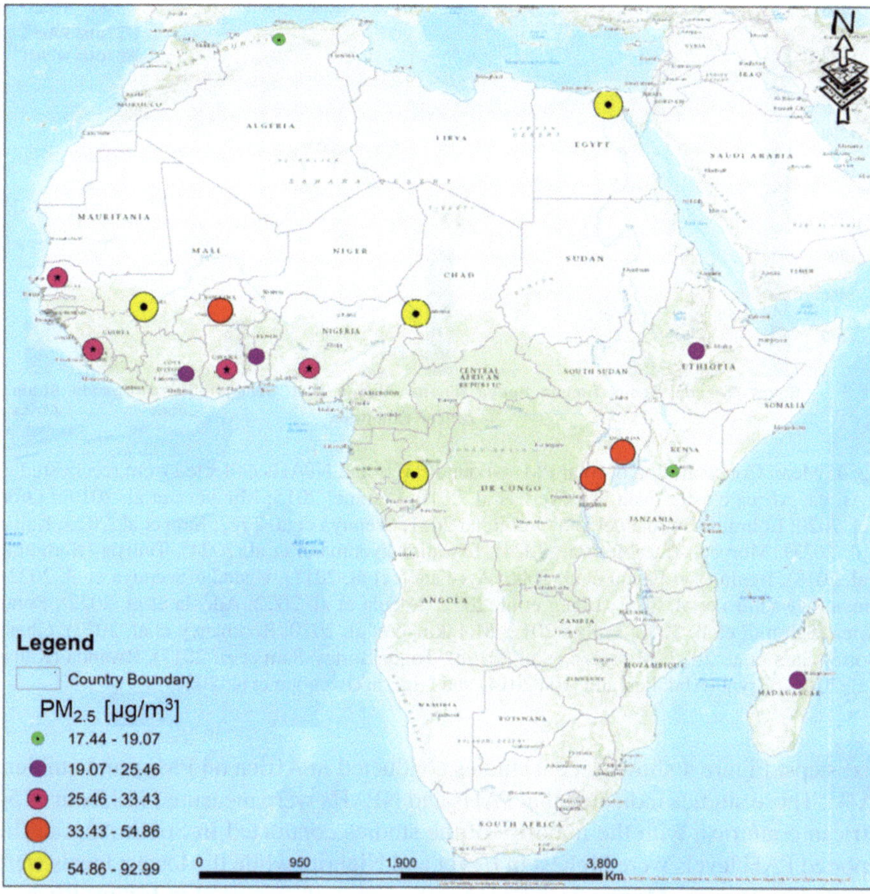

Fig. 3 Annual mean concentrations of $PM_{2.5}$ were obtained from near-reference stations operated by the US Embassy in Africa (2016–2024). Hourly Data is available from AirNow: https://www.airnow.gov/

and decision-makers to make informed decisions about intervening. However, they do indicate that immediate attention is required to deal with these toxic pollutants. There is a dearth of evidence on the extent of both exposure and impacts (health and environmental) of PAHs and NPAHs. This is largely due to resource constraints, as there are limited funding, tools and proficient human resources to conduct comprehensive monitoring. Additionally, despite several studies conducted in Africa on this subject, the resulting data have proven to be disjointed and incoherent (Ofori et al. 2020), thereby exacerbating the process of obtaining a comprehensive perspective of the problem for evidence-based decisions and policy formulation. Although countries like Nigeria, Angola, Egypt, South Africa and Kenya have commenced the application of remedial actions for PAHs (Ofori et al. 2020), robust evidence of the scope of this plague in the continent is required to determine and guide actionable

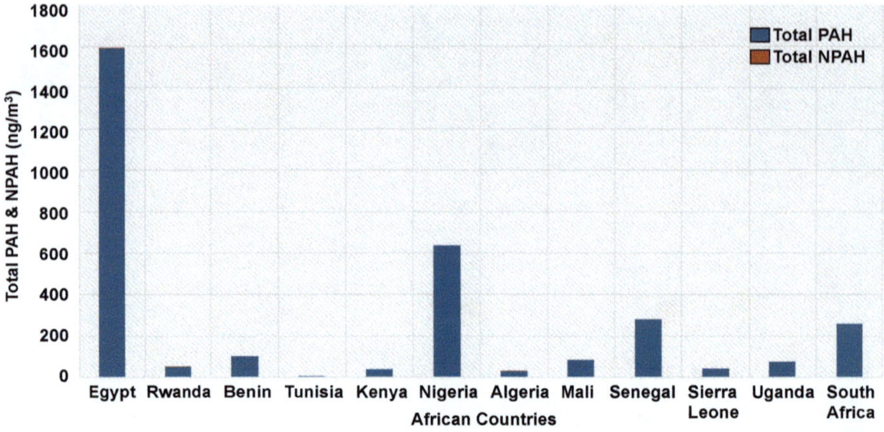

Fig. 4 Mean Concentration of total $PM_{2.5}$-bound PAHs and NPAHs collected from recent studies in Africa. Africa cluster (Van Der Werf et al. 2010; Lammel 2015; Munyeza et al. 2019b; Ofori et al. 2020; Ephraim-Emmanuel and Ordinioha 2021; Ssepuya et al. 2022; Sam et al. 2023; Kalisa et al. 2023), Morocco (Deabji et al. 2024), Eswatini (Neumann et al. 2024), Tunisia (Barhoumi et al. 2018), Benin (Ayi Fanou et al. 2006; Ayi-Fanou et al. 2011), Uganda (Ssepuya et al. 2022), Kenya (Mwachiro et al. 2021; Liang et al. 2022; Adeola et al. 2022; Adeola et al. 2022), South Africa (Schmidl et al. 2008; Forbes 2012; Morakinyo et al. 2020; Roomaney et al. 2022), Ghana (Bortey-Sam et al. 2017; Munyeza et al. 2019b), Togo (Bortey-Sam et al. 2017), Rwanda (Kalisa et al. 2018), Egypt (Abdallah and Atia 2014) and Nigeria (Iwegbue et al. 2022)

next steps. Figure 4 shows recent studies conducted in Africa on $PM_{2.5}$ and ambient PAHs. These studies indicated that PAHs and NPAHs were measured in 12 out of 54 African countries, with the majority of the studies conducted in South Africa. The reported PAH levels were highest in Egypt and Nigeria, while the lowest levels were reported in Tunisia. Moreover, NPAH levels were only reported in Egypt, Rwanda and Algeria. This shows that a significant gap still exists in the exploration and understanding of NPAHs and their impacts in Africa.

In Africa, PAHs are generated and released into the atmosphere and environment from myriads of natural and anthropogenic (human-induced) sources. These PAH sources are mainly due to Africa's rapid growth in industrialization, commercialization and urbanization (in urban settings), over-reliance on traditional biomass cooking fuels (in rural and remote areas) due to socioeconomic factors, use of poor-quality vehicular fuels, and wildfires. Analysis from studies that have reported sources of PAHs in Africa showed that biomass burning is the dominant source, accounting for 48%, while industrial emissions are the least significant source of PAHs, accounting for only 8% (Fig. 5).

Using the PAH Molecular Diagnostic Ratios, a few studies further characterized the sources of PAH reported (Fig. 6). The application of molecular diagnostic ratios in 6 studies showed ratios such as fluoranthene to fluoranthene + pyrene ([Fla]/[Fla + Pyr]), benz[*a*]anthracene to benz[*a*]anthracene + chrysene ([BaA]/[BaA + Chr]), indeno[1,2,3-*cd*]pyrene to indeno[1,2,3-*cd*]pyrene + benzo[*ghi*]perylene ([IcdP]/

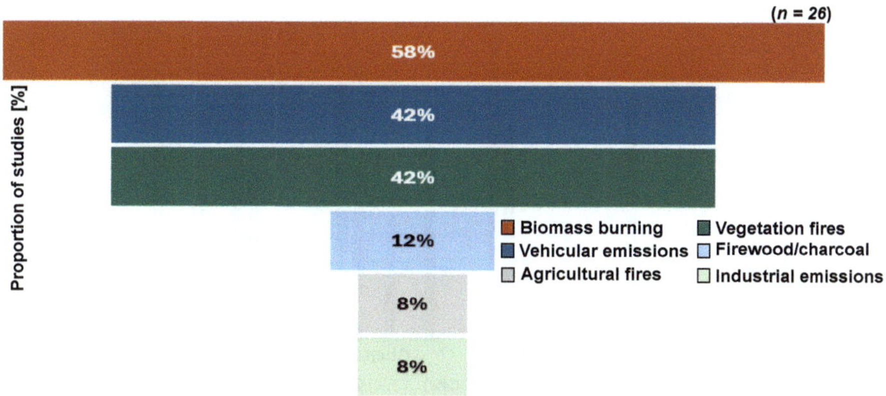

Fig. 5 PAHs and NPAHs sources reported in Africa from 26 studies. Biomass burning (Shi et al. 2015; Bortey-Sam et al. 2017; Munyeza et al. 2019a, b; Mwachiro et al. 2021; Thompson et al. 2022; Iwegbue et al. 2022; Wu et al. 2022; Adeola et al. 2022; Roomaney et al. 2022; Onaiwu and Ifijen 2024; Neumann et al. 2024; Bai et al. 2024), vehicular emission (Ayi Fanou et al. 2006; Abdallah and Atia 2014; Bortey-Sam et al. 2017; Barhoumi et al. 2018; Kalisa et al. 2018; Munyeza et al. 2019b; Iwegbue et al. 2022; Raheja et al. 2022; Onaiwu and Ifijen 2024; Deabji et al. 2024), firewood/charcoal (Ayi Fanou et al. 2006; Forbes and Rohwer 2010; Forbes 2012; Bortey-Sam et al. 2017; Kalisa et al. 2018; Munyeza et al. 2019a, b; Mwachiro et al. 2021; Thompson et al. 2022; Adeola et al. 2022; Ssepuya et al. 2022; Bai et al. 2024), vegetation fires (Lammel et al. 2013; Shi et al. 2015; Wu et al. 2022), agricultural fires (Forbes and Rohwer 2010; Bortey-Sam et al. 2017; Deabji et al. 2024), and industrial emissions (Forbes and Rohwer 2010; Munyeza et al. 2019b; Morakinyo et al. 2020)

[[IcdP + BghiP]), Anthracene to Anthracene + Phenanthrene ([Ant]/([Ant + Phe]), benz[*a*]pyrene to benz[*a*]pyrene + chrysene ([BaP]/[BaP + Chr]), fluorene to fluorene + pyrene ([Flo]/[Flo + Pyr]) are often identified based on PAHs markers. The value from these ratios showed that the prevalent sources of PAHs in Africa countries are petrogenic (vehicular emissions, combustion engines) and biomass burning.

(a) **Vegetation Fires and Agricultural Practices**

Forest fires and savanna burning release significant amounts of PAHs into the atmosphere (Lammel et al. 2013). These can take the form of wildfires, controlled burns or seasonal fires. Wildfires are uncontrolled fires ignited by lightning strikes that spread quickly across vegetation, including savannas, forests, and grasslands. While during the dry season, many regions in Africa experience seasonal fires, controlled burns resulting from agricultural practices are prevalent, including slash-and-burn agriculture for land preparation ahead of cultivation. Notably, Africa accounts for about 52% of the total global fire emissions resulting from the increasing rate of forest, savannah, grassland and cropland burning (Van Der Werf et al. 2010). Slash-and-burn farming practices are commonplace, contributing to PAH emissions and air pollution, with more than 90% of the total biomass burning in Africa attributable to vegetation fires, while fires due to agricultural practices (crop residue burning) have

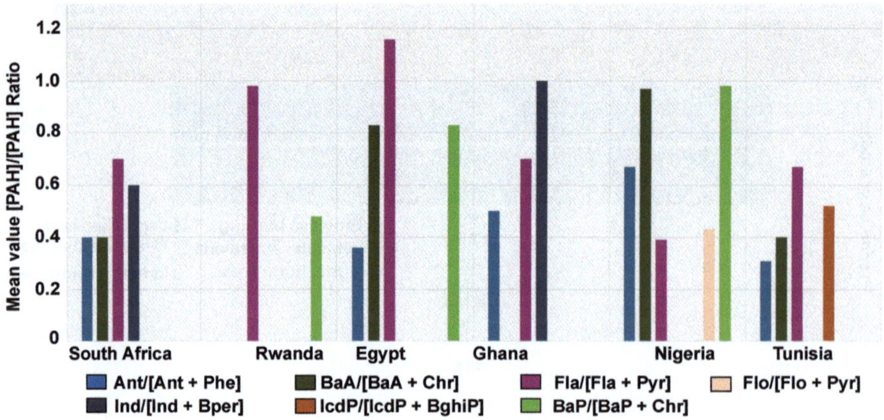

Fig. 6 Mean value of PAH molecular diagnostic ratios reported in six countries in Africa (Thompson et al. 2022; Onaiwu and Ifijen 2024; Abdallah and Atia 2014; Barhoumi et al. 2018; Kalisa et al. 2018; Morakinyo et al. 2020)

contributed less than 10% (Wu et al. 2022). This rapid deforestation is attributable to the susceptibility of the tropics to an increasing risk of wildfires (Shi et al. 2015).

PAHs released from fires are long-range pollutants that affect both rural and urban populations. Lammel (2015) reported from field observations and modeling that PAHs undergo long-range atmospheric transport, moving across remote continental regions, the oceans and the Arctic. According to Wu et al. (2022), approximately 17% of the total levels of benzo(a)pyrene (BaP)—the most toxic and carcinogenic congener among the 16 parent PAHs—in southern Europe, and to some degree in South America, Western and Northern Europe, can be attributed to fires originating from Africa.

(b) **Vehicle Emissions and Transportation Growth**

Africa, in the past decade, has experienced a rapid expansion in its transportation network and infrastructure. The key drivers of this mushrooming are urbanization, economic development, and increased mobility, including intra- and inter-region trade and movements. The policy environment has been conducive to this: for example, the implementation of the African Continental Free Trade Area (AfCFTA) promotes easy movement of goods and labor within the West Africa region and is expected to double road freight from 201 to 403 million tonnes, according to the United Nations Economic Commission for Africa (ECA) (ECA 2022). This increase in vehicular movements, particularly older models and used vehicles from developed countries, has increased emissions of PAHs, exacerbating ambient air pollution. Public transport systems in many African cities also rely on these older-model diesel and petrol-powered vehicles, including taxis and buses, which lack strict emission controls, worsening emissions. Between 2005 and 2015, car ownership increased by up to 6%, 17%, 26% and 27% in Nigeria, Egypt, Kenya and Algeria respectively.

Correspondingly, fuel consumption grew by up to 123%, 21%, 86% and 44% in these countries Sustainable Low carbon Transport (SLOCAT) (SLOCAT 2021).

These data indicate that there was a consequential increase in fuel consumption as car ownership and vehicle sales increased in these countries over the period from 2010 to 2019. This further exacerbated pollution through increased emissions of PAHs from transportation due to widespread poor-quality gasoline in this region. According to the SLOCAT Transport, Climate and Sustainability Global Status Report, transport emissions grew by approximately 27% from a low baseline between 2010 and 2019, second only to Asia at 41% (SLOCAT 2021). Bortey-Sam et al. (2017), in their analysis of children's urine in Kumasi (an urban city in Ghana), found high concentrations of hydroxylated PAHs ($\sum$OHPAHs) resulting from exposure to vehicular emissions compared to samples collected from the control site. Using Principal Component Analysis (PCA), a method that has been widely applied in determining the sources of PAHs and their derivatives, including NPAHs in aerosol studies, Kalisa et al. (2018), in their study of particulate-bound PAHs and NPAHs in Rwanda, reported varying means for $\sum$15PAH in $PM_{2.5}$ and in PM_{10} at three different locations. They found urban roadside locations had a higher mean level (52.4 ng/m^3 and 105.0 ng/m^3) compared to rural area sites (31.5 ng/m^3 and 64.0 ng/m^3) and urban background sites (20.8 ng/m^3 and 23.9 ng/m^3), with these high concentration levels attributable to the vehicular emissions from diesel and gasoline-powered vehicles.

According to Abdallah and Atia (2014), Assiut, considered the commercial capital of Upper Egypt, is ranked among the most PAH-contaminated cities in the world, with levels of target PHAs in the atmosphere averaging 1590 ng/m^3. Compared to suburban sites, they determined the major sources of PAHs to be vehicular combustion and traffic exhaust emissions (gasoline and diesel vehicles), using diagnostic PAH concentration ratios. In their application of Laser Induced Fluorescence (LIF) in monitoring atmospheric PAHs in South Africa, Forbes and Rohwer (2010) identified naphthalene as the prevalent PAH, with gaseous concentrations monitored in the primary trap reaching 1.8 µg.m^{-3} from diesel vehicular emissions. Similarly, Ayi Fanou et al. (2006) measured exposure to PAHs using personal monitoring and biomarkers in Cotonou and a village and found the total level of PAHs, notably benzene, ranged from 76.21 to 103.23 ng/m^3 in Cotonou versus 1.55 ng/m^3 for the village. This high benzene concentration is linked to higher traffic density and the use of low-quality gasoline in Cotonou compared to the village. In most European cities, despite higher traffic densities, the ambient concentration of benzene is considerably lower compared to Cotonou because good quality gasoline is in use (Ayi Fanou et al. 2006).

Onaiwu and Ifijen (2024) analyzed, using molecular diagnostic ratios, $PM_{2.5}$-bound PAHs in the ambient air of automobile workshops in Benin City, Nigeria. The analysis showed annual average concentrations of $PM_{2.5}$-bound PAHs in Benin City varied seasonally and ranged from 269.87 ± 249.32 ng/m^3 (dry season) to 216.30 ± 204.89 ng/m^3 (wet season) and identified gasoline combustion and diesel combustion to be the primary sources of PAHs near vehicle workshops. Likewise, in Nigeria's megacity, Lagos, Iwegbue et al. (2022) determined the concentrations and distribution patterns of PAHs using gas chromatography–mass spectrometry and found the

PAH concentrations in urban street dust ranged from 545 to 10,785 μg kg^{-1}, 289 to 17,943 μg kg^{-1} and 616 to 13,174 μg kg^{-1} for industrial, commercial, and residential areas, respectively. Further analysis, including PAH isomeric ratios and multivariate analysis, showed that PAHs in urban street dust originated primarily from vehicular traffic emissions. Also, Barhoumi et al. (2018) identified diesel vehicle emissions as the dominant source of PAHs while assessing the concentrations, sources, dry deposition and human health risks associated with exposure to PAHs and other pollutants in Bizerte City, Tunisia.

(c) **Household Air Pollution from Cooking Methods and Practices**

Traditional cooking practices, including open fires and traditional stoves (three-stone stoves, clay/mud stoves, traditional metal stoves, rocket stoves, charcoal stoves, wood-burning cookstoves, pit fires/mud pit stoves, and other varieties), involving the use of solid fuels including charcoal, wood, and biomass (crop residues, animal dung) are commonplace in developing countries, especially in Africa. With more than 600 million people depending on traditional cooking methods and energy sources, these practices are prevalent in rural and remote areas because they are affordable and easily accessible, while cultural norms, customs, and beliefs are also factors that determine the type of cooking method. However, these solid fuels undergo incomplete combustion due to the combustion inefficiency of the different stoves. This results in the release of harmful pollutants, notably PAHs. To further worsen this, most of the houses in these rural settings are built with poor ventilation, thereby concentrating these pollutants indoors and increasing the exposure of households to numerous health risks. Women and children are particularly vulnerable to PAH exposure from indoor and inefficient cooking fumes, leading to respiratory diseases and skin cancers, among other illnesses.

Globally, the average concentration of PAHs in cooking areas is highest in Africa at 14.74 $\mu g/m^3$. This is greater than the levels recorded in other continents by a factor of up to 160 (Bai et al. 2024). Mwachiro et al. (2021) found exposure to PAHs to be extremely high among sampled participants, mostly women (up to 28 times higher compared to sampled populations in other climes) in rural Kenya to be strongly associated with the type of fuel used (wood) and cooking method (cooking in an enclosed room on open, unvented stoves, without chimneys or adequate room ventilation).

Correspondingly, Munyeza et al. (2019a) analyzed the effects of fuel type and cooking device on the release of PAHs in households in rural and urban Kenya and found that wood, charcoal, kerosene, and gas stoves produced high concentrations of PAHs (average of $46.23 \pm 3.24\ \mu g/m^3$). Emissions were highest from wood-burning in three-stone stoves. Additionally, an analysis and comparison of PAH emissions from cooking devices in rural and urban Kenya showed that wood combustion using three-stone cooking devices produced the highest total PAH concentrations of up to 71 $\mu g/m^3$ (Adeola et al. 2022). The findings of Munyeza et al. (2019a) and Adeola et al. (2022) in rural Kenya were consistent, indicating that wood-fueled three-stone stoves contributed the largest total PAH concentrations among cooking devices analyzed. However, further studies would be required to determine the level of contribution of Jiko and kerosene stove types to total PAH concentrations.

In Eswatini, where more than 60% of households still rely on solid fuel (largely wood) for cooking, particulate PAH intake concentrations from biomass fuel users are extremely high (Neumann et al. 2024). Ssepuya et al. (2022) studied the association between measured levels of PAHs in breast milk of nursing mothers and exposure to household fuels and cooking methods in Uganda, where they found varying levels of PAHs with nursing mothers who used charcoal for cooking accumulating higher levels of $\sum$13PAHs in their breast milk samples compared to those who used firewood. These all indicate the high risk of exposure of the rural population to these harmful pollutants.

(d) **Industrial Emissions and Informal Activities**

In the developing regions of the world, particularly in Africa, the activities of industries and the informal economy are two major contributors to pollution, especially PAHs. This is largely due to poor regulations in these sectors. Industrial activities that focus on energy generation, including coal and biomass combustion and petroleum refining processes, involve incomplete combustion of organic materials that release significant quantities of PAHs into the atmosphere and environment. Additionally, manufacturing processes such as steel and iron, cement production, chemicals production, and paper production, are all high-temperature reactions that generate large amounts of PAHs.

The informal economy, which is dominated by informal workers who are poorly skilled and often operate small-scale activities that involve largely unregulated processes, is a key driver of increased exposure to PAHs. They largely operate informal recycling businesses (recovery of metals or plastics, burning waste materials, etc.), artisanal mining (small-scale mining, notably gold and other minerals, involving burning of fuels or other organic materials in makeshift furnaces), household waste burning in open fires (including plastics, textiles, and electronic waste (e-waste)), and small-scale brick kilns using low-grade coal or biomass for firing the bricks. Most of these activities are operated proximal to residential areas and further exacerbate human exposure to PAHs. Morakinyo et al. measured seasonal concentrations of $PM_{2.5}$-bound PAHs in an industrial area in Pretoria, South Africa (with 13 industrial facilities and two power plants and metallurgical industries) and found total PAHs to be 10.97 ng/m^3, with concentrations highest during the winter season (Morakinyo et al. 2020).

4 Health Impacts of PAHs and NPAHs in Africa

The health risks linked with exposure (through inhalation, ingestion and dermal routes) to PAHs in Africa are quite alarming. This is due to the vast majority of the African population still relying on solid fuels and biomass for energy and domestic use (cooking) and the rapid urbanization and industrialization transition.

In sub-Saharan Africa, a broad range of cardiovascular and respiratory diseases and premature deaths (Gordon et al. 2023) have been associated with PAHs. The inhalation of PAHs from household air pollution resulting from the use of solid fuels/biomass for cooking and urban air pollution have caused inflammation and oxidative stress, resulting in increased risks of heart attacks and strokes. The increased risks for respiratory diseases, including asthma, chronic obstructive pulmonary disease (COPD), and bronchitis, which are prevalent among children and the elderly, have been strongly associated with exposure to PAHs (WHO 2021; Ofori et al. 2021).

Long-term exposure to some PAH groups classified as carcinogens by the International Agency for Research on Cancer (IARC), especially benzo(a)pyrene (BaP), significantly increases the risk of developing different types of cancers. The most common types of cancer linked to PAH exposure include lung cancer, skin cancer, and bladder cancer. Risks are heightened in rural settings for households that predominantly rely on open fire burning or traditional cooking practices and urban households situated proximal to industrial areas or commercial areas with high vehicular traffic. Most of the studies in Africa have measured the levels of BaP, while only a few have quantified lifetime cancer risk (LCR) due to exposure (Table 1). BaP levels were highest in Kenya and Eswatini while the lowest BaP levels were reported in South Africa. The LCRs reported in South Africa, Rwanda, and Eswatini exceed the recommended minimum threshold, indicating significant risk to the population.

The majority of studies in Africa (as shown in Fig. 7) have measured and reported high levels of mutagenicity and carcinogenicity resulting from exposure to PAHs, especially BaP. Lung cancer, DNA adducts, leukemia, esophageal squamous dysplasia (ESD), and esophageal squamous cell cancer are common cancer types associated with exposure to BaP.

Carcinogenicity (Ayi Fanou et al. 2006; Van Der Werf et al. 2010; Ayi-Fanou et al. 2011; Abdallah and Atia 2014; Barhoumi et al. 2018; Kalisa et al. 2018; IHME 2019; Munyeza et al. 2019a; Morakinyo et al. 2020; Mwachiro et al. 2021; Liang et al. 2022; Iwegbue et al. 2022; Adeola et al. 2022; Onaiwu and Ifijen 2024; Neumann et al. 2024; Bai et al. 2024), premature deaths (Roomaney et al. 2022; Gordon et al.

Table 1 BaP levels reported in Africa

Country	BaP (ng/m^3)	Lifetime cancer risk	References
South Africa	0.54	1.43×10^{-5}	Morakinyo et al. (2020)
Rwanda	11.28	1.15×10^{-3}	Kalisa et al. (2018)
Kenya	453.88		Munyeza et al. (2019a)
Egypt	73.6		Abdallah and Atia (2014)
Benin	3.81		Ayi Fanou et al. (2006)
Ghana	2.26		Thompson et al. (2022)
Nigeria	32.47		Onaiwu and Ifijen (2024)
Eswatini	341.61	7.72×10^{-5}	Neumann et al. (2024)
Tunisia	1.49	1.60×10^{-6}	Barhoumi et al. (2018)

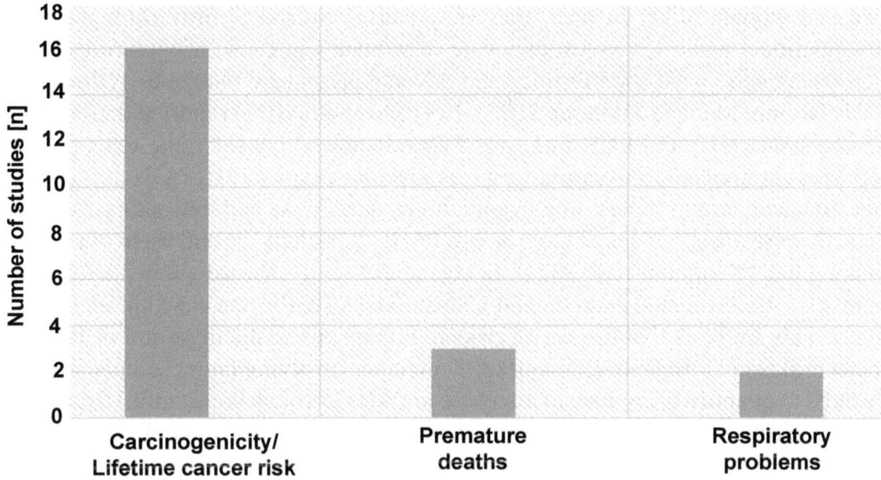

Fig. 7 Number of studies that have commonly reported major health risk impacts of particulate matter associated with PAHs and NPAHs in Africa from 21 studies

2023; Neumann et al. 2024), respiratory problems (Bortey-Sam et al. 2017; Iwegbue et al. 2022).

Exposure resulting from inhalation of BaP from wildfires has shown that the LCR levels in many African countries are greater than 1×10^{-6}, exceeding the WHO standard (Wu et al. 2022). Morakinyo et al. (2020) reported the potential risk of developing cancer over a lifetime of 70 years among people exposed (through inhalation/ ingestion and dermal contact) to PAHs in South Africa to be 1 to 8 per million persons. Additionally, Bai et al. (2024) measured the Total Incremental Lifetime Cancer Risk (TILCR) due to exposure to inefficient cooking methods in Africa and found it to exceed acceptable levels. Concurrently, evidence from Rwanda showed that LCRs emanating from exposure due to inhalation of PM-bound PAHs and NPAHs exceeded World Health Organization safe limits (Kalisa et al. 2018). Elevated DNA-adducts levels were found to be significantly associated with exposure to PAHs from vehicular emissions (worsened by poor quality gasoline) among city residents in Benin (Ayi Fanou et al. 2006, 2011).

Additionally, exposure to PAHs in early life has been linked to developmental and cognitive impairments in infants, and increased risk of childhood cancers. Children who are exposed to PAH-laden smoke from cooking fires are at higher risk of respiratory problems and developmental delays. PAHs' neurotoxicity effects can impact the development and function of the brain in children, and prenatal exposure can also adversely impact neurological development in children, causing a range of learning disabilities, cognitive impairments and behavioral disorders (Hisamuddin and Jalaludin 2023). PAHs, besides worsening pre-existing health conditions through comorbidity, especially diabetes and respiratory diseases, and making management more challenging, have also been linked with fertility and reproductive issues in men and women alike. In men, they may cause reduced fertility through sperm abnormalities, while long-term exposure in women can cause irregularities in the menstrual cycle, weakened fertility and adverse pregnancy outcomes (Ofori et al. 2021; Hisamuddin and Jalaludin 2023). Iwegbue et al. (2022) reported that both the Hazard Index (HI) and LCR in Lagos-Nigeria showed considerable risks that are both non-carcinogenic and carcinogenic as a result of exposure to PAHs through oral ingestion and dermal routes. In Eswatini, high cancer risks and low embryo survival chances were observed due to exposure to PAHs, especially among users of biomass cooking fuel (Neumann et al. 2024). In Uganda, although hazard quotients for PAHs were < 1, the estimated Incremental Cancer Risks (ICRs) were within acceptable levels. This suggests that the health risks to infants due to the ingestion of PAHs in breast milk were tolerable, although further studies involving larger sample sizes are required to quantify the impact of exposure to PAHs (through breast milk) from traditional cooking practices on infants to establish validity (Ssepuya et al. 2022). Also, in Ghana, exposure to PAHs, especially naphthalene, through inhalation was found to significantly increase the occurrence of diverse respiratory problems, including incessant cough, tachycardia, inflammation, nasal congestion, dyspnea and headache, especially among women (Bortey-Sam et al. 2017). Mwachiro et al. (2021) reported a strong association between exposure to PAH, notably 2-hydroxynaphthalene, and advanced ESD among sampled participants (who used wood as cooking fuel) in rural Kenya, especially women.

5 Air Quality Monitoring and Regulatory Landscape in Africa

Air quality monitoring in Africa is critical due to the plethora of exposure pathways, including industrial activities and vehicular emissions (resulting from rapid urbanization, industrialization and expansion of transportation network), biomass burning, and traditional cooking practices. Assessing the levels of exposure, formulating public health policies, and implementing environmental protection policies that advance the population's health, especially among the vulnerable, requires effective monitoring of fine particulate matter ($PM_{2.5}$) and PAHs is major component of $PM_{2.5}$. However, Africa faces significant hurdles to effectively monitor air quality, including determining/tracking long-term $PM_{2.5}$ levels due to several factors, including resource constraints, methodological inconsistencies, a dearth of data and limited technical capacity. Most African countries lack the necessary infrastructure and funding required to conduct ongoing monitoring of air quality and pollution. The paucity of data on PAH levels in most regions remains a big challenge due to limited studies on PAHs from the different contexts and regions of the continent. Even where data exist, comparability has been plagued by inconsistencies in data resulting from methodological differences in data collection, sampling and analytical techniques.

While air quality research has been conducted in several countries in Africa, only South Africa, Senegal, Morrocco and Rwanda have programs in place to monitor air quality and $PM_{2.5}$ (Fig. 8) (HEI 2024) As shown in Fig. 8, limited $PM_{2.5}$ research and no forms of air quality monitoring are prevalent in Africa, while most of the countries that have conducted some $PM_{2.5}$ bound PAH research are yet to take action to improve the air pollution menace.

Moreover, the extant regulatory landscape for PAH monitoring in Africa is largely underdeveloped, further exacerbating the prevalence of the menace. Comprehensive and standardized regulations targeted at controlling PAH emissions are lacking. A few countries, including South Africa, Rwanda, Morocco, and Senegal, have put in place measures to regulate air pollution; however, such environmental laws do not make clear provisions for monitoring and controlling PAHs. This weakness in the regulatory framework can be attributed to dominating informal sector activities,

Fig. 8 Air quality monitoring status in Africa with and without national regulations (HEI 2024)

sparse data and research and limited funding, among other issues. The activities of the informal economy, which contributes largely to PAH emissions, make it difficult to enforce regulations, and the limited available data due to insufficient research on PAH levels and their corresponding impacts (health and environmental) limits evidence-driven regulations.

6 Potential Opportunities to Mitigate PAHs and NPAH Impacts and Policy Implications for Africa

The level of risk posed to the vast population of Africa by perpetual exposure to PAHs is huge, and prompt actions need to be taken to reverse current trends. Addressing this issue requires concerted efforts and a multifaceted approach, including improved PAH monitoring, investing in/transitioning to renewable energy (countries like South Africa, Kenya, Uganda and Morrocco are already investing in EVs), infrastructural upgrades, strengthened regulatory and policy frameworks, partnerships with the international community, creating demand for efficient energy, and public education. Described below are options that country governments in Africa can explore with support from the international community:

(a) **Aggressive monitoring of PAHs and NPAHs**

Monitoring PAHs and NPAHs, as a key action step in Africa, requires aggressive effort. Effectively monitoring these pollutants will involve surmounting obvious hurdles, notably scarce resources, lack of infrastructure and technical capacity, and data gaps. However, several strategies, including but not limited to the following, can be explored to subordinate these issues and improve PAH and NPAH monitoring.

First, Africa needs to standardize and harmonize clear methods and protocols for measuring (sampling, analyzing and reporting) PAH and NPAH concentrations in the different media in which they exist, including air, water, soil and food. This will foster consistency in data management methods and eliminate the problem of results incomparability across regions in the continent. To achieve this, monitoring stations need to be set up strategically, both nationally and regionally, to collect data for a holistic overview of PAH and NPAH levels. In specific areas categorized as high risk due to elevated PAH levels resulting from industrial emissions, high traffic density and prevalent traditional cooking methods, monitoring hotspots can be set up. Frugal (cost-effective or low-cost) PAH and NPAH monitoring tools, including portable samplers and air quality sensors, should be deployed in these stations and hotspots and used widely for real-time monitoring. Additionally, remote sensing technologies and satellite imagery should be used to track PAH sources and emission levels in especially hard-to-reach areas.

Moreover, PAH and NPAH monitoring data should be appropriately stored using online platforms or government portals for ease of accessibility (to the public, researchers and policymakers to raise awareness, advance research, and inform transparent and evidence-driven policy directions, respectively), analysis, and sharing using centralized databases at country and regional levels. These efforts should be implemented in alignment with the global monitoring programs of the United Nations Environment Programme and the World Health Organization to promote data comparability, validity and reliability for planning and decision-making.

(b) **Transition to Cleaner Energy and Technologies**

Africa, being rich in renewable resources, requires transitioning to cleaner energy sources, leveraging technologies, to meet and sustain its rapid growth and energy demand sustainably. Investment in clean energy technologies, including solar power and energy-efficient appliances such as cookstoves, is critical to reducing indoor exposure to PAHs and NPAHs. Electrifying public transportation systems, including the use of electric vehicles (EVs) in the form of buses, taxis and cars, can significantly cut PAH emissions from urban traffic and improve ambient air quality. This pivot will deliver collective benefits, including improving public and population health and reducing environmental degradation.

This shift will improve air quality by drastically reducing the level of ambient, indoor and environmental pollutants, including PAHs and NPAHs. Some regions in Africa receive high solar insolation, making them viable locations to harness solar energy. Solar energy has the potential to make power available to remote and off-grid areas. Small-scale solar solutions or technologies, including mobile solar systems or solar-powered mini-grids, can be invested in to provide off-grid power to rural communities, providing clean electricity and reducing the use and reliance of these communities on biomass for their cooking and heating needs. The North-Eastern region of Africa (the Horn of Africa) is known to have very high wind energy potential. This wind energy can generate electricity for household and industrial use if investments can be made in wind farms (large, medium or small scale). Additionally, geothermal energy, which is one of the most stable renewable energy sources, is heavily available in parts of East Africa (Kenya, Ethiopia, and Tanzania) and expanding geothermal plants in these regions can expand access to reliable electricity to rural dwellers.

In addition, hydropower, which is widely used in Africa, can also be expanded responsibly to sustainably provide power to millions of people. Agricultural wastes (crop residue, animal waste) and organic waste, which are abundant due to the agrarian nature of Africa, can also be converted to biogas as a cleaner energy source, replacing charcoal and wood for cooking in rural settings. Adopting new technologies and upgrading existing infrastructure and energy architecture is also a crucial step in Africa's transition. The adoption of batteries to store and transmit energy, especially in seasons when there is limited sunlight or water to service solar and hydropower plants, would ensure energy is accessible all year round. Smart grid technologies, which aid the efficient distribution of energy, thereby eliminating wastage, should be considered and integrated into infrastructures for renewable energy. Investing in energy-efficient appliances and creating demand for them, including efficient cookstoves, will contribute immensely to cutting back on PAH emissions and enhance population health.

(c) Strengthening Regulatory Frameworks and Policies

To strengthen regulatory frameworks and policy environments, governments need to explore innovative approaches to developing national policies and incorporating PAHs into existing policies and implementing them for a more robust environmental monitoring strategy. Policies need to be clear and targeted, including PAHs as part of routine air quality monitoring programs, especially in regions with high levels of vehicular emissions or industrial pollution, by expanding air quality monitoring stations to include sampling for PAHs and NPAHs alongside other air pollutants. PAH and NPAHs exposure data should also be integrated into public health surveillance systems to better understand the association between PAH and NPAHs exposure and disease outcomes including cancer and respiratory problems. Such policies must be evidence-based and must focus on transition to renewable and efficient energy sources.

Additionally, the policies must include incentives such as subsidies and tax breaks to create demand and buy-in for the transition from both individuals and the private sector. Policies must also include carbon pricing and emissions standards that will spur industries to embrace the transition to cleaner energy sources. These regulatory frameworks should also include green policies, in which governments can encourage car-free days and green urban planning to reduce vehicular emissions and adopt sustainable agricultural practices that can minimize PAHs from vegetation fires.

(d) Establishing Partnerships

Establishing partnerships and collaborations to attract catalytic foreign and domestic investments from entities such as the World Bank (WB), the African Development Bank (AfDB) and private investors will be critical to funding this pivot without it being a catastrophic investment expenditure for individual governments. This will also foster a seamless transfer of technology. Collaborations should be established with non-governmental organizations and civil society groups to assist in mobilizing financing and technical resources for PAH monitoring efforts and in disseminating monitoring data and raising public awareness of the risks of exposure to PAHs through the currently known diverse sources.

International institutions such as the World Health Organization (WHO), the United Nations Environment Programme (UNEP), and the United Nations Development Programme (UNDP), which are deeply invested in global environment efforts, should be partnered with to assist with strengthening local monitoring capacities and mobilizing resources for PAH and NPAHs monitoring initiatives. Collaborations with universities, research institutions and laboratories in developed countries with vast expertise and experience in PAH monitoring should be pursued as these will likely expose Africa to the use of advanced research methods, analytical tools, and potentially, additional funding. Additionally, Africa needs to be better represented and participate in global air quality initiatives and dialogues, as this will help the continent better align national policies with international standards for PAH and NPAH monitoring and reduction.

(e) **Capacity building and citizen awareness**

Investing in capacity-building initiatives to develop local expertise is crucial. Training, coaching and mentoring initiatives for local scientists, researchers, and technicians in the management—including installation, operationalization, and maintenance—of innovative systems, PAH and NPAH sampling, laboratory analysis, and data interpretation will empower and localize institutional capacity to lead monitoring efforts and reduce dependency on external experts. Workshops, conferences, and collaborative research and learning initiatives can be leveraged to promote knowledge exchange, experience and best practices sharing to improve monitoring capacities in the continent.

Citizen education and awareness are equally critical to the success of this transitional journey for Africa and can be funded through the partnership vehicle. Raising citizen awareness about PAH and NPAH pollution and its associated health risks is instrumental in achieving effective monitoring and long-term solutions. Hence, the government needs to invest in creating public awareness for buy-in to understand how transitioning to cleaner energy technologies can improve their quality of life, protect the environment and preclude sabotage. Strategies, including community-based public health campaigns about the health risks of PAH and NPAH exposure, particularly in households using solid fuels for cooking, can prove effective and foster buy-in for pivoting to cleaner technologies, such as improved cookstoves and solar-powered cooking solutions. Incorporating indigenous knowledge by considering culturally relevant practices, norms, customs and beliefs and how they impact local environmental pollution can also be effective in monitoring and deploying more sustainable options.

7 Conclusion

The exacerbation of ambient particulate matter pollution in Africa has been largely attributed to economic growth and urbanization. With the worsening impact of PAHs and NPAH on human health, economic gains in the continent may soon be eroded if prompt actions are not taken to intervene. Viable interventions will include a mix of policy shifts (car-free days, regulations on vehicle importation, air quality management, energy and fuel types,), adoption/adaption of technology (investment in renewable energy, clean cooking devices, solar technology, compressed natural gas) and citizen awareness through education (health education and promotion, and community-based participatory programmes).

References

Abdallah MA-E, Atia NN (2014) Atmospheric concentrations, gaseous–particulate distribution, and carcinogenic potential of polycyclic aromatic hydrocarbons in Assiut, Egypt. Environ Sci Pollut Res 21(13):8059–8069. https://doi.org/10.1007/s11356-014-2746-6

Adeola AO, Nsibande SA, Osano AM, Maghanga JK, Naudé Y, Forbes PBC (2022) Analysis of gaseous polycyclic aromatic hydrocarbon emissions from cooking devices in selected rural and urban kitchens in Bomet and Narok counties of Kenya. Environ Monit Assess 194(6):435. https://doi.org/10.1007/s10661-022-10062-3

Agbo KE, Walgraeve C, Eze JI, Ugwoke PE, Ukoha PO, Van Langenhove H (2021) A review on ambient and indoor air pollution status in Africa. Atmospheric Pollut Res 12(2):243–260. https://doi.org/10.1016/j.apr.2020.11.006

Ayi Fanou L, Mobio TA, Creppy EE, Fayomi B, Fustoni S, Møller P, Kyrtopoulos S, Georgiades P, Loft S, Sanni A, Skov H, Øvrebø S, Autrup H (2006) Survey of air pollution in Cotonou, Benin—air monitoring and biomarkers. Sci Total Environ 358(1–3):85–96. https://doi.org/10.1016/j.scitotenv.2005.03.025

Ayi-Fanou L, Avogbe PH, Fayomi B, Keith G, Hountondji C, Creppy EE, Autrup H, Rihn BH, Sanni A (2011) DNA-adducts in subjects exposed to urban air pollution by benzene and polycyclic aromatic hydrocarbons (PAHs) in Cotonou, Benin. Environ Toxicol 26(1):93–102. https://doi.org/10.1002/tox.20533

Bai L, Geng X, Liu X (2024) Review of polycyclic aromatic hydrocarbons pollution characteristics and carcinogenic risk assessment in global cooking environments. Environ Pollut 361:124816. https://doi.org/10.1016/j.envpol.2024.124816

Barhoumi B, Castro-Jiménez J, Guigue C, Goutx M, Sempéré R, Derouiche A, Achour A, Touil S, Driss MR, Tedetti M (2018) Levels and risk assessment of hydrocarbons and organochlorines in aerosols from a North African coastal city (Bizerte, Tunisia). Environ Pollut 240:422–431. https://doi.org/10.1016/j.envpol.2018.04.109

Bililign S, Brown SS, Westervelt DM, Kumar R, Tang W, Flocke F, Vizuete W, Ture K, Pope FD, Demoz B, Asa-Awuku A, Levelt PF, Kalisa E, Raheja G, Ndyabakira A, Gatari MJ (2024) East African Megacity Air Quality: Rationale and Framework for a Measurement and Modeling Program. Bull Am Meteorol Soc. https://doi.org/10.1175/BAMS-D-23-0098.1

Bortey-Sam N, Ikenaka Y, Akoto O, Nakayama SMM, Asante KA, Baidoo E, Obirikorang C, Saengtienchai A, Isoda N, Nimako C, Mizukawa H, Ishizuka M (2017) Oxidative stress and respiratory symptoms due to human exposure to polycyclic aromatic hydrocarbons (PAHs) in Kumasi, Ghana. Environ Pollut 228:311–320. https://doi.org/10.1016/j.envpol.2017.05.036

Deabji N, Fomba KW, Dos Santos Souza EJ, Mellouki A, Herrmann H (2024) Influence of anthropogenic activities on metals, sugars and PAHs in PM10 in the city of Fez, Morocco: Implications on air quality. Environ Sci Pollut Res 31(17):25238–25257. https://doi.org/10.1007/s11356-024-32740-0

ECA (2022, February 9) Africa's transport sector to strongly benefit from African Continental Free Trade Area (AfCFTA). https://www.uneca.org/stories/africa%E2%80%99s-transport-sector-to-strongly-benefit-from-african-continental-free-trade-area-#:~:text=Addis%20Ababa%2C%209%20February%202022,Commission%20for%20Africa%20(ECA)

Ephraim-Emmanuel BC, Ordinioha B (2021) Exposure and Public Health Effects of Polycyclic Aromatic Hydrocarbon Compounds in Sub-Saharan Africa: A Systematic Review. Int J Toxicol 40(3):250–269. https://doi.org/10.1177/10915818211002487

Forbes PBC (2012) Particle emissions from household fires in South Africa. WIT Trans Ecol Environ157: 445–456.

Forbes PBC, Rohwer ER (2010) Application of laser induced fluorescence to monitor atmospheric polycyclic aromatic hydrocarbons. WIT Trans Ecol Environ 136: 367-377.

Gordon JND, Bilsback KR, Fiddler MN, Pokhrel RP, Fischer EV, Pierce JR, Bililign S (2023) The Effects of Trash, Residential Biofuel, and Open Biomass Burning Emissions on Local and Transported $PM_{2.5}$ and Its Attributed Mortality in Africa. GeoHealth 7(2):e2022GH000673. https://doi.org/10.1029/2022GH000673

Health Effects Institute (HEI). 2024. State of Global Air 2024. Available: www.stateofglobalair.org [accessed 02/15/2015]
Health Effects Institute (HEI). 2022. State of Global Air 2024. Available: www.stateofglobalair.org [accessed 02/15/2015]
Health Effects Institute (HEI). 2020. State of Global Air 2024. Available: www.stateofglobalair.org [accessed 02/15/2015].
Hisamuddin NH, Jalaludin J (2023) Children's exposure to polycyclic aromatic hydrocarbon (PAHs): a review on urinary 1-hydroxypyrene and associated health effects. Rev Environ Health 38(1):151–168. https://doi.org/10.1515/reveh-2021-0013
IHME I (2019) State of Global Air 2019 Report. In: Inst. Health Metr. Eval. https://www.healthdata.org/news-release/state-global-air-2019-report. Accessed 26 Aug 2022
Iwegbue CMA, Ehigbor MJ, Tesi GO, Eguavoen OI, Martincigh BS (2022) Occurrence, Sources and Exposure Risk of Polycyclic Aromatic Hydrocarbons (PAHs) in Street Dusts from the Nigerian Megacity, Lagos. Polycycl Aromat Compd 42(1):49–69. https://doi.org/10.1080/10406638.2020.1716027
Kalisa E, Clark ML, Ntakirutimana T, Amani M, Volckens J (2023) Exposure to indoor and outdoor air pollution in schools in Africa: Current status, knowledge gaps, and a call to action. Heliyon 9(8):e18450. https://doi.org/10.1016/j.heliyon.2023.e18450
Kalisa E, Nagato EG, Bizuru E, Lee KC, Tang N, Pointing SB, Hayakawa K, Archer SDJ, Lacap-Bugler DC (2018) Characterization and Risk Assessment of Atmospheric PM2.5 and PM10 Particulate-Bound PAHs and NPAHs in Rwanda, Central-East Africa. Environ Sci Technol 52(21):12179–12187. https://doi.org/10.1021/acs.est.8b03219
Lammel G (2015) Long-range atmospheric transport of polycyclic aromatic hydrocarbons is worldwide problem - results from measurements at remote sites and modelling. Acta Chim Slov :729–735. https://doi.org/10.17344/acsi.2015.1387
Lammel G, Heil A, Stemmler I, Dvorská A, Klánová J (2013) On the Contribution of Biomass Burning to POPs (PAHs and PCDDs) in Air in Africa. Environ Sci Technol 47(20):11616–11624. https://doi.org/10.1021/es401499q
Liang S, Yan X, An B, Yang Y, Liu M (2022) Distribution and ecological risk assessment of polycyclic aromatic hydrocarbons in different compartments in Kenya: A review. Emerg Contam 8:351–359. https://doi.org/10.1016/j.emcon.2022.08.001
Morakinyo OM, Mokgobu MI, Mukhola MS, Hunter RP (2016) Health Outcomes of Exposure to Biological and Chemical Components of Inhalable and Respirable Particulate Matter. Int J Environ Res Public Health 13(6):592. https://doi.org/10.3390/ijerph13060592
Morakinyo OM, Mukhola MS, Mokgobu MI (2020) Concentration levels and carcinogenic and mutagenic risks of PM2.5-bound polycyclic aromatic hydrocarbons in an urban–industrial area in South Africa. Environ Geochem Health 42(7):2163–2178. https://doi.org/10.1007/s10653-019-00493-2
Munyeza CF, Osano AM, Maghanga JK, Forbes PBC (2019a) Polycyclic Aromatic Hydrocarbon Gaseous Emissions from Household Cooking Devices: A Kenyan Case Study. Environ Toxicol Chem 39(3):538–547. https://doi.org/10.1002/etc.4648
Munyeza CF, Rohwer ER, Forbes PBC (2019b) A review of monitoring of airborne polycyclic aromatic hydrocarbons: An African perspective. Trends Environ Anal Chem 24:e00070. https://doi.org/10.1016/j.teac.2019.e00070
Mwachiro MM, Pritchett N, Calafat AM, Parker RK, Lando JO, Murphy G, Chepkwony R, Burgert SL, Abnet CC, Topazian MD, White RE, Dawsey SM, Etemadi A (2021) Indoor wood combustion, carcinogenic exposure and esophageal cancer in southwest Kenya. Environ Int 152:106485. https://doi.org/10.1016/j.envint.2021.106485
Neumann M, Dlamini WN, Sallah-Ud-Din R, Berekute AK, Siregar S, Getnet ME, Maulana M, Pan W-C, Lung S-CC, Yu K-P (2024) Assessment of air pollution emitted during cooking using biomass and cleaner fuels in the Shiselweni region of Eswatini (Swaziland). Clean Technol Environ Policy 26(9):3003–3020. https://doi.org/10.1007/s10098-024-02786-2

Ofori SA, Cobbina SJ, Doke DA (2020) The occurrence and levels of polycyclic aromatic hydrocarbons (PAHs) in African environments—a systematic review. Environ Sci Pollut Res 27(26):32389–32431. https://doi.org/10.1007/s11356-020-09428-2

Ofori SA, Cobbina SJ, Imoro AZ, Doke DA, Gaiser T (2021) Polycyclic Aromatic Hydrocarbon (PAH) Pollution and its Associated Human Health Risks in the Niger Delta Region of Nigeria: a Systematic Review. Environ Process 8(2):455–482. https://doi.org/10.1007/s40710-021-00507-1

Onaiwu GE, Ifijen IH (2024) PM2.5-bound polycyclic aromatic hydrocarbons (PAHs): quantification and source prediction studies in the ambient air of automobile workshop using the molecular diagnostic ratio. Asian J Atmospheric Environ 18(1):6. https://doi.org/10.1007/s44273-024-00027-y

Raheja G, Sabi K, Sonla H, Gbedjangni EK, McFarlane CM, Hodoli CG, Westervelt DM (2022) A Network of Field-Calibrated Low-Cost Sensor Measurements of $PM_{2.5}$ in Lomé, Togo, Over One to Two Years. ACS Earth Space Chem 6(4):1011–1021. https://doi.org/10.1021/acsearthspacechem.1c00391

Roomaney RA, Cairncross E, Tesfaye M, Kapwata T, Abdulatif N, Olivier C, Mathibela K, Cois A, Neethling I, Botai J, Turawa EB, Awotiwon OF, Chetty K, Nojilana B, Wright CY, Pacella R, Bradshaw D, Pillay-van Wyk V (2022) Estimating the burden of disease attributable to ambient air pollution (ambient PM2.5 and ambient ozone) in South Africa for 2000, 2006 and 2012. S Afr Med J :705–717. https://doi.org/10.7196/SAMJ.2022.v112i8b.16483

Sam K, Onyena AP, Zabbey N, Odoh CK, Nwipie GN, Nkeeh DK, Osuji LC, Little DI (2023) Prospects of emerging PAH sources and remediation technologies: insights from Africa. Environ Sci Pollut Res 30(14):39451–39473. https://doi.org/10.1007/s11356-023-25833-9

Schmidl C, Marr IL, Caseiro A, Kotianová P, Berner A, Bauer H, Kasper-Giebl A, Puxbaum H (2008) Chemical characterisation of fine particle emissions from wood stove combustion of common woods growing in mid-European Alpine regions. Atmos Environ 42(1):126–141. https://doi.org/10.1016/j.atmosenv.2007.09.028

Shi Y, Matsunaga T, Yamaguchi Y (2015) High-Resolution Mapping of Biomass Burning Emissions in Three Tropical Regions. Environ Sci Technol 49(18):10806–10814. https://doi.org/10.1021/acs.est.5b01598

SLOCAT (2021) SLOCAT Transport, Climate and Sustainability Global Status Report. https://tcc-gsr.com/

Ssepuya F, Odongo S, Musa Bandowe BA, Abayi JJM, Olisah C, Matovu H, Mubiru E, Sillanpää M, Karume I, Kato CD, Shikuku VO, Ssebugere P (2022) Polycyclic aromatic hydrocarbons in breast milk of nursing mothers: Correlates with household fuel and cooking methods used in Uganda, East Africa. Sci Total Environ 842:156892. https://doi.org/10.1016/j.scitotenv.2022.156892

Thompson N, Adjei JK, Bentum JK, Essumang DK, Duodu GO, Hadzi G, Adjei GA (2022) Vehicular influence on atmospheric concentrations and source apportionment of polycyclic aromatic hydrocarbons in some major cities in three regions of Ghana using epiphytic lichens. Toxicol Rep 9:1691–1699. https://doi.org/10.1016/j.toxrep.2022.08.002

Van Der Werf GR, Randerson JT, Giglio L, Collatz GJ, Mu M, Kasibhatla PS, Morton DC, DeFries RS, Jin Y, Van Leeuwen TT (2010) Global fire emissions and the contribution of deforestation, savanna, forest, agricultural, and peat fires (1997–2009). Atmospheric Chem Phys 10(23):11707–11735. https://doi.org/10.5194/acp-10-11707-2010

WHO (2021) Human health effects of polycyclic aromatic hydrocarbons as ambient air pollutants - Report of the Working Group on Polycyclic Aromatic Hydrocarbons of the Joint Task Force on the Health Aspects of Air Pollution. World Health Organization

World Health Organization (2021) WHO global air quality guidelines: particulate matter (PM2.5 and PM10), ozone, nitrogen dioxide, sulfur dioxide and carbon monoxide. World Health Organization https://www.who.int/publications/i/item/9789240034228

Wu M, Luo J, Huang T, Lian L, Chen T, Song S, Wang Z, Ma S, Xie C, Zhao Y, Mao X, Gao H, Ma J (2022) Effects of African BaP emission from wildfire biomass burning on regional and global environment and human health. Environ Int 162:107162. https://doi.org/10.1016/j.envint.2022.107162

Environmental Reaction

Formation of a Quinoid Polycyclic Aromatic Hydrocarbon Promoted by Mineral Dust Aerosols

Takayuki Kameda and Daiki Nojo

Abstract Quinoid polycyclic aromatic hydrocarbons (PAH quinones; PAHQs), which are known to cause adverse health effects, are partly formed by the atmospheric oxidation of their parent compounds, PAHs. However, the heterogeneous oxidation of PAHs by gaseous O_3 on natural mineral substrates, such as desert dust aerosols, is poorly understood. This study examines the potential for the secondary formation of a representative PAHQ, 9,10-anthraquinone (AntQ), by the oxidation of anthracene (Ant) on mineral dust aerosols. Experiments with a flow reactor and analysis of the water content on the substrate surfaces showed that the decay rate of Ant with O_3 was typically high when the surface water content on the substrates was increased. We showed through surface analysis by temperature-programed desorption using NH_3 that the Ant–O_2 reaction was accelerated on the acidic surfaces of the mineral dust, particularly on the clay minerals. Both processes led to AntQ formation, suggesting that chemical reactions on the surface of the mineral dust aerosols, including clay minerals, result in the unintended formation of toxic organic compounds in the atmosphere.

Keywords PAH-quinones · Heterogeneous reaction · Oxidation · Secondary formation · Atmospheric reaction

1 Introduction

Polycyclic aromatic hydrocarbon quinones (PAHQs), a major class of hazardous substances. For example, in addition to disrupting signaling pathways by arylating cellular proteins, PAHQs can transfer electrons from reducing agents to molecular oxygen, generating reactive oxygen species (ROS) that cause oxidative stress

T. Kameda (✉)
Graduate School of Energy Science, Kyoto University, Yoshida-Honmachi, Sakyo-Ku, Kyoto, Japan
e-mail: tkameda@energy.kyoto-u.ac.jp

D. Nojo
Faculty of Engineering, Kyoto University, Yoshida-Honmachi, Sakyo-Ku, Kyoto, Japan

K. Hayakawa and Y. Aoki (eds.), *A New Era of Polycyclic Aromatic Hydrocarbons*,
https://doi.org/10.1007/978-981-96-7695-8_4

(Kumagai 2009). Furthermore, PAHQs can bind to aryl hydrocarbon receptors (AhR) and thyroid hormone receptors (TR), potentially causing endocrine disruption (Bekki et al. 2009). PAHQs are commonly detected in airborne particulate matter (Cho et al. 2004; Barradas-Gimate et al. 2017; Yang et al. 2018). These quinones are formed through atmospheric chemical reactions involving polycyclic aromatic hydrocarbons (PAHs) (Lee et al. 2015) as well as through human activities, including fuel combustion (Cho et al. 2004; Jakober et al. 2007). Some PAHQ types are formed via the gas-phase reactions of semi-volatile PAHs and are subsequently deposited on airborne particulates. For example, 9,10-phenanthrenequinone is formed from the gas-phase reaction of phenanthrene with OH or NO_3 radicals in the presence of O_2 and NO_2 (Helmig and Harger 1994; Kwok et al. 1994; Wang et al. 2007). One of the most abundant PAHQs is 9,10-anthraquinone (AntQ), which is formed through the combustion of fossil fuels such as gasoline and diesel fuel (Cho et al. 2004; Jakober et al. 2007; Wang et al. 2023). AntQ can also be formed from gas-phase radical-initiated reactions (Lee et al. 2015) or gas-particle phase heterogeneous reactions (Kwamena et al. 2006; Nájera et al. 2011; Chen and Zhu 2014). It is formed by the reaction of anthracene (Ant) with gaseous O_3 on various substrates such as glass surfaces (Kwamena et al. 2006), octanol thin films (Kahan et al. 2006), silica particles (Perraudin et al. 2007), and phenylsiloxane oil or azelaic acid aerosols (Kwamena et al. 2007). Although previous studies on heterogeneous PAHQ formation used simple, representative materials to simulate atmospheric aerosols, these substances fall short of replicating the complexity of real-world aerosols and are thus less effective for studying heterogeneous PAHQ formation. Studies exploring the heterogeneous formation of atmospheric 1-nitropyrene (1-NP), a nitro-polycyclic aromatic hydrocarbon, showed that the reaction rate and yield of 1-NP on natural mineral substrates markedly exceeded those on conventional substrates such as silica and metal oxides (Kameda et al. 2016). Mineral dust is a major component of airborne particulates on a global scale (Cwiertny et al. 2008). It is transported by winds from deserts or semiarid regions (Tanaka and Chiba 2006), which account for 40% of the total world land area (Fernández 2002). Organic compounds absorbed onto mineral dust surfaces can have notable health effects (Falkovich et al. 2004). Nonetheless, the heterogeneous oxidation of PAHs by gaseous O_3 on natural mineral substrates, such as desert dust aerosols, remains unexplored.

Our hypothesis is that the heterogeneous formation of PAHQs on mineral dust may have been underestimated in significance, as natural mineral aerosols possess a more reactive surface than the model materials that were utilized in previous studies. To test this hypothesis, we investigated the impact of authentic mineral dust on the formation of AntQ from Ant and O_3 under conditions involving O_2. The findings suggest that mineral dust aerosols substantially influence the Ant-to-AntQ conversion process.

2 Methods

2.1 Heterogeneous Reaction of Ant with O_3

The surface of the substrates was coated with Ant by a standard procedure based on liquid–solid adsorption. In brief, 10 mL of 30 μmol L^{-1} acetonitrile solution of Ant was well mixed with 300 mg of a substrate in a Pyrex recovery flask, after which acetonitrile was gently removed using a rotary evaporator at 323 K under reduced pressure in the dark. When all the Ant molecules were adsorbed on the surface, the amount of supported Ant per mg of the substrate was calculated as 1 nmol mg^{-1}. At this concentration, we estimated that the surface coverage of Ant was less than 0.7 for all substrates. Hence, Ant was considered a monolayer in the reactions, assuming uniform adsorption.

Ant was heterogeneously reacted with 1 ppm of gaseous O_3/air in a Pyrex flow reactor (6 cm ID × 18 cm height, ca. 500 mL) under constant reaction conditions at 298 ± 1 K and < 5% relative humidity in the dark. A portion of the Ant-coated substrate (about 10 mg) was spread on a Pyrex Petri dish (1.9 cm ID × 1.2 cm depth). The Petri dish was placed in the reactor and subsequently exposed to O_3/air supplied by passing pure air through an ozone generator (EcoDesign, ED-OG-L7) at a constant flow rate. The relative humidity and the O_3 concentration in the reaction system were measured with a humidity probe (Vaisala, HMI41) and an ozone analyzer (Dylec, Model 1200). The reaction products and residual Ant after the prescribed reaction time were extracted with acetonitrile. The extracted chemicals were identified and quantified by gas chromatographic–mass spectrometry (GC/MS) by comparing their retention times, fragmentation patterns, and peak areas with those of authentic standards, as described below.

2.2 GC/MS Analysis of Soluble Organic Fractions from Laboratory Experiment Samples

GC/MS analysis of the reaction products and the residual reactant was performed using a mass spectrometer (Shimadzu, GC-2010 connected to QP-2010 Plus) with the electron ionization (EI) method in the selected ion monitoring (SIM) mode: ionization voltage, 70 eV; interface temperature, 573 K; ion source temperature, 573 K. The GC column used was a ZB-5MS (5% phenylmethylsilicone, 0.25 mm ID × 30 m, Phenomenex). The temperature program for the GC/MS analysis was as follows: 323 K for 4 min, to 583 K with a ramping rate of 15 K min^{-1}, isothermal for 20 min. The monitored ions were *m/z* 179, 178 and 176 for Ant; *m/z* 188, 187 and 187 for Ant-d_{10}; *m/z* 208, 180 and 152 for AntQ; and *m/z* 216, 188 and 160 for AntQ-d_8. Ant-d_{10} and AntQ-d_8 were added as internal standards for analysis.

2.3 Curve Fitting

To quantitatively evaluate the degradation rate of Ant on each substrate, the kinetics of the reactions of Ant adsorbed on the substrates tested in this study were determined by observing Ant decay as a function of the reaction time. The Ant degradation showed an exponential pattern, suggesting that the reactions could be reasonably described by pseudo-first-order kinetics. In our experiments, the reactions of Ant adsorbed on some substrates were not complete and finally reached a plateau independent of the reaction time. In such cases, the experimental data were fitted by a plateau-shifted first-order exponential function, as shown in Eq. (1):

$$[\mathrm{Ant}]_t = [\mathrm{Ant}]_{\mathrm{plateau}} + \left([\mathrm{Ant}]_0 - [\mathrm{Ant}]_{\mathrm{plateau}}\right)e^{-k_{\mathrm{obs}}t}, \quad (1)$$

where $[\mathrm{Ant}]_t$ is the concentration of adsorbed Ant at a given time, $[\mathrm{Ant}]_0$ is the initial concentration of adsorbed Ant, $[\mathrm{Ant}]_{\mathrm{plateau}}$ is the concentration of adsorbed Ant at the plateau, and k_{obs} is the apparent rate constant of the pseudo-first-order reaction. In Eq. (1), both k_{obs} and $[\mathrm{Ant}]_{\mathrm{plateau}}$ are fitting parameters. In the case with no plateau, the experimental data were fitted with a simple first-order exponential function.

The heterogeneous reaction results were expressed as the mean from three independent experiments $\pm$ standard deviation (SD). Nonlinear least-squares curve fitting based on the Levenberg–Marquardt algorithm was performed using KaleidaGraph 5.0 J (Hulinks). The errors in the fitting parameters represent one standard-error value.

2.4 Characterization of the Substrates

The surface acidity of the substrates was evaluated by temperature-programed desorption (TPD) using NH_3 as a probe. TPD analysis was performed using a multiple adsorption apparatus (Microtrac BEL, TPD-1-ATw). Approximately 50 mg of the substrate was heated at 573 K for 1 h under a He flow at 50 mL min^{-1}, followed by NH_3 adsorption at 373 K by passing 0.5% NH_3/He through the substrate at a flow rate of 100 mL min^{-1} for 1 h and further He exposure at 373 K for 0.5 h in order to remove residual NH_3 from the sample surface. Subsequently, the sample was heated from 373 to 873 K under flowing He at a rate of 10 K min^{-1} for NH_3 desorption. The desorbed gas was monitored using a mass spectrometer by tracking the m/z = 16.

The water content of the substrates was measured according to the Japanese Industrial Standard (JIS) C 2142. Briefly, the samples were placed under an atmosphere of 298 K and 30% relative humidity for 2 days, after which they were dried in a desiccator set at 383 K for 24 h. The mass of the samples after humidity conditioning and drying was measured. Sample water amount and sample water content were calculated by subtracting the sample mass after drying from the sample mass after humidity conditioning.

2.5 *Materials*

Arizona Test Dust (ATD; standard test dust made from Arizona desert sand, ISO 12103-1, A2 Fine Test Dust) was obtained from Power Technology. Quartz (Silica) was purchased from Kanto Chemical. Montmorillonite (Mont M) (JCSS3102) was obtained from the Clay Science Society of Japan. Kaolin (JCRM R605) was obtained from the Ceramic Society of Japan. Feldspar (JF-1) and limestone (JLs-1) were obtained from the Geological Survey of Japan. Montmorillonite K10 (Mont K) was purchased from Sigma-Aldrich.

3 Results and Discussion

3.1 *O_3 Exposure Experiments of the Particle-Bound Ant*

Ant degradation was measured under 1 ppmv O_3-air in the dark. In Figs. 1, 2, 3, 4, 5 and 6, the amounts of formed AntQ and the remaining Ant were normalized to the sum of the initial amounts of Ant and AntQ on each substrate. On the ATD particles, more than 80% of the initial amount of Ant was degraded, and the yield of AntQ reached 36% after a reaction time of 1 h (Fig. 1). On feldspar particles, Ant was slowly degraded, reaching a ~ 15% yield of AntQ in 1 h (Fig. 2). Desert dust often comprises various minerals, such as quartz, carbonates, feldspars, and clay minerals (Usher et al. 2003). To quantify the Ant degradation rate on each substrate, the kinetics of the heterogeneous reaction between O_3 and Ant adsorbed on the substrates tested in this study were determined by following the Ant consumption as a function of O_3 exposure time. We found that the Ant degradation rate on each substrate could be fitted to a pseudo-first-order exponential function (see Methods) using nonlinear least-squares fitting. The apparent rate constant values of the pseudo-first-order reaction, k_{obs}, were 2.01 and 0.65 h^{-1} on ATD and feldspar, respectively, at an O_3 concentration of 1 ppmv (Table 1). To determine which components contribute to the rapid degradation of Ant, we compared k_{obs} on various substrates that generally constitute desert dust. The most active components were natural kaolin, followed by ATD (Table 1). In these cases, the conversion of Ant to AntQ was completed within 2 h. The least active component was limestone, for which k_{obs} was 0.02 h^{-1}, 100 times lower than that for ATD.

In addition to k_{obs}, Table 1 shows the concentrations of Ant and AntQ at $t = 0$ ($[Ant]_0$ and $[AntQ]_0$, respectively), and the water and acid site concentrations on the substrate surface ($[H_2O]_{surf}$/wt% and $[NH_3]_{TPD}$/μmol g^{-1}, respectively). Although

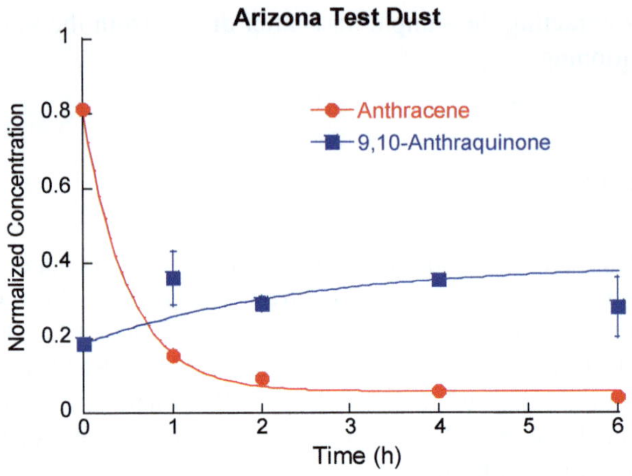

Fig. 1 Normalized concentrations of anthracene (Ant) and 9,10-anthraquinone (AntQ) on Arizona Test Dust (ATD) after exposure to 1 ppmv O_3 for the indicated times. The data points represent the mean values (± 1 SD) of triplicate experiments: circles, Ant; squares, AntQ. The amounts of formed AntQ and the remaining Ant were normalized to the sum of the initial amounts of Ant and AntQ. The curve for the Ant decay is an exponential nonlinear least-squares fit assuming a first-order reaction. See Methods for details. The curve for AntQ formation is only for illustrative purposes

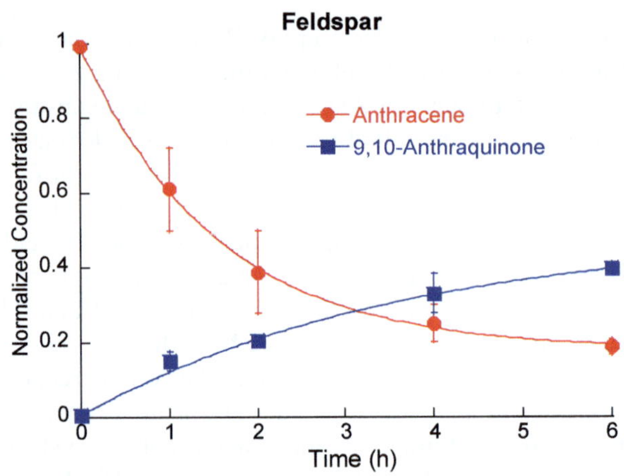

Fig. 2 Normalized concentrations of anthracene (Ant) and 9,10-anthraquinone (AntQ) on feldspar after exposure to 1 ppmv O_3 for the indicated times. The data points represent the mean values (± 1 SD) of triplicate experiments: circles, Ant; squares, AntQ. The amounts of formed AntQ and the remaining Ant were normalized to the sum of the initial amounts of Ant and AntQ. The curve for the Ant decay is an exponential nonlinear least-squares fit assuming a first-order reaction. See Methods for details. The curve for AntQ formation is only for illustrative purposes

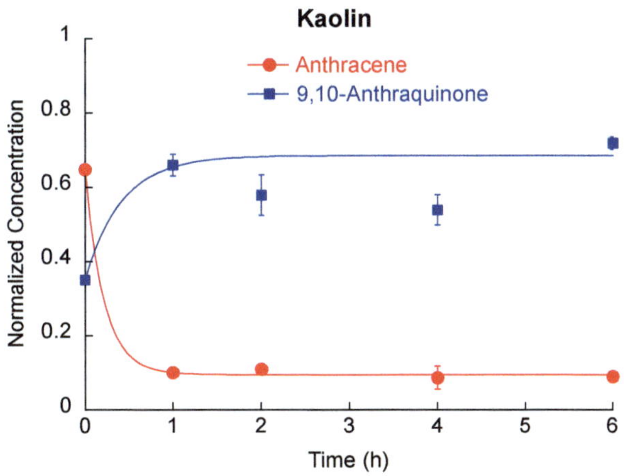

Fig. 3 Normalized concentrations of anthracene (Ant) and 9,10-anthraquinone (AntQ) on kaolin after exposure to 1 ppmv O_3 for the indicated times. The data points represent the mean values (± 1 SD) of triplicate experiments: circles, Ant; squares, AntQ. The amounts of formed AntQ and the remaining Ant were normalized to the sum of the initial amounts of Ant and AntQ. The curve for the Ant decay is an exponential nonlinear least-squares fit assuming a first-order reaction. See Methods for details. The curve for AntQ formation is only for illustrative purposes

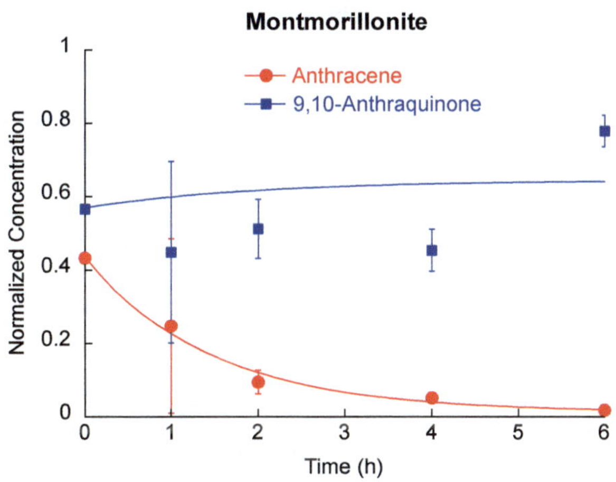

Fig. 4 Normalized concentrations of anthracene (Ant) and 9,10-anthraquinone (AntQ) on montmorillonite (Mont M) after exposure to 1 ppmv O_3 for the indicated times. The data points represent the mean values (± 1 SD) of triplicate experiments: circles, Ant; squares, AntQ. The amounts of formed AntQ and the remaining Ant were normalized to the sum of the initial amounts of Ant and AntQ. The curve for the Ant decay is an exponential nonlinear least-squares fit assuming a first-order reaction. See Methods for details. The curve for AntQ formation is only for illustrative purposes

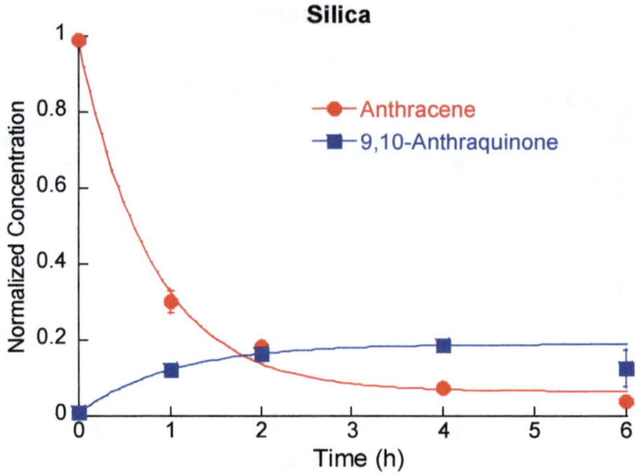

Fig. 5 Normalized concentrations of anthracene (Ant) and 9,10-anthraquinone (AntQ) on silica after exposure to 1 ppmv O_3 for the indicated times. The data points represent the mean values (± 1 SD) of triplicate experiments: circles, Ant; squares, AntQ. The amounts of formed AntQ and the remaining Ant were normalized to the sum of the initial amounts of Ant and AntQ. The curve for the Ant decay is an exponential nonlinear least-squares fit assuming a first-order reaction. See Methods for details. The curve for AntQ formation is only for illustrative purposes

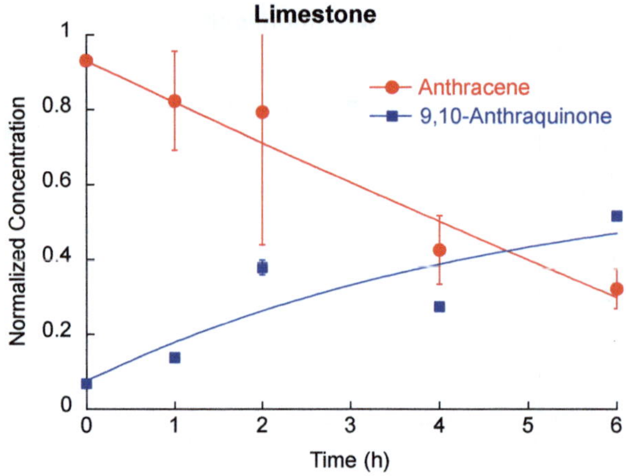

Fig. 6 Normalized concentrations of anthracene (Ant) and 9,10-anthraquinone (AntQ) on limestone after exposure to 1 ppmv O_3 for the indicated times. The data points represent the mean values (± 1 SD) of triplicate experiments: circles, Ant; squares, AntQ. The amounts of formed AntQ and the remaining Ant were normalized to the sum of the initial amounts of Ant and AntQ. The curve for the Ant decay is an exponential nonlinear least-squares fit assuming a first-order reaction. See Methods for details. The curve for AntQ formation is only for illustrative purposes

Table 1 Observed pseudo-first-order rate constants for the reaction of Ant on the substrates examined in this study with 1 ppmv O_3 (k_{obs}), observed initial concentrations of AntQ ($[AntQ]_0$) and Ant ($[Ant]_0$), water content on the substrates ($[H_2O]_{surf}$), and surface acidity of the substrates ($[NH_3]_{TPD}$)

Substrates	k_{obs}/h^{-1}	$[AntQ]_0/\mu M$[a]	$[Ant]_0/\mu M$[b]	$[H_2O]_{surf}/\%$[c]	$[NH_3]_{TPD}/\mu mol\ g^{-1}$[f]
Kaolin[d]	4.65	3.32	6.13	1.35	167
Arizona test dust (ATD)	2.01	1.57	6.82	1.32	NA[†]
Silica	1.04	0.09	9.56	0.27	10
Montmorillonite (Mont M)	0.85	5.66	4.34	6.62	78
Feldspar	0.65	0.08	10.55	0.09	14
Limestone	0.02	0.64	8.46	0.07	10
Montmorillonite K10 (Mont K)[e]	NA	5.16	NA	6.39	152

NA, not available
[a]Concentration in the sample solution for GC/MS measurement; the expected initial concentrations of AntQ and Ant are 0 and 10 μM, respectively
[b]Not measured due to the carbonate-derived interference peaks
[c]On the basis of weight
[d]Note that kaolin consists largely of kaolinite
[e]Acid-activated montmorillonite
[f]NH_3 adsorption per unit weight of the substrate

the $[AntQ]_0$ should be zero, we found that nonnegligible amounts of AntQ were generated on ATD, Kaolin, Mont M, and Mont K at the initial stage of the reactions (Table 1; Figs. 1, 3, and 4). In particular, on Mont K at $t = 0$, no Ant was detected, while AntQ was formed in amounts corresponding to 52% of the initial amount of Ant (Table 1). Notably, these substrates may have acted as catalysts when Ant was oxidized by O_2 in the air or in the solvents used for pretreatment.

AntQ formation from the reaction of Ant with O_2 on solid-state catalysts has been reported previously (Cepeda and Diaz 1987). A mechanism has been proposed for AntQ formation by adding O_2 to the 9,10-carbon following the formation of the radical cation of Ant at the Lewis acid sites on the catalyst surface (Fig. 7). Similar mechanisms for radical cation formation on natural clay or cation-modified clay minerals have also been proposed (Li et al. 2015; Jia et al. 2018; González Henao et al. 2022). Therefore, the relationship between the acidic properties of the substrate surface and the $[AntQ]_0$ was examined, and the $[AntQ]_0$ tended to be higher on the substrate with a higher surface acidity ($[NH_3]_{TPD}/mmol\ g^{-1}$) (Fig. 8). Kaolin, Mont M, and Mont K, where the reaction proceeded rapidly, are all clay minerals. Clay minerals are also present in ATD, which is natural desert dust (Kameda et al. 2016). Clay minerals generally exhibit Lewis acidity, which is derived from the electron-accepting sites in structures (Kameda et al. 2016). Based on the findings shown in

Fig. 8, the Lewis acid sites on ATD, Kaolin, Mont M, and Mont K may have acted as catalysts, with the reaction proceeding via a mechanism similar to the one in Fig. 7.

The reaction rate constant k_{obs} was the highest for kaolin ($k_{obs} = 4.65\ h^{-1}$), followed by ATD ($k_{obs} = 2.01\ h^{-1}$), Silica ($k_{obs} = 1.04\ h^{-1}$), Mont M ($k_{obs} =$

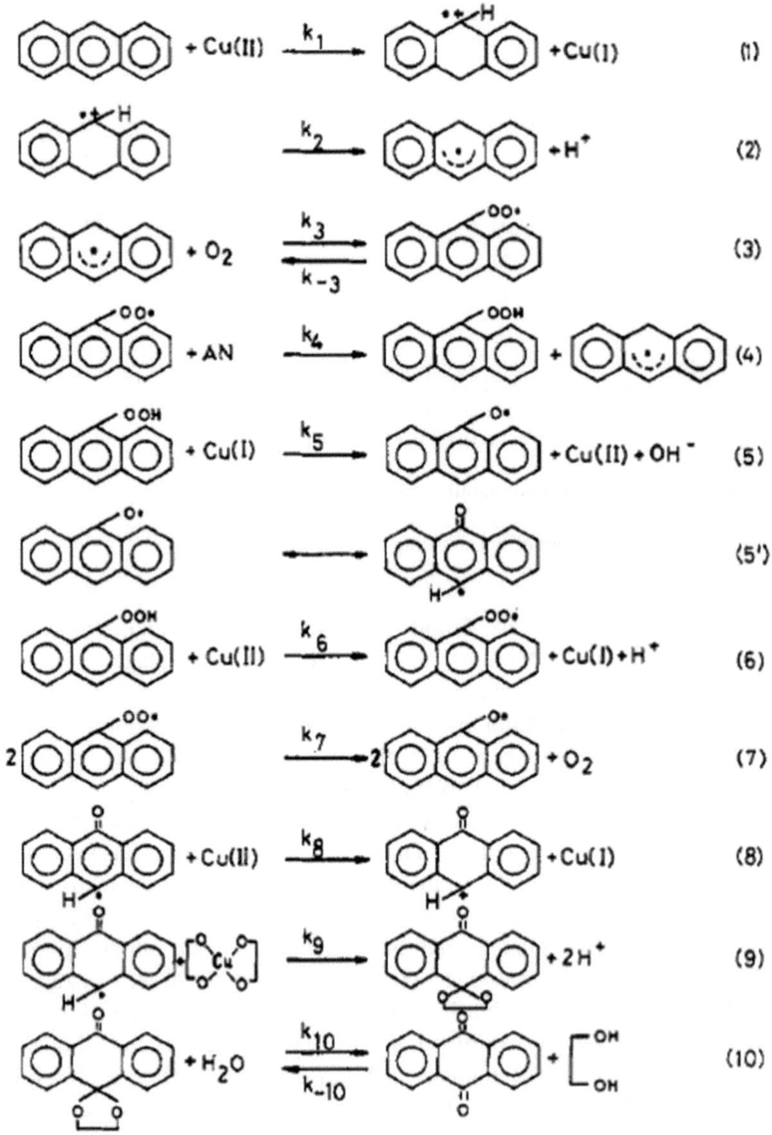

Fig. 7 Proposed mechanisms for the formation of 9,10-anthraquinone via the oxidation of anthracene by O_2 with Cu (II) catalysis. Adapted with permission from Cepeda and Diaz (1987). Copyright 1987 American Chemical Society

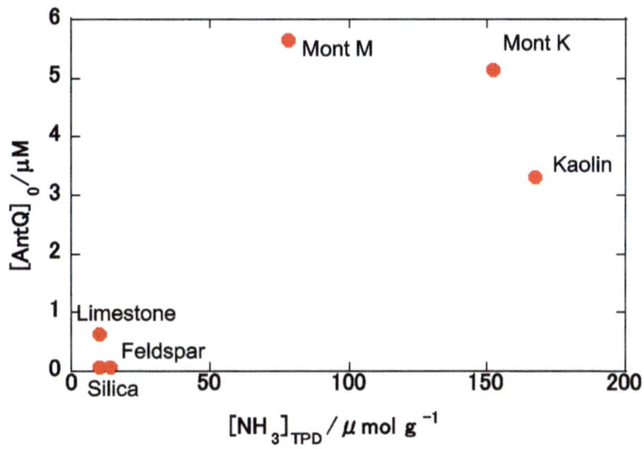

Fig. 8 Relationship between $[AntQ]_0$ and the surface acidity ($[NH_3]_{TPD}$) of the substrates

0.85 h^{-1}), feldspar ($k_{obs} = 0.65\ h^{-1}$), and limestone ($k_{obs} = 0.02\ h^{-1}$) (Table 1). As described above, O_2 oxidation proceeded rapidly on Mont K. Since Ant had already disappeared in the initial stage of the reaction by O_3, evaluating the reaction rate of Ant on Mont K was not feasible.

Bailey et al. (1964) proposed a pathway for AntQ formation by reacting Ant with O_3 (Fig. 9). The most O_3-sensitive sites of Ant are the 9- and 10-carbon atoms, and the following two pathways are possible. In the first pathway, only one terminal oxygen atom of the O_3 molecule attacks the carbons at positions 9 and 10 of Ant one by one to release oxygen molecules, followed by Anthracen-9,10-diol (2OHAnt) to form quinones (direct reactions; solid and dashed arrows in Fig. 9). The other pathway involves adding O_3 across the 9- and 10-carbon of Ant to form a primary ozonide (Ant_POZ), and the ketone hydroperoxide (Ant_KHP) formed by the intramolecular rearrangement dehydrates to form a quinone (non-direct reaction; dotted arrow in Fig. 9). As shown in Fig. 9, water molecules are involved in the progression of the reaction in both pathways, and the quantity of coexisting water is expected to affect the AntQ formation rate.

Therefore, the relationship between the surface water content ($[H_2O]_{surf}$/wt%) and the Ant decay rate on each substrate was examined. The relationship between $[H_2O]_{surf}$ and k_{obs} for each substrate is shown in Fig. 10a. Except for Mont M, which had the highest water content, a good positive correlation (correlation coefficient $R =$ 0.86) was found between $[H_2O]_{surf}$ and k_{obs} (Fig. 10b). This result suggests that the water on the substrate surface plays a key role in the progression of Ant-ozonation.

As described above, Kaolin, Mont M, and Mont K, on which the reaction proceeded rapidly, are clay minerals. Clay minerals comprise layered aluminosilicates, which consist of tetrahedral silicate (T) and octahedral aluminate (O) sheets (Laszlo 1987). Clays are classified according to the relative number of T and O layers. For example, kaolinite comprises alternating O and T layers (OT structure). In contrast, in montmorillonite, one O layer is sandwiched between two T layers (TOT structure). Montmorillonites contain water molecules (interlayer water) between their layers. The Mont M used in this study showed a high water content, suggesting that

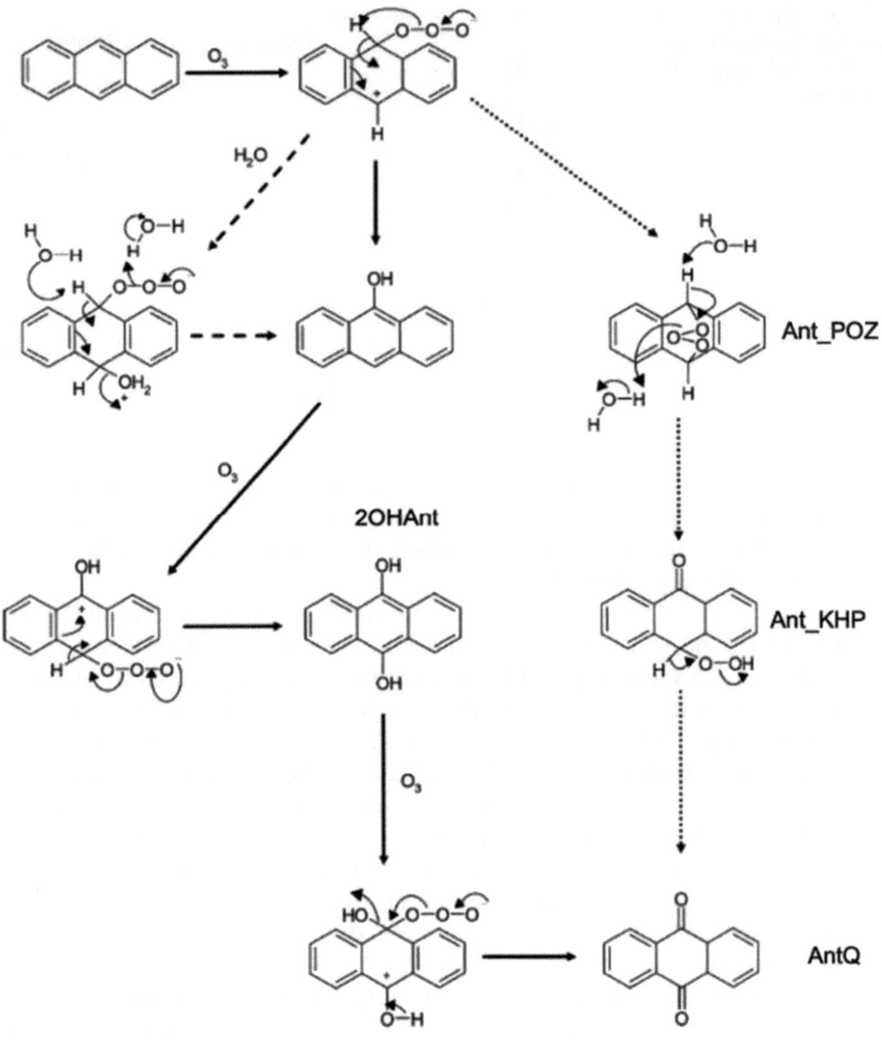

Fig. 9 The proposed mechanisms for the formation of 9,10-anthraquinone (AntQ) via the ozonation of anthracene. Adapted with permission from Kwamena et al. (2006). Copyright 2006 American Chemical Society

water was retained between the TOT layers. The amount of interlaminar water in the clay minerals varies markedly with humidity, and the basal spacing (thickness of the unit structure) varies. The ability of water to enter the interlaminar space and to extend the basal spacing is called swelling. Because the reaction experiments were conducted under dry conditions (at less than 5% RH) in this study, a narrow basal spacing was adopted, and organic molecules, such as hydrophobic Ant, might not have efficiently filled in-between the layers, which were the reaction active sites.

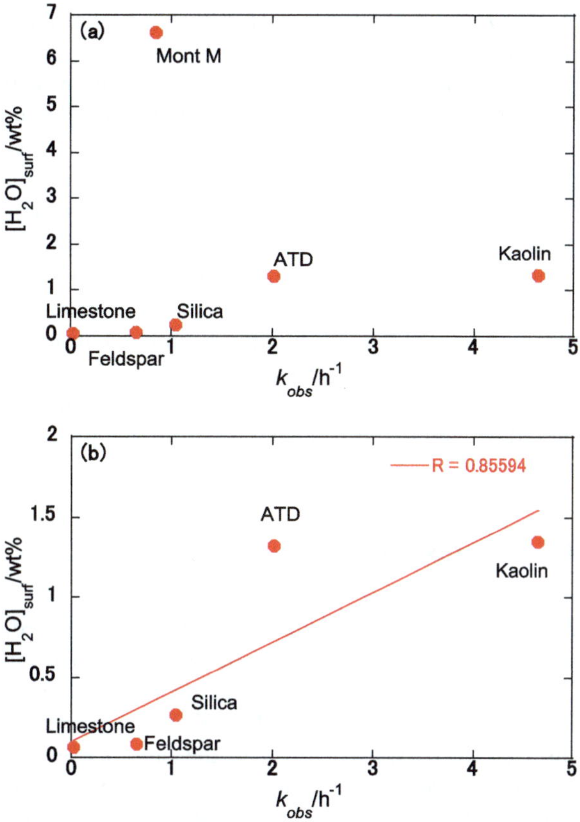

Fig. 10 Relationship between $[H_2O]_{surf}$ and k_{obs} on the substrates tested in this study. **a** plotting on all substrates for which k_{obs} was obtained; **b** plotting without Mont M

These factors likely reduced the expected Ant-ozonation reaction rate on Mont M, despite the high water content in Mont M.

Kaolin has OT structures, with one face covered by oxygen (oxygen shared by two adjacent tetrahedra) and the other by hydroxyl groups (−OH). Because kaolin has no interlaminar water, nonpolar organic molecules, such as Ant, face notable difficulty penetrating the interlayer. Therefore, Ant adsorption and subsequent reactions could efficiently occur on the external surface of kaolin rather than in the interlayer. ATD was also confirmed to contain clay minerals, and its relatively high water content was presumed to be derived from clay minerals.

Conversely, silica (SiO_2) and limestone ($CaCO_3$) exhibit hydrophobicity and therefore lower water content. In addition, feldspar, which is an aluminosilicate mineral with no layered structure such as a clay mineral, has a low water content; therefore, the reaction on feldspar is considered slow similar to that in the case of silica and limestone.

Another possible reason for promoting the Ant-O_3 reaction with water is the enhanced affinity of O_3 to the substrates. For example, Kwamena et al. (2006) observed that the pseudo first-order rate coefficient for the reaction of Ant with

O_3 on azelaic acid at high RH was 25% higher than the value observed under dry conditions.

These findings suggest that the contribution to the formation of AntQ on the mineral dust can be primarily attributed to the clay minerals. In addition, the Lewis acidity of the clay minerals contributes to the O_2-induced oxidation of Ant, whereas the higher water content of the clay minerals can be ascribed to O_3-induced oxidation. Higher concentrations of AntQ have been observed in the ambient airborne particles relative to the parent Ant (Schnelle-Kreis et al. 2005; Ringuet et al. 2012). This might be a result of the secondary formation of AntQ in the atmosphere, as well as the effect of direct emission from combustion sources. While this study elucidates the promotion of AntQ formation on mineral dust particles, our understanding of the atmospheric formation processes of PAHQs remains insufficient. To further clarify them, experimental and observational approaches are needed, and more attention should be paid to the atmospheric formation process of PAHQs.

4 Conclusion

In this study, AntQ, a typical PAHQ, was selected, and the possibility of AntQ secondary formation on the mineral dust surface was verified via laboratory reaction experiments. Seven mineral aerosol model materials (feldspar, silica, limestone, Mont M, kaolin, ATD, and Mont K) were selected, and the Ant deposited on their surfaces were exposed to O_3 in the presence of O_2. The results showed that the reaction on the clay minerals kaolin, Mont M, and Mont K and on ATD proceeded rapidly, while the reaction on feldspar, silica, and limestone was slower. In particular, on Mont K, oxidation proceeded with O_2 in air or solvent before exposure to O_3, resulting in the loss of all Ant during the reaction's initial stage. The reaction experimental results suggest that the contribution of the clay minerals to the promotion of AntQ formation on the mineral dust is significant and that the Lewis acidity of the clay minerals in the oxidation of Ant by O_2 and the high water content of the clay minerals in the oxidation by O_3 are responsible for the accelerated reaction. The study results suggest that harmful organic compounds are unintentionally formed in the atmosphere by chemical reactions on mineral aerosol surfaces. This finding provides useful information for estimating the health hazards of atmospheric aerosols. Future investigations should be focused on assessing the possibility of the formation of more toxic compounds through similar or analogous reactions.

References

Bailey, P. S., Kolsaker, P., Sinha, B., Ashton, J. B., Dobinson, F., & Batterbee, J.E. (1964). Competing reactions in the ozonation of anthracene, *The Journal of Organic Chemistry, 29*(6), 1400–1409 (1964). https://doi.org/10.1021/jo01029a034

Barradas-Gimate, A., Murillo-Tovar, M.A., Díaz-Torres, J.D., Hernández-Mena, L., Saldarriaga-Noreña, H.A., Delgado-Saborit, J.M., & López-López, A. (2017). Occurrence and Potential Sources of Quinones Associated with PM2.5 in Guadalajara, Mexico. *Atmosphere, 8*, 140. https://doi.org/10.3390/atmos8080140

Bekki, K., Takigami, H., Suzuki, G., Tang, N., & Hayakawa, K. (2009). Evaluation of Toxic Activities of Polycyclic Aromatic Hydrocarbon Derivatives Using In Vitro Bioassays, *Journal of Health Science, 55*(4), 601–610.

Cepeda, E.A. & Diaz, M. (1987). Catalyzed Oxidation Kinetics of Anthracene with Oxygen in Ethylene Glycol. *Industrial and Engineering Chemistry Research, 26*(12), 2401–2403. https://doi.org/10.1021/ie00072a001

Chen, W., & Zhu, T. (2014). Formation of nitroanthracene and anthraquinone from the heterogeneous reaction between NO_2 and anthracene adsorbed on NaCl particles. *Environmental science & technology, 48*(15), 8671–8678. https://doi.org/10.1021/es501543g

Cho, A.K., Di Stefano, E., You, Y., Rodriguez, C.E., Schmitz, D.A., Kumagai, Y., Miguel, A.H., Eiguren-Fernandez, A., Kobayashi, T., Avol, E.L., & Froines, J.R. (2004). Determination of Four Quinones in Diesel Exhaust Particles, SRM 1649a, and Atmospheric $PM_{2.5}$. *Aerosol Science and Technology, 38*, 68–81. https://doi.org/10.1080/02786820390229471

Cwiertny, D. M., Young, M. A. & Grassian, V. H. (2008). Chemistry and photochemistry of mineral dust aerosol. *Annual Review of Physical Chemistry, 59*, 27–51. https://doi.org/10.1146/annurev.physchem.59.032607.093630

Falkovich, A. H., Schkolnik, G., Ganor, E. & Rudich, Y. (2004). Adsorption of organic compounds pertinent to urban environments onto mineral dust particles. *Journal of Geophysical Research Atmospheres, 109*, D02208. https://doi.org/10.1029/2003jd003919.

Fernández, R. J. (2002). Do humans create deserts? *Trends in Ecology and Evolution, 17*, 6–7. https://doi.org/10.1016/S0169-5347(01)02366-7

González Henao, S., Karanauskas, V., Drummond, S. M., Dewitt, L. R., Maloney, C. M., Mulu, C., Weber, J. M., Barge, L. M., Videau, P., & Gaylor, M. O. (2022). Planetary minerals catalyze conversion of a polycyclic aromatic hydrocarbon to a prebiotic quinone: Implications for origins of life. *Astrobiology, 22*(2), 197–209. https://doi.org/10.1089/ast.2021.0024

Helmig, D., & Harger, W.P. (1994). OH radical-initiated gas-phase reaction products of phenanthrene. *Science of The Total Environment, 148*, 11–21. https://doi.org/10.1016/0048-9697(94)90368-9

Jakober, C. A., Riddle, S. G., Robert, M. A., Destaillats, H., Charles, M. J., Green, P. G., & Kleeman, M. J. (2007). Quinone emissions from gasoline and diesel motor vehicles. *Environmental Science & Technology, 41*(13), 4548–4554. https://doi.org/10.1021/es062967u

Jia, H., Zhao, S., Shi, Y., Zhu, L., Wang, C., & Sharma, V. K. (2018). Transformation of Polycyclic Aromatic Hydrocarbons and Formation of Environmentally Persistent Free Radicals on Modified Montmorillonite: The Role of Surface Metal Ions and Polycyclic Aromatic Hydrocarbon Molecular Properties. *Environmental Science & Technology, 52*(10), 5725–5733. https://doi.org/10.1021/acs.est.8b00425

Kahan, T.F., Kwamena, N.O., & Donaldson, D.J. (2006). Heterogeneous ozonation kinetics of polycyclic aromatic hydrocarbons on organic films. *Atmospheric Environment, 40*(19), 3448–3459. https://doi.org/10.1016/j.atmosenv.2006.02.004

Kameda, T., Azumi, E., Fukushima, A. Tang, N., Matsuki, A., Kamiya, Y., Toriba, A., & Hayakawa, K. (2016). Mineral dust aerosols promote the formation of toxic nitropolycyclic aromatic compounds. *Scientific Reports, 6*, 24427. https://doi.org/10.1038/srep24427

Kumagai, Y.. (2009). Polycyclic Aromatic Hydrocarbon Quinones as Redox and Electrophilic Chemicals Contaminated in the Atmosphere. *Journal of Health Science, 55*(6), 887–894.

Kwamena, N. O., Earp, M. E., Young, C. J., & Abbatt, J. P. (2006). Kinetic and product yield study of the heterogeneous gas-surface reaction of anthracene and ozone. *The Journal of Physical Chemistry A, 110*(10), 3638–3646. https://doi.org/10.1021/jp056125d

Kwamena, N. O., Staikova, M. G., Donaldson, D. J., George, I. J., & Abbatt, J. P. (2007). Role of the aerosol substrate in the heterogeneous ozonation reactions of surface-bound PAHs. *The Journal of Physical Chemistry A, 111*(43), 11050–11058. https://doi.org/10.1021/jp075300i

Kwok, E. S., Harger, W. P., Arey, J., & Atkinson, R. (1994). Reactions of Gas-Phase Phenanthrene under Simulated Atmospheric Conditions. *Environmental Science & Technology, 28*(3), 521–527. https://doi.org/10.1021/es00052a027

Laszlo P. (1987). Chemical reactions on clays. *Science (New York, N.Y.), 235*(4795), 1473–1477. https://doi.org/10.1126/science.235.4795.1473

Lee, J.Y., Lane, D.A., & Kim, Y.P. (2015). Formation of polyaromatic hydrocarbon (PAH)-quinones during the gas phase reactions of PAHs with the OH radical in the atmosphere. *Environmental Chemistry, 12*, 307–315. https://doi.org/10.1071/EN14150

Li, L., Jia, H., Li, X., & Wang, C. (2015). Transformation of anthracene on various cation-modified clay minerals. *Environmental Science and Pollution Research International, 22*(2), 1261–1269. https://doi.org/10.1007/s11356-014-3424-4

Nájera, J.J., Wamsley, R., Leather, K.E., Percival, C.J., & Horn, A.B. (2011). Heterogeneous oxidation reaction of gas-phase ozone with anthracene in thin films and on aerosols by infrared spectroscopic methods. *International Journal of Chemical Kinetics, 43*, 694–707. https://doi.org/10.1002/kin.20602

Perraudin, E., Budzinski, H., & Villenave, E. (2007). Identification and quantification of ozonation products of anthracene and phenanthrene adsorbed on silica particles. *Atmospheric Environment, 41*(28), 6005–6017. https://doi.org/10.1016/j.atmosenv.2007.03.010

Ringuet, J., Albinet, A., Leoz-Garziandia, E., Budzinski, H., & Villenave, E. (2012). Diurnal/nocturnal concentrations and sources of particulate-bound PAHs, OPAHs and NPAHs at traffic and suburban sites in the region of Paris (France). *The Science of the total environment, 437*(15), 297–305. https://doi.org/10.1016/j.scitotenv.2012.07.072

Schnelle-Kreis, J., Sklorz, M., Peters, A., Cyrys, J., & Zimmermann, R. (2005). Analysis of particle-associated semi-volatile aromatic and aliphatic hydrocarbons in urban particulate matter on a daily basis. *Atmospheric Environment, 39*(40), 7702–7714. https://doi.org/10.1016/j.atmosenv.2005.04.001

Tanaka, T.Y., & Chiba, M. (2006). A numerical study of the contributions of dust source regions to the global dust budget. *Global and Planetary Change, 52*, 88–104. https://doi.org/10.1016/j.gloplacha.2006.02.002

Usher, C.R., Michel, A.E., & Grassian, V.H. (2003). Reactions on mineral dust. *Chemical Reviews, 103*(12), 4883–4940. https://doi.org/10.1021/cr020657y

Wang, L., Atkinson, R., & Arey, J. (2007). Formation of 9,10-phenanthrenequinone by atmospheric gas-phase reactions of phenanthrene. *Atmospheric Environment, 41*(10), 2025–2035. https://doi.org/10.1016/j.atmosenv.2006.11.008

Wang, L., Wen, W., Yan, J., Zhang, R., Li, C., Jiang, H., Chen, S., Pardo, M., Zhu, K., Jia, B., Zhang, W., Bai, Z., Shi, L., Cheng, Y., Rudich, Y., Morawska, L., & Chen, J. (2023). Influence of polycyclic aromatic compounds and oxidation states of soot organics on the metabolome of human-lung cells (A549): Implications for vehicle fuel selection. *Environmental Science & Technology, 57*(51), 21593–21604. https://doi.org/10.1021/acs.est.3c05228

Yang, M., Ahmed, H., Wu, W., Jiang, B., & Jia, Z. (2018). Cytotoxicity of air pollutant 9,10-phenanthrenequinone: Role of reactive oxygen species and redox signaling. *BioMed Research International, 2018*. https://doi.org/10.1155/2018/9523968

Polycyclic Aromatic Hydrocarbons from Biomass Burning: Source Contribution Analysis and Risk Assessment

Fumikazu Ikemori

Abstract Polycyclic aromatic hydrocarbons (PAHs) are ubiquitous environmental pollutants that are considered strongly carcinogenic. A source contribution analysis was conducted by applying positive matrix factorization to measurement data on fine particulate matter in three Japanese cities to identify PAH sources, and a risk assessment was conducted by estimating the toxicity equivalent concentration (TEQ) of each source. Five sources were identified: the major sources of biomass burning, semi-volatile PAHs, and ship and industrial emissions and the minor sources of road traffic and plastic combustion. Biomass burning was a significant PAH source during the cold season. Moreover, biomass burning made a significant contribution to the estimated TEQ for each city. These results suggest that biomass burning is an important source of toxic PAHs. The presented source contribution analysis and risk assessment can help other countries and regions understand the sources and risks of PAHs and collect essential data for control and management.

Keywords Biomass burning · Polycyclic aromatic hydrocarbon · Positive matrix factorization · Toxicity equivalent concentration · Organic tracer

1 Introduction

Biomass burning due to forest fires, the open burning of agricultural residue, and burning of wood for residential use is a major emission source of pollutants in the troposphere (Andreae and Merlet 2001; van der Werf et al. 2010). The gases and particulate matter (PM) emitted from biomass burning influence cloud formation, the radiation balance, ozone formation, and ultimately human health and the global climate (Andreae and Merlet 2001; Chen et al. 2017; Huang et al. 2011; Petetin et al. 2018). In addition, biomass burning can cause severe haze that affects air quality at the national, regional, and local scales. Numerous studies utilizing diverse

F. Ikemori (✉)
Nagasaki University, Bunkyo-machi, Nagasaki, Japan
e-mail: f-ikemori@nagasaki-u.ac.jp

K. Hayakawa and Y. Aoki (eds.), *A New Era of Polycyclic Aromatic Hydrocarbons*,
https://doi.org/10.1007/978-981-96-7695-8_5

approaches including observational data, laboratory experiments, and simulation models have been conducted to understand the effects of biomass burning on atmospheric aerosol characteristics (Tian et al. 2015; Chuang et al. 2015; Zhang et al. 2010). Biomass burning has been shown to emit harmful levels of various pollutants such as fine particulate matter less than 2.5 μm in diameter ($PM_{2.5}$), methane, sulfur oxides, nitrogen oxides, carbon monoxide, carbon dioxide, and black carbon (Chanana et al. 2023; Ravindra et al. 2019). In particular, biomass burning emits polycyclic aromatic hydrocarbons (PAHs) such as benzo[*a*]pyrene (BaP), which are considered hazardous (Fushimi et al. 2017; Ichikawa and Naito 2019). Fushimi et al. (2021) conducted a source contribution analysis on organic aerosols using chemical and in vitro biological assays and a chemical mass balance (CMB) model and reported that open biomass burning was the second-largest contributor to oxidative and inflammatory potentials after naphthalene-derived secondary organic aerosols. They also demonstrated that open biomass burning had very strong aryl hydrocarbon receptor activity. Specific compounds or families of compounds present in the PM of biomass burning such as PAHs and their derivatives have been identified as key contributors to observed health effects (Hayakawa 2024a; Pardo et al. 2024). Thus, there is considerable interest in the health effects of PAHs derived from biomass burning.

Previous studies have used chemical transport and receptor models to investigate the impact of biomass burning on aerosols at multiple scales. Chemical transport models such as the community multiscale air quality model use emission inventory data to evaluate the impact of emission sectors such as ships, domestic vehicles, and biomass burning on air quality (Chuang et al. 2015; Uranishi et al. 2017). Meanwhile, receptor models such as the CMB model and positive matrix factorization (PMF) use aerosol composition data such as cations, anions, inorganic elements, organic carbon (OC), elemental carbon (EC), and organic tracers (Bressi et al. 2014, Ikemori et al. 2021a, Zhang et al. 2010). Levoglucosan is a sugar anhydride produced from the thermal decomposition of cellulose (Simoneit et al. 1999) and is known as a useful organic tracer of biomass burning (Bhattarai et al. 2019). Additionally, several analytical methods have been proposed to identify PAH sources. Tobiszewski and Namieśnik (2012) used the concentration ratios of several PAH isomer pairs to quickly and easily identify PAH sources such as automobiles, coal, and biomass burning. However, the disadvantage of this method is that it is not quantitative, and the identified sources may differ depending on which PAH pair is used (Hayakawa 2024b). Hayakawa et al. (2018) proposed an isomer-based analytical method based on the experimental theory that the concentration ratio of nitrated PAHs (NPAHs) to their parent PAHs is highly dependent on the temperature, which they applied to quantify the contributions of automobiles and coal (Hayakawa 2024b). However, a more detailed source contribution analysis is needed where the results can be compared with the results of other methods and analyses of biomass burning (Ikemori et al. 2019, 2021a) and plastic combustion (Ikemori et al. 2022). However, a detailed source contribution analysis is needed, including comparison with other methods and analytical results, and quantification of other PAH sources such as biomass burning (Ikemori et al. 2019, 2021a) and plastic burning (Ikemori et al. 2022).

In Japan, the annual average $PM_{2.5}$ concentration is below the domestic environmental standard of 15 μg/m^3 but exceeds the standard of 5 μg/m^3 recommended by the World Health Organization. Previous studies using organic tracer measurements and models have shown that biomass burning is a major source of $PM_{2.5}$ and OC in Japan, especially from fall to winter. Ikemori et al. (2021a) conducted a PMF analysis on 13 sites in central Japan using levoglucosan as an organic tracer for biomass burning and demonstrated that biomass burning made larger contributions to $PM_{2.5}$, OC, and EC in the fall of 2014 than in the following winter. In terms of location, we found that biomass burning made a larger contribution to $PM_{2.5}$ in agricultural areas (Fig. 1). Uranishi et al. (2017) demonstrated that biomass burning in Japan is characterized by small-scale biomass burning to treat agricultural residue, which is difficult to detect by satellite remote sensing. They observed that biomass burning contributed to OC and $PM_{2.5}$ from fall to winter in the suburban and urban areas of Japan. Yamagami et al. (2019) reported that EC produced by incomplete combustion as well as PAHs decreased significantly in Nagoya in the 2000s. However, they noted that the decrease in EC was slower in the 2010s and that the main sources became more complex. Because PAHs are also produced by incomplete combustion, their results suggest that the major sources of PAHs in Japan may also be becoming more complex. At present, the actual contribution of biomass burning to PAHs in Japan is not well understood and has not been sufficiently distinguished from the contribution of automobiles, which are another major source of PAHs.

This chapter presents a source contribution analysis on PAHs in $PM_{2.5}$ measurement data from three Japanese cities. PMF was used to identify PAH sources and estimate their contributions. The source characteristics in the warm and cold seasons were investigated under the assumption that the primary anthropogenic emission sources differed markedly according to the ambient conditions, particularly those of biomass burning. Finally, a carcinogenic risk assessment was carried out on the PAH sources by estimating the toxicity equivalent concentration (TEQ).

2 Methods

2.1 Sample Collection and Chemical Analyses

$PM_{2.5}$ was collected on quartz filters by a high-volume air sampler equipped with an impactor-type particle separator in three cities in Japan (Fig. 2): Nagoya in Aichi Prefecture, which is an urban city near a port; Habikino in Osaka Prefecture, which is an urban city near the mountains; and Sakurai in Nara Prefecture, which is a suburban city surrounded by mountains. Measurements were taken during the warm and cold seasons of 2021–2022 during both the daytime (9:00–17:00) and nighttime (17:00–9:00), and the amounts of PAHs, PAH quinones (PAHQs), S-containing PAHs (PASHs), and organic tracer compounds were quantified.

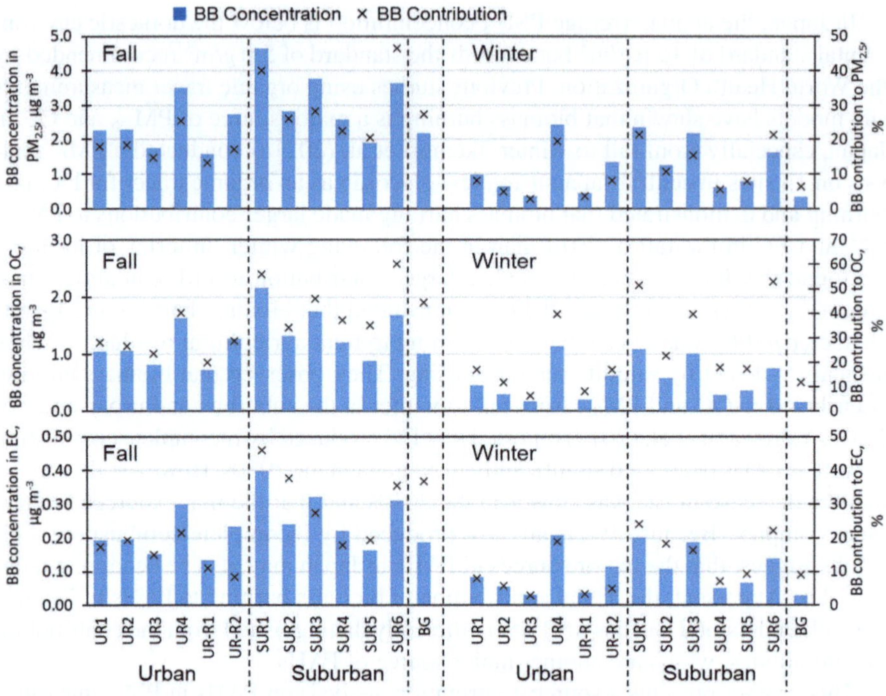

Fig. 1 Seasonal mean concentrations and contributions of biomass burning (BB) to $PM_{2.5}$, OC, and EC during fall and winter at urban (UR), suburban (SUR), and background (BG) sites in central Japan (Fig. 5 in Ikemori et al. 2021a)

The measurement of PAHs, PAH analogs, hopanes, and polar organics has been described in a previous report (Ikemori et al. 2025). In brief, organics were extracted by using dichloromethane/methanol (2:1, v/v, 5 mL, Wako Pure Chemical Industries) under ultrasonication for 20 min. One of the solvent extracts was concentrated with pure nitrogen gas before drying. The resulting extracts were then diluted with an organic solvent mixture of toluene, dichloromethane, and isooctane (2:1:2), and the PAHs, PAHQs, PASHs, and organic tracers (hopanes) were measured by gas chromatography–tandem mass spectrometry (GC–MS/MS). Another solvent extract was concentrated by using pure nitrogen gas. The dried extracts were then reacted with 35 µL of N,O-bis(trimethylsilyl)trifluoroacetamide containing 1% trimethylsilyl chloride (BSTFA; GL-sciences) and 15 µL of pyridine (Wako Pure Chemical Industries) at 70 °C for 1.5 h to convert the compounds with alcohol and carboxylic bases into trimethylsilyl derivatives (Ikemori et al. 2021b). The resulting derivatives were then diluted with 100 µL of isooctane containing the recovery standard (n-pentacosan-d52; C/D/N Isotopes), and organic tracers were measured by GC–MS/MS. All organics were quantified by using authentic standards. The multiple reaction monitoring results were normalized to corresponding internal standards and quantified by using a linear calibration curve ($R^2 > 0.995$).

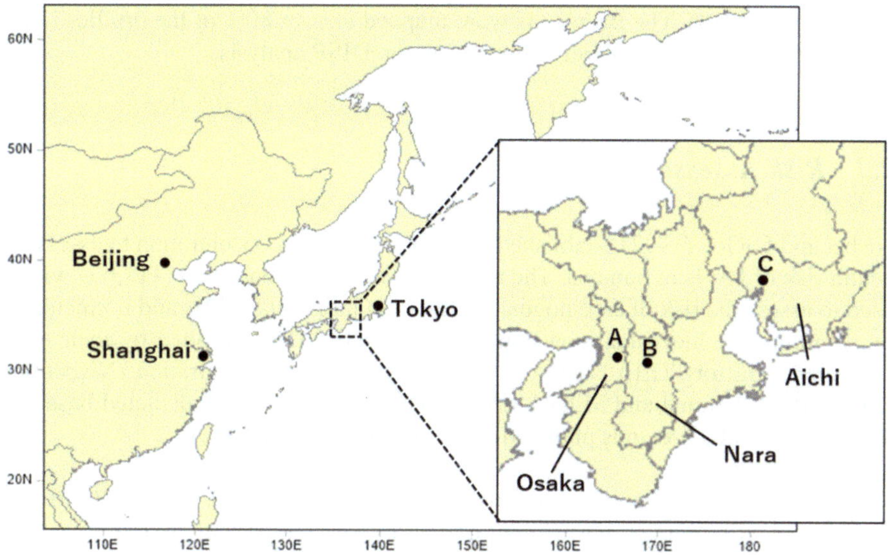

Fig. 2 Locations of $PM_{2.5}$ observation sites: **a** Habikino, **b** Sakurai, and **c** Nagoya (Fig. 1 in Ikemori et al. 2025)

2.2 *Source Contribution Analysis*

The technical procedure for PAH source contribution analysis using PMF has been described in a previous report (Ikemori et al. 2025). The EPA PMF5.0 model was applied to datasets containing the total PAH mass and chemical components to identify possible sources of PAHs and estimate their contributions to the PAH concentration. The dataset comprised measurements of 35 PAHs, three PASHs, and seven organic tracers for combustion sources including biomass, fossil fuels, and plastic at the three observation sites. Naphthalene, acenaphthene, and acenaphthylene were excluded from the PMF analysis because they are mostly gaseous even during the cold season when the distribution rate to particles increases owing to lower temperatures (Pratt et al. 2018). In addition, 2-methylanthracene, benz[*b*]anthracene, 7,12-dimethylbenz[*a*]anthracene, 3-methylcholanthrene, benzo[*c*]chrysene, dibenzo[*a,l*]pyrene, dibenzo[*a,i*]pyrene, and dibenzo[*a,h*]pyrene were excluded from the PMF analysis due to low detection numbers. PAHQs were excluded from the PMF analysis because a secondary generation tracer was needed to isolate them (Kojima et al. 2010; Eiguren-Fernandez et al. 2008). Data from all sites and both seasons were entered into the PMF model, which was run 100 times with different numbers of factors (3–7) using the species category settings. The calculations with five factors gave stable base runs and meaningful factor profiles. The solutions of the five-factor calculations were then evaluated using the displacements of the factor elements (DISP) and bootstrap (BS) analysis

(Norris et al. 2014). The BS results were mapped to over 80% of the profiles in the calculations, and no swaps were observed in the DISP analysis.

2.3 *Risk Assessment*

Nisbet and LaGoy (1992) established the BaP equivalent concentration to assess the health risk of PAHs to humans. The toxic equivalent concentration (TEQ) is widely used to assess the risk of carcinogenic potency of individual PAHs and is calculated by summing the individual concentrations of each PAH multiplied by their toxic equivalency factor (TEF) based on the BaP equivalent concentration (Nisbet and LaGoy 1992; Křůmal and Mikuška 2020). Here, the TEQ was calculated based on TEFs provided by previous publication (Nisbet and LaGoy 1992):

$$\begin{aligned} \mathrm{TEQ} = {} & 0.001 \times (\mathrm{Flu} + \mathrm{Phe} + \mathrm{Fluor} + \mathrm{Py}) + 0.01 \times (\mathrm{Ant} + \mathrm{Chry} + \mathrm{BghiP}) \\ & + 0.1 \times (\mathrm{BaA} + \mathrm{BbF} + \mathrm{BkF} + \mathrm{IcdP}) + 1 \times \mathrm{BaP} + 5 \times \mathrm{DBahA} \end{aligned} \tag{1}$$

where Flu is the concentration of fluorene, Phe is the concentration of phenanthrene, Fluor is the concentration of fluoranthene, Py is the concentration of pyrene, Ant is the concentration of anthracene, Chry is the concentration of chrysene, BghiP is the concentration of benzo[*ghi*]perylene, BaA is the concentration of benz[*a*]anthracene, BbF is the concentration of benzo[*b*]fluoranthene, BkF is the concentration of benzo[k]fluoranthene, IcdP is the concentration of indeno[1,2,3-*cd*]pyrene, BaP is the concentration of benzo[*a*]pyrene, and DBahA is the concentration of dibenz[*a,h*]anthracene. The TEQ was calculated for each source during the warm and cold seasons at each site using the profiles isolated by the PMF analysis.

3 Results and Discussion

3.1 *Sample Characterization*

Seasonal mean concentrations were obtained of individual PAHs, sums of PAHs according to the number of rings (3–6), total PAHs, PAHQs, and PASHs during the warm and cold seasons at all sites (Tables 1 and 2). In the warm season, the mean concentrations of the total PAHs, PAHQs, and PASHs were higher in Nagoya (3.5, 0.64, and 0.078 ng/m^3, respectively; $n = 19$) than in Habikino (0.95, 0.28, and 0.018 ng/m^3, respectively; $n = 18$) and Sakurai (0.72, 0.23, and 0.012 ng/m^3, respectively; $n = 19$). In the cold season, the mean concentration of the total PAHs was lower in Nagoya (2.1 ng/m^3; $n = 18$) than in Habikino (2.9 ng/m^3; $n = 9$) and Sakurai (3.2 ng/m^3; $n = 18$) while the mean concentrations of PAHQ and PASH were comparable in Nagoya (1.1 and 0.035 ng/m^3, respectively; $n = 18$) and Sakurai (1.3 and 0.033

ng/m^3, respectively; $n = 18$). The day–night variations in concentrations of the total PAHs, PAHQs, and PASHs were also observed at each site during the warm and cold seasons (Fig. 3). Specific high-concentration events with a total PAH concentration exceeding 15 ng/m^3 were observed in Nagoya during the daytime on August 5, 2021 and August 23, 2022, which were accompanied by high concentrations of PAHQs and PASHs. Consequently, these events were attributed to the high average concentrations of PAHs and PAH analogs during the warm season in Nagoya.

Seven organic tracers were measured and were used for classification into three groups according to their origin: 17a(H),21b(H)-norhopane and 17a(H),21b(H)-hopane for fossil fuel burning (Rogge et al. 1993); levoglucosan, mannosan, and galactosan for biomass burning (Simoneit et al. 1999), and 1,3,5-triphenylbenzene and terephthalic acid for plastic burning (Simoneit et al. 2005). Their concentrations were comparable to previous findings in Japan at Nagoya, Osaka, and Nara (Ikemori et al. 2018, 2019, 2021a, b). The organic tracers generally had higher mean concentrations in the cold season than in the warm season with the exception of terephthalic acid in Nagoya. A high concentration of terephthalic acid was observed in Nagoya during the daytime on August 21, 2021, which did not correspond to a day with a high PAH concentration (Fig. 4). These results suggest that variations in PAH concentrations in Nagoya may be attributed to another source that does not correspond to the measured organic tracers.

3.2 *Source Contribution Analysis*

PMF was used to divide the sources of PAHs, and these sources were assigned on the basis of their physical properties and organic tracers. Five sources were identified: biomass burning, road traffic, plastic burning, semi-volatile PAHs, and ship and industrial emissions. Profiles for the chemical components of the five sources were obtained (Fig. 5). A detailed interpretation of the sources is given by Ikemori et al. (2025).

The average estimated concentrations of PAHs from each source were obtained for each season at the three sites (Figs. 6 and 7). The contribution from biomass burning increased during the cold season at all sites and was particularly high at Sakurai at over 50%. Biomass burning was the biggest source at Habikino during the daytime and at Nagoya during the daytime and nighttime. Other studies have shown that biomass burning makes a small contribution during the warm season and a large contribution during the cold season in Japan (Hoshi and Saito 2020; Ikemori et al. 2019, 2021a; Kumagai et al. 2010). A previous study on the contributions of biomass burning at 13 sites in central Japan (Ikemori et al. 2021a) also showed that the biomass burning contribution was highest near Sakurai.

Biomass burning includes the open burning of agricultural residues, wood burning for residential use, and forest fires. More specific biomass burning sources can be

Table 1 PAH concentrations (ng/m^3) in $PM_{2.5}$ during the warm and cold seasons at each site. (Table S7 in Ikemori et al. (2025) in supplementary materials)

(a) Warm season

Compound	Abbreviation	Ring size	HAB ($n = 18$)		SAK ($n = 19$)		NGY ($n = 19$)	
			Av	(Min–Max, n)	Av	(Min–Max, n)	Av	(Min–Max, n)
Naphthalene	Nap	2	0.021	(N.D. – 0.060, 5)	0.019	(N.D. – 0.077, 13)	0.17	(N.D. – 1.1, 16)
Acenaphthylene	Acy	3	0.0050	(N.D. – 0.013, 17)	0.0032	(N.D. – 0.0060, 8)	0.013	(0.0010 – 0.073, 19)
Acenaphthene	Ace	3	0.0052	(N.D. – 0.0074, 4)	0.0036	(N.D. – 0.0055, 6)	0.014	(N.D. – 0.050, 10)
Fluorene	Flu	3	0.0034	(N.D. – 0.013, 13)	0.0025	(N.D. – 0.0076, 9)	0.023	(N.D. – 0.18, 18)
Phenanthrene	Phe	3	0.029	(0.0090 – 0.10, 18)	0.022	(N.D. – 0.065, 14)	0.15	(0.0058 – 1.1, 19)
Anthracene	Ant	3	0.0054	(N.D. – 0.019, 14)	0.0056	(N.D. – 0.015, 12)	0.026	(N.D. – 0.16, 18)
2-Methylphenanthrene	2MPhe	3	0.0094	(0.0038 – 0.028, 18)	0.0083	(N.D. – 0.019, 15)	0.021	(0.0014 – 0.11, 19)
2-Methylanthracene	2MAnt	3	0.0026	(N.D. – 0.0046, 7)	0.0017	(N.D. – 0.0026, 7)	0.0049	(N.D. – 0.019, 13)
1-Methylphenanthrene	1MPhe	3	0.0071	(0.0025 – 0.021, 18)	0.0063	(N.D. – 0.015, 16)	0.015	(0.0013 – 0.067, 19)
Retene	Ret	3	0.0077	(0.0026 – 0.019, 18)	0.0070	(0.0014 – 0.020, 19)	0.011	(0.0021 – 0.050, 19)
Fluoranthene	Fluor	4	0.066	(0.021 – 0.22, 18)	0.044	(0.0019 – 0.12, 19)	0.31	(0.010 – 1.9, 19)

(continued)

Table 1 (continued)

(a) Warm season								
Compound	Abbreviation	Ring size	HAB ($n = 18$)		SAK ($n = 19$)		NGY ($n = 19$)	
			Av	(Min–Max, *n*)	Av	(Min–Max, *n*)	Av	(Min–Max, *n*)
Pyrene	Py	4	0.059	(0.022 – 0.17, 18)	0.043	(0.0020 – 0.11, 19)	0.28	(0.0087 – 1.7, 19)
11H-Benzo[*a*]fluorene	BaFlu	4	0.0098	(0.0026 – 0.032, 18)	0.0077	(N.D. – 0.019, 17)	0.048	(N.D. – 0.29, 16)
11H-Benzo[*b*]fluorene	BbFlu	4	0.0061	(0.0014 – 0.020, 18)	0.0044	(N.D. – 0.013, 18)	0.031	(N.D. – 0.20, 17)
7H-Benzo[*c*]fluorene	BcFlu	4	0.0045	(N.D. – 0.0079, 8)	0.0028	(N.D. – 0.0048, 10)	0.018	(N.D. – 0.083, 11)
Benzo[*c*]phenanthrene	BcPhe	4	0.012	(N.D. – 0.020, 7)	0.0065	(N.D. – 0.014, 13)	0.032	(N.D. – 0.15, 14)
Benz[*a*]anthracene	BaA	4	0.025	(0.0066 – 0.088, 18)	0.019	(0.00082 – 0.046, 19)	0.12	(0.0057 – 0.76, 19)
Triphenylene	Tri	4	0.022	(0.0063 – 0.065,18)	0.015	(0.00086 – 0.038, 19)	0.077	(0.0046 – 0.43,19)
Chrysene	Chry	4	0.051	(0.014 – 0.14, 18)	0.036	(N.D. – 0.089, 18)	0.20	(0.0051 – 1.2, 19)
Benz[*b*]anthracene	BbA	4	N.D	(N.D., 0)	N.D.	(N.D., 0)	0.019	(N.D. – 0.027, 2)

(continued)

Table 1 (continued)

(a) Warm season

Compound	Abbreviation	Ring size	HAB ($n = 18$)		SAK ($n = 19$)		NGY ($n = 19$)	
			Av	(Min–Max, n)	Av	(Min–Max, n)	Av	(Min–Max, n)
7,12-Dimethylbenz[*a*]anthracene	7,12DMBaA	4	N.D	(N.D., 0)	N.D.	(N.D., 0)	0.0092	(N.D. – 0.0092, 1)
Benzo[*ghi*]fluoranthene	BghiF	5	0.012	(0.0021 – 0.034, 18)	0.012	(N.D. – 0.029, 18)	0.043	(N.D. – 0.21, 17)
Cyclopenta[*cd*]pyrene	CycdP	5	0.015	(0.0069 – 0.031, 18)	0.017	(0.00086 – 0.033, 19)	0.042	(0.0033 – 0.26, 19)
Benzo[*b*]fluoranthene	BbF	5	0.090	(0.033 – 0.21, 18)	0.063	(0.0042 – 0.14, 19)	0.30	(0.017 – 1.7, 19)
Benzo[*k*]fluoranthene	BkF	5	0.034	(0.0094 – 0.084, 18)	0.026	(0.0015 – 0.058, 19)	0.11	(0.0054 – 0.62, 19)
Benzo[*j*]fluoranthene	BjF	5	0.036	(0.012 – 0.085, 18)	0.030	(0.0020 – 0.064, 19)	0.11	(0.0075 – 0.62, 19)
Benzo[*a*]fluoranthene	BaF	5	0.0071	(0.0025 – 0.027, 18)	0.0073	(N.D. – 0.018, 18)	0.022	(0.0018 – 0.11, 19)
Benzo[*e*]pyrene	BeP	5	0.071	(0.022 – 0.17, 18)	0.048	(0.0013 – 0.11, 19)	0.20	(0.0087 – 1.3, 19)
Benzo[*a*]pyrene	BaP	5	0.053	(0.020 – 0.16, 18)	0.043	(0.0017 – 0.088, 19)	0.22	(0.011 – 1.3, 19)

(continued)

Table 1 (continued)

(a) Warm season

Compound	Abbreviation	Ring size	HAB ($n = 18$)		SAK ($n = 19$)		NGY ($n = 19$)	
			Av	(Min–Max, n)	Av	(Min–Max, n)	Av	(Min–Max, n)
Perylene	Pery	5	0.0079	(N.D. – 0.032, 15)	0.0063	(N.D. – 0.012, 18)	0.037	(N.D. – 0.21, 18)
3-Methylcholanthrene	3MChA	5	N.D.	(N.D., 0)	N.D.	(N.D., 0)	N.D.	(N.D., 0)
Benzo[*c*]chrysene	BcC	5	0.0023	(N.D. – 0.0036, 7)	0.0025	(N.D. – 0.0045, 13)	0.0059	(N.D. – 0.017, 9)
Dibenz[*a,j*]anthracene	DBajA	5	0.0069	(0.0016 – 0.026, 18)	0.0049	(N.D. – 0.0098, 17)	0.028	(N.D. – 0.14, 14)
Dibenz[*a,c*]anthracene	DBacA	5	0.0058	(0.0017 – 0.023, 18)	0.0042	(0.00042 – 0.0087, 19)	0.021	(N.D. – 0.12, 18)
Dibenz[*a,h*]anthracene	DBahA	5	0.012	(N.D. – 0.034, 11)	0.0073	(N.D. – 0.011, 17)	0.042	(N.D. – 0.23, 14)
Benzo[*b*]chrysene	BbC	5	0.0042	(0.0014 – 0.022, 18)	0.0036	(0.00037 – 0.0086, 19)	0.013	(0.00093 – 0.094, 19)
Picene	Pic	5	0.058	(0.014 – 0.20, 18)	0.041	(0.0020 – 0.083, 19)	0.15	(N.D.-1.1, 18)
Indeno[1,2,3-*cd*]pyrene	IcdP	6	0.066	(N.D. – 0.18, 17)	0.053	(0.0048 – 0.11, 19)	0.19	(0.012 – 1.0, 19)

(continued)

Table 1 (continued)

(a) Warm season

Compound	Abbreviation	Ring size	HAB (n = 18)		SAK (n = 19)		NGY (n = 19)	
			Av	(Min–Max, *n*)	Av	(Min–Max, *n*)	Av	(Min–Max, *n*)
Benzo[*ghi*]perylene	BghiP	6	0.080	(0.026 – 0.20 18)	0.061	(0.0068 – 0.11, 19)	0.24	(0.025 – 1.3, 19)
Anthanethrene	ANTH	6	0.0079	(0.0031 – 0.027, 18)	0.0062	(0.00054 – 0.012, 19)	0.030	(0.0022 – 0.19, 19)
Dibenzo[*a,e*]fluoranthene	DBaeF	6	0.013	(0.0031 – 0.044, 18)	0.010	(0.0010 – 0.020, 19)	0.038	(0.0019 – 0.25, 19)
Dibenzo[*a,l*]pyrene	DBalP	6	0.0053	(N.D. – 0.0053, 1)	0.0089	(N.D. – 0.017, 8)	0.050	(N.D. – 0.19, 7)
Dibenzo[*a,e*]pyrene	DBaeP	6	0.015	(N.D. – 0.037, 10)	0.010	(N.D. – 0.019, 16)	0.042	(N.D. – 0.24, 16)
Coronene	Cor	6	0.042	(0.015 – 0.094, 18)	0.029	(0.0049 – 0.054, 19)	0.085	(0.014 – 0.43,19)
Dibenzo[*a,i*]pyrene	DBaiP	6	0.0099	(N.D. – 0.034, 9)	0.0058	(N.D. – 0.012, 13)	0.049	(N.D. – 0.20, 9)
Dibenzo[*a,h*]pyrene	DBahP	6	0.0053	(N.D. – 0.0053, 1)	0.0042	(N.D. – 0.0042, 1)	0.011	(N.D. – 0.024, 4)
Sum PAHs in each ring size								

(continued)

Table 1 (continued)

(a) Warm season

Compound	Abbreviation	Ring size	HAB ($n = 18$)		SAK ($n = 19$)		NGY ($n = 19$)	
			Av	(Min–Max, n)	Av	(Min–Max, n)	Av	(Min–Max, n)
Sum of 3 ring PAHs	3 ring	3	0.067	(0.021 – 0.23, 18)	0.043	(0.0014 – 0.14, 19)	0.27	(0.014 – 1.8, 19)
Sum of 4 ring PAHs	4 ring	4	0.25	(0.076 – 0.74, 18)	0.17	(0.0055 – 0.45, 19)	1.1	(0.035 – 6.6, 19)
Sum of 5 ring PAHs	5 ring	5	0.41	(0.13 – 1.1, 18)	0.31	(0.014 – 0.64, 19)	1.3	(0.063 – 7.3, 19)
Sum of 6 ring PAHs	6 ring	6	0.22	(0.070 – 0.59, 18)	0.18	(0.018 – 0.33, 19)	0.66	(0.059 – 3.7, 19)
Total PAHs	2 to 6 ring		0.95	(0.35 – 2.2, 18)	0.72	(0.039 – 1.6, 19)	3.5	(0.21 – 20, 19)

(b) Cold season

Compound	Abbreviation	Ring size	HAB ($n = 9$)		SAK ($n = 18$)		NGY ($n = 18$)	
			Av	(Min–Max, n)	Av	(Min–Max, n)	Av	(Min–Max, n)
Naphthalene	Nap	2	0.12	(0.011 – 0.27, 9)	0.077	(0.019 – 0.15, 18)	0.10	(0.024 – 0.47, 18)
Acenaphthylene	Acy	3	0.014	(0.0033 – 0.023, 9)	0.019	(0.0039 – 0.058, 18)	0.016	(0.0030 – 0.050, 18)

(continued)

Table 1 (continued)

(b) Cold season

Compound	Abbreviation	Ring size	HAB ($n = 9$)		SAK ($n = 18$)		NGY ($n = 18$)	
			Av	(Min–Max, n)	Av	(Min–Max, n)	Av	(Min–Max, n)
Acenaphthene	Ace	3	0.0098	(N.D. – 0.013, 7)	0.0057	(N.D. – 0.0098, 12)	0.012	(N.D. – 0.030, 9)
Fluorene	Flu	3	0.011	(0.0034 – 0.020, 9)	0.014	(0.0028 – 0.034, 18)	0.013	(0.0019 – 0.035, 18)
Phenanthrene	Phe	3	0.16	(0.036 – 0.25, 9)	0.20	(0.033 – 0.45, 18)	0.16	(0.027 – 0.46, 18)
Anthracene	Ant	3	0.017	(0.0056 – 0.024, 9)	0.019	(0.0039 – 0.041, 18)	0.015	(0.0033 – 0.043, 18)
2-Methylphenanthrene	2MPhe	3	0.036	(0.0089 – 0.056, 9)	0.030	(0.0084 – 0.044, 18)	0.027	(0.0079 – 0.069, 18)
2-Methylanthracene	2MAnt	3	0.0050	(N.D. – 0.0073, 8)	0.0047	(N.D. – 0.0093, 17)	0.0038	(N.D. – 0.0085, 14)
1-Methylphenanthrene	1MPhe	3	0.028	(0.0097 – 0.038, 9)	0.026	(0.0088 – 0.048, 18)	0.022	(0.0085 – 0.053, 18)
Retene	Ret	3	0.034	(0.013 – 0.057, 9)	0.058	(N.D. – 0.27, 17)	0.036	(0.0087 – 0.13, 18)
Fluoranthene	Fluor	4	0.28	(0.060 – 0.47, 9)	0.32	(0.049 – 0.61, 18)	0.23	(0.048 – 0.69, 18)

(continued)

Table 1 (continued)

(b) Cold season

Compound	Abbreviation	Ring size	HAB ($n = 9$)		SAK ($n = 18$)		NGY ($n = 18$)	
			Av	(Min–Max, *n*)	Av	(Min–Max, *n*)	Av	(Min–Max, *n*)
Pyrene	Py	4	0.21	(0.042 – 0.34, 9)	0.26	(0.039 – 0.55, 18)	0.17	(0.036 – 0.57, 18)
11H-Benzo[*a*]fluorene	BaFlu	4	0.040	(0.0066 – 0.077, 9)	0.043	(0.0077 – 0.11, 18)	0.020	(0.0031 – 0.063, 18)
11H-Benzo[*b*]fluorene	BbFlu	4	0.025	(0.0045 – 0.052, 9)	0.025	(0.0051 – 0.068, 18)	0.012	(0.0024 – 0.038, 18)
7H-Benzo[*c*]fluorene	BcFlu	4	0.013	(0.0023 – 0.023, 9)	0.016	(0.0029 – 0.044, 18)	0.0078	(N.D. – 0.023, 16)
Benzo[*c*]phenanthrene	BcPhe	4	0.029	(0.0067 – 0.052, 9)	0.035	(0.0068 – 0.076, 18)	0.019	(0.0035 – 0.063, 18)
Benz[*a*]anthracene	BaA	4	0.088	(0.019 – 0.17, 9)	0.12	(0.024 – 0.27, 18)	0.057	(0.010 – 0.18, 18)
Triphenylene	Tri	4	0.078	(0.022 – 0.11, 9)	0.073	(0.017 – 0.14, 18)	0.059	(0.013 – 0.17, 18)
Chrysene	Chry	4	0.15	(0.033 – 0.26, 9)	0.17	(0.035 – 0.32, 18)	0.11	(0.021 – 0.34, 18)
Benz[*b*]anthracene	BbA	4	0.0092	(N.D. – 0.010, 3)	0.0064	(N.D. – 0.0078, 4)	N.D.	(N.D., 0)

(continued)

Table 1 (continued)

(b) Cold season

Compound	Abbreviation	Ring size	HAB ($n = 9$)		SAK ($n = 18$)		NGY ($n = 18$)	
			Av	(Min–Max, *n*)	Av	(Min–Max, *n*)	Av	(Min–Max, *n*)
7,12-Dimethylbenz[*a*]anthracene	7,12DMBaA	4	N.D.	(N.D., 0)	0.0052	(N.D. – 0.0082, 3)	N.D.	(N.D., 0)
Benzo[*ghi*]fluoranthene	BghiF	5	0.077	(0.025 – 0.13, 9)	0.11	(0.023 – 0.21, 18)	0.072	(0.017 – 0.24, 18)
Cyclopenta[*cd*]pyrene	CycdP	5	0.062	(0.026 – 0.10, 9)	0.095	(0.025 – 0.21, 18)	0.035	(0.0082 – 0.092, 18)
Benzo[*b*]fluoranthene	BbF	5	0.21	(0.057 – 0.35, 9)	0.22	(0.046 – 0.40, 18)	0.14	(0.021 – 0.37, 18)
Benzo[*k*]fluoranthene	BkF	5	0.093	(0.024 – 0.16, 9)	0.097	(0.019 – 0.18 18)	0.059	(0.0081 – 0.16, 18)
Benzo[*j*]fluoranthene	BjF	5	0.11	(0.036 – 0.20, 9)	0.13	(0.029 – 0.28, 18)	0.074	(0.014 – 0.22, 18)
Benzo[*a*]fluoranthene	BaF	5	0.026	(0.0095 – 0.047, 9)	0.035	(0.011 – 0.090, 18)	0.015	(0.0021 – 0.046, 18)
Benzo[*e*]pyrene	BeP	5	0.16	(0.042 – 0.27, 9)	0.16	(0.035 – 0.29, 18)	0.11	(0.017 – 0.29, 18)
Benzo[*a*]pyrene	BaP	5	0.13	(0.035 – 0.21, 9)	0.15	(0.034 – 0.32, 18)	0.077	(0.016 – 0.27, 18)

(continued)

Table 1 (continued)

(b) Cold season								
Compound	Abbreviation	Ring size	HAB ($n = 9$)		SAK ($n = 18$)		NGY ($n = 18$)	
			Av	(Min–Max, n)	Av	(Min–Max, n)	Av	(Min–Max, n)
Perylene	Pery	5	0.017	(0.0061 – 0.029, 9)	0.020	(0.0051 – 0.044, 18)	0.011	(N.D. – 0.036, 17)
3-Methylcholanthrene	3MChA	5	N.D.	(N.D.,0)	N.D.	(N.D.,0)	N.D.	(N.D., 0)
Benzo[*c*]chrysene	BcC	5	0.0075	(0.0019 – 0.014, 9)	0.0082	(0.0028 – 0.017, 18)	0.0057	(N.D. – 0.014, 12)
Dibenz[*a,j*]anthracene	DBajA	5	0.017	(0.0041 – 0.030, 9)	0.014	(0.0036 – 0.028, 18)	0.0075	(0.00094 – 0.022, 18)
Dibenz[*a,c*]anthracene	DBacA	5	0.014	(0.0029 – 0.026, 9)	0.012	(0.0038 – 0.026, 18)	0.0069	(0.0012 – 0.021, 18)
Dibenz[*a,h*]anthracene	DBahA	5	0.023	(N.D. – 0.040, 8)	0.015	(0.0046 – 0.035, 18)	0.013	(N.D. – 0.030, 12)
Benzo[*b*]chrysene	BbC	5	0.013	(0.0056 – 0.027, 9)	0.011	(0.0040 – 0.029, 18)	0.0066	(N.D. – 0.021, 16)
Picene	Pic	5	0.12	(0.031 – 0.21, 9)	0.087	(0.024 – 0.19, 18)	0.056	(0.0047 – 0.18, 18)
Indeno[1,2,3-*cd*]pyrene	IcdP	6	0.17	(0.053 – 0.28, 9)	0.16	(0.043 – 0.34, 18)	0.096	(0.015 – 0.33, 18)

(continued)

Table 1 (continued)

(b) Cold season

Compound	Abbreviation	Ring size	HAB ($n = 9$)		SAK ($n = 18$)		NGY ($n = 18$)	
			Av	(Min–Max, *n*)	Av	(Min–Max, *n*)	Av	(Min–Max, *n*)
Benzo[*ghi*]perylene	BghiP	6	0.18	(0.061 – 0.31, 9)	0.18	(0.049 – 0.35, 18)	0.12	(0.022 – 0.40, 18)
Anthanethrene	ANTH	6	0.019	(0.011 – 0.025, 9)	0.020	(0.0068 – 0.050, 18)	0.011	(0.0018 – 0.029, 18)
Dibenzo[*a,e*]fluoranthene	DBaeF	6	0.029	(0.0078 – 0.051, 9)	0.023	(0.0068 – 0.052, 18)	0.014	(N.D. – 0.040, 17)
Dibenzo[*a,l*]pyrene	DBalP	6	N.D.	(N.D., 0)	0.013	(N.D. – 0.013, 1)	N.D.	(N.D., 0)
Dibenzo[*a,e*]pyrene	DBaeP	6	0.022	(0.0055 – 0.039, 9)	0.018	(0.0053 – 0.040, 18)	0.015	(N.D. – 0.032, 11)
Coronene	Cor	6	0.093	(0.038 – 0.15, 9)	0.090	(0.026 – 0.18, 18)	0.071	(0.018 – 0.20, 18)
Dibenzo[*a,i*]pyrene	DBaiP	6	0.012	(N.D. – 0.020, 8)	0.0095	(N.D. – 0.027, 9)	0.0054	(N.D. – 0.0098, 9)
Dibenzo[*a,h*]pyrene	DBahP	6	N.D.	(N.D., 0)	0.0023	(N.D. – 0.0023, 1)	N.D.	(N.D., 0)
Sum PAHs in each ring size								

(continued)

Table 1 (continued)

(b) Cold season

Compound	Abbreviation	Ring size	HAB ($n = 9$)		SAK ($n = 18$)		NGY ($n = 18$)	
			Av	(Min–Max, n)	Av	(Min–Max, n)	Av	(Min–Max, n)
Sum of 3 ring PAHs	3 ring	3	0.31	(0.079 – 0.45, 9)	0.37	(0.082 – 0.77, 18)	0.30	(0.069 – 0.87, 18)
Sum of 4 ring PAHs	4 ring	4	0.91	(0.20 – 1.5, 9)	1.1	(0.19 – 1.9, 18)	0.69	(0.15 – 2.1, 18)
Sum of 5 ring PAHs	5 ring	5	1.1	(0.31 – 1.8, 9)	1.2	(0.27 – 2.3, 18)	0.68	(0.12 – 2.0, 18)
Sum of 6 ring PAHs	6 ring	6	0.52	(0.18 – 0.87, 9)	0.50	(0.14 – 1.0, 18)	0.32	(0.060 – 1.0, 18)
Total PAHs			2.9	(0.79 – 4.5, 9)	3.2	(0.73 – 5, 18)	2.1	(0.44 – 6.5, 18)

HAB Habikino; *SAK* Sakurai; *NGY* Nagoya; *N.D.* Not detected

Table 2 PAHQ and PASH concentrations (ng/m^3) in $PM_{2.5}$ during the warm and cold seasons at each site (Table S8 in Ikemori et al. 2025 in Supplementary Materials)

(a) Warm season

Compound	Abbreviation	Ring size	HAB ($n = 18$)		SAK ($n = 19$)		NGY ($n = 19$)	
			Av.	(Min–Max, *n*)	Av.	(Min–Max, *n*)	Av.	(Min–Max, *n*)
PAHQs								
9H-Fluoren-9-one	Flu-9-on	3	0.015	(0.0041 – 0.035, 18)	0.017	(0.0018 – 0.043, 19)	0.037	(0.0064 – 0.12, 19)
9,10-Anthracenedione	9,10Ant-on	3	0.14	(0.030 – 0.45, 18)	0.095	(0.022 – 0.26, 19)	0.34	(0.043–1.2, 19)
2-Methylanthracenequinone	2MAnt-on	3	0.048	(0.0074 – 0.13, 18)	0.043	(0.0082 – 0.081, 19)	0.096	(0.014 – 0.32, 19)
11H-Benzo[*a*]fluoren-11-one	11H-BaFlu-11-on	4	0.016	(0.0053 – 0.051, 18)	0.014	(0.0013 – 0.034, 19)	0.037	(0.0049 – 0.17, 19)
Benzanthrone	Bant-q	4	0.045	(0.026 – 0.076, 18)	0.048	(0.0090 – 0.11, 19)	0.091	(0.015 – 0.45, 19)
5,12-Naphthacenequinone	5,12Nap-q	4	0.016	(N.D. – 0.038, 15)	0.011	(0.0028 – 0.020, 19)	0.034	(N.D. – 0.20, 17)
Total PAHQs			0.28	(0.083 – 0.78, 18)	0.23	(0.046 – 0.53, 19)	0.64	(0.12 – 2.1, 19)
PASHs								
Dibenzothiophene	DBT	3	0.0025	(0.00091 – 0.0052, 18)	0.0025	(N.D. – 0.0041, 18)	0.011	(0.0010 – 0.059 19)
Benzo[*b*]naphtho[2,1-*d*]thiophene	BbN21T	4	0.014	(0.0028 – 0.040, 18)	0.0084	(N.D. – 0.023, 18)	0.058	(0.0020 – 0.35, 19)

(continued)

Table 2 (continued)

(a) Warm season

Compound	Abbreviation	Ring size	HAB ($n = 18$)		SAK ($n = 19$)		NGY ($n = 19$)	
			Av.	(Min–Max, *n*)	Av.	(Min–Max, *n*)	Av.	(Min–Max, *n*)
Benzo[*b*]naphtho[1,2-*d*]thiophene	BbN12T	4	0.0025	(N.D. – 0.0063, 14)	0.0015	(N.D. – 0.0035, 15)	0.010	(N.D. – 0.052, 17)
Total PASHs			0.018	(0.0043 – 0.051, 18)	0.012	(N.D. – 0.030, 19)	0.078	(0.0032 – 0.44, 19)

(continued)

Table 2 (continued)

(b) Cold season

Compound	Abbreviation	Ring size	HAB ($n = 9$)		SAK ($n = 18$)		NGY ($n = 18$)	
			Av.	(Min–Max, *n*)	Av.	(Min–Max, *n*)	Av.	(Min–Max, *n*)
PAHQs								
9H-Fluoren-9-one	Flu-9-on	3	0.29	(0.13 – 0.57, 9)	0.29	(0.063 – 0.77, 18)	0.45	(0.086 – 1.3, 18)
9,10-Anthracenedione	9,10Ant-on	3	0.61	(0.26 – 1.1, 9)	0.53	(0.21 – 1.1, 18)	0.42	(0.13 – 1.1, 18)
2-Methylanthracenequinone	2MAnt-on	3	0.10	(0.041 – 0.15, 9)	0.095	(0.044 – 0.17, 18)	0.076	(0.019 – 0.24, 18)
11H-Benzo[*a*]fluoren-11-one	11H-BaFlu-11-on	4	0.069	(0.019 – 0.10, 9)	0.079	(0.017 – 0.18, 18)	0.055	(0.011 – 0.18, 18)
Benzanthrone	Bant-q	4	0.16	(0.058 – 0.31, 9)	0.23	(0.055 – 0.54, 18)	0.13	(0.023 – 0.42, 18)
5,12-Naphthacenequinone	5,12Nap-q	4	0.028	(0.011 – 0.046, 9)	0.027	(0.0066 – 0.051, 18)	0.017	(N.D. – 0.033,16)
Total PAHQs			1.3	(0.64 – 2.1, 9)	1.3	(0.47 – 2.3, 18)	1.1	(0.38 – 2.2, 18)
PASHs								

(continued)

Table 2 (continued)

(b) Cold season

Compound	Abbreviation	Ring size	HAB ($n = 9$)		SAK ($n = 18$)		NGY ($n = 18$)	
			Av.	(Min–Max, n)	Av.	(Min–Max, n)	Av.	(Min–Max, n)
Dibenzothiophene	DBT	3	0.011	(0.0042 – 0.018, 9)	0.0085	(0.0020 – 0.019, 18)	0.012	(0.0050 – 0.025, 18)
Benzo[*b*]naphtho[2,1-*d*]thiophene	BbN21T	4	0.037	(0.0035 – 0.070, 9)	0.021	(0.0014 – 0.048, 18)	0.020	(0.0058 – 0.046, 18)
Benzo[*b*]naphtho[1,2-*d*]thiophene	BbN12T	4	0.0084	(N.D. – 0.012, 7)	0.0041	(N.D. – 0.0083, 16)	0.0038	(N.D. – 0.0085, 17)
Total PASHs			0.054	(0.0078 – 0.094, 9)	0.033	(0.0036 – 0.075, 18)	0.035	(0.014 – 0.079, 18)

HAB Habikino; *SAK* Sakurai; *NGY* Nagoya; *N.D.* Not detected

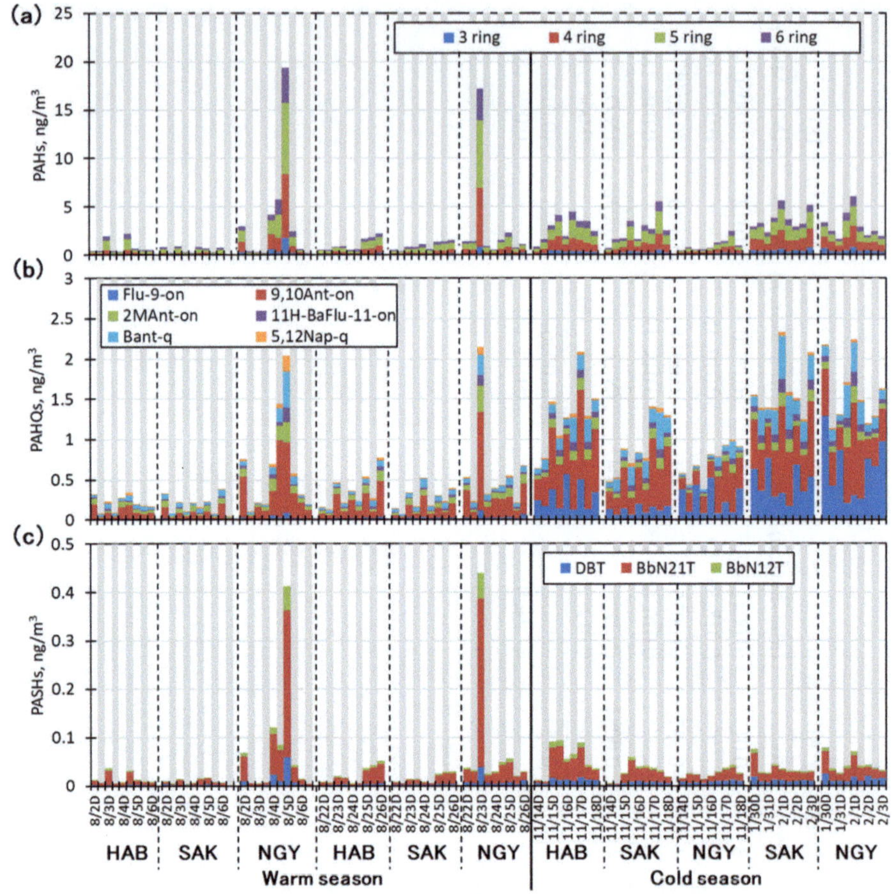

Fig. 3 Chemical compositions of **a** PAHs, **b** PAHQs, and **c** PASHs detected in $PM_{2.5}$ collected at each site (HAB: Habikino; SAK: Sakurai; NGY: Nagoya). Gray hatches mean nighttime samples. In (a), PAHs are classified according to the number of rings. The abbreviations for the PAHQs and PASHs are listed in Table 2 (Fig. 2 in Ikemori et al. 2025)

determined by measuring the ratio of levoglucosan to mannosan and lignin combustion decomposition products (Bhattarai et al. 2019; Wan et al. 2019). Such components have been used in some cases to evaluate the transition between biomass burning sources (Ikemori et al. 2021c; Kawamura et al. 2024). Including such components in the PMF analysis would help investigate the biomass burning sources of PAHs in more detail.

In Japan, PAHs from automobile emissions have been significantly reduced because of the automobile NO_x and PM Law (MOE 2020), which was aimed at controlling emissions based on long-term observational data for PAHs and NPAHs (Hayakawa et al. 2018, 2020). In Nagoya, EC was reduced by 90% from 2003 to 2018 when automobile emissions were regulated (Yamagami et al. 2019), which suggests

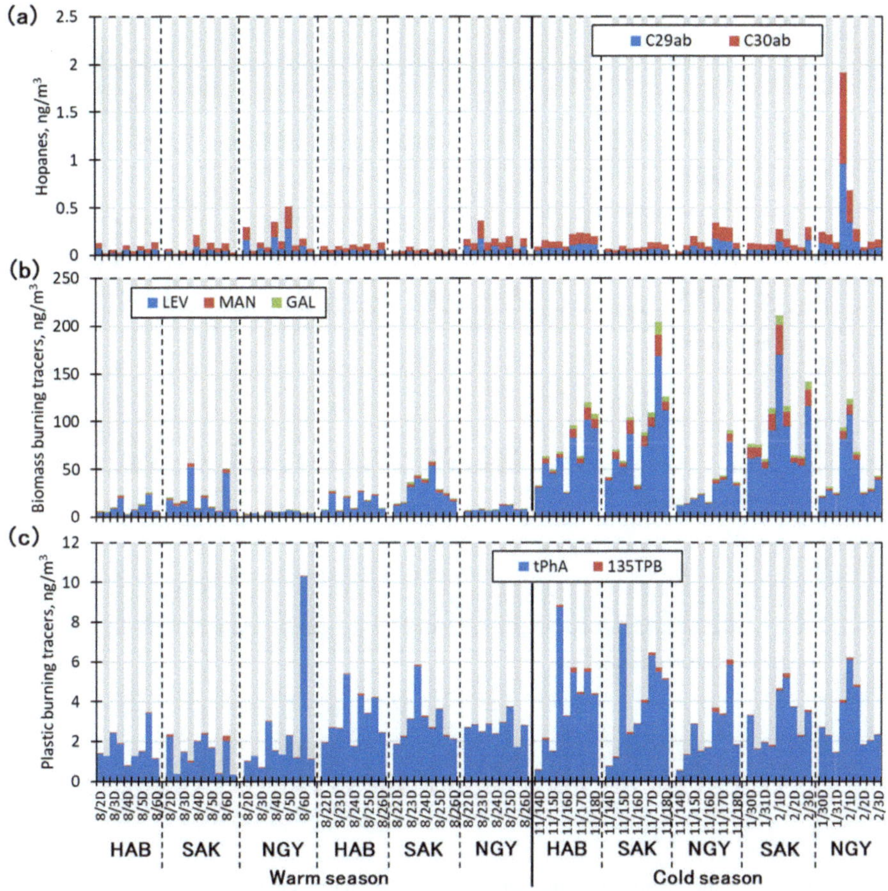

Fig. 4 Chemical compositions of organic tracers for **a** fossil fuel burning (C29ab: 17a(H),21b(H)-norhopane, C30ab: 17a(H),21b(H)-hopane), **b** biomass burning (LEV: levoglucosan, MAN: mannosan, GAL: galactosan), and **c** plastic burning (tPhA: terephthalic acid, 135TPB: 1,3,5-triphenylbenzene) detected in $PM_{2.5}$ collected at each site (HAB: Habikino; SAK: Sakurai; NGY: Nagoya). Gray hatches mean nighttime samples (Fig. 3 in Ikemori et al. 2025)

that traffic-derived PAHs have also been significantly reduced. Therefore, the contribution of this source was considered small at < 5% for each location. Nagoya is heavily urbanized with many roads, so the contribution of automobile emissions was twice that of Sakurai, which is mostly suburban.

Plastic burning was a relatively large contributor, especially during the warm season in Habikino and Sakurai. In Japan, the Dioxin Control Law prohibits plastic burning in open fields. However, terephthalic acid, which is considered an indicator of plastic combustion, also has been detected around Nagoya and in inland areas (Ikemori et al. 2022). These results suggest that plastic burning needs to be carefully monitored to reduce PAH emissions.

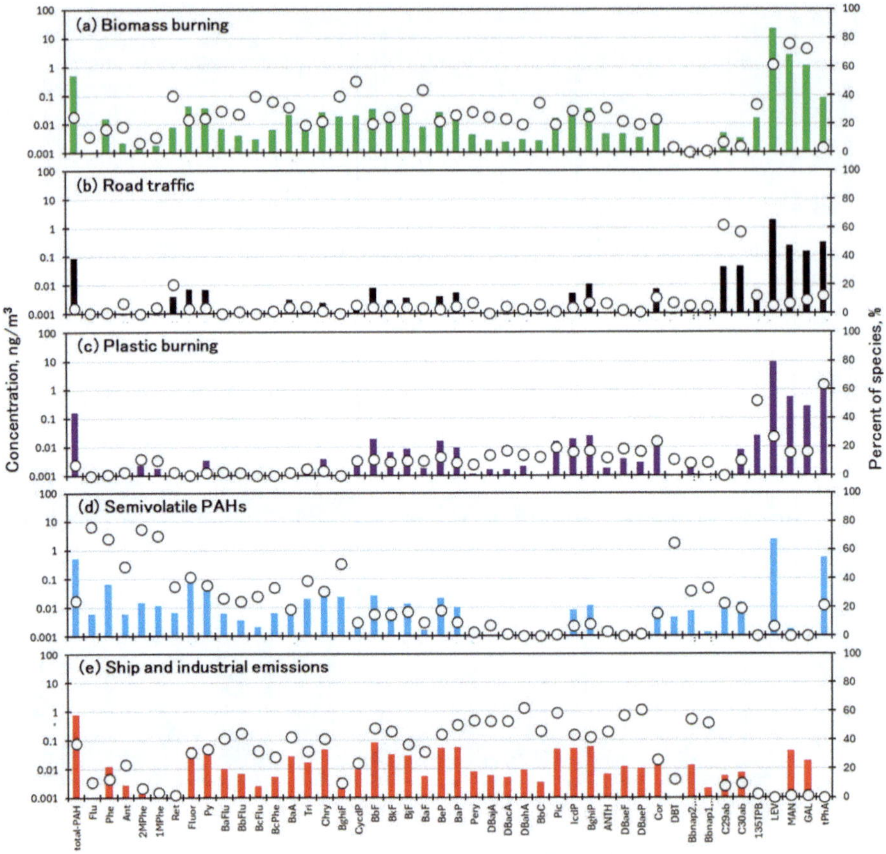

Fig. 5 Profiles of the chemical components of the five factors calculated using positive matrix factorization. Bars show the concentration of each component (left axis), and the circles show the contribution of each component to each factor as a percentage (right axis). The abbreviations for the PAHQs and PASHs are listed in Table 1 (Fig. 4 in Ikemori et al. 2025)

The contribution of semi-volatile PAHs was higher in the cold season than in the warm season for Sakurai and Habikino. The lower temperatures of the cold season mean that semi-volatile PAHs are generally in the particle phase. However, the contribution of semi-volatile PAHs was higher during the warm season in Nagoya, which increased the average concentration of PAHs at the same time. The origin of semi-volatile PAHs needs to be clarified by combining observations of gaseous PAHs and particulate PAHs with conducting source analysis. Isolating semi-volatile PAHs is an important issue because they are also emitted by biomass burning (Xue et al. 2024).

Similar to semi-volatile PAHs, the contribution of ship and industrial emissions was particularly high during the daytime for the warm season in Nagoya at over 70%. The daytime concentration exceeded 15 ng/m^3 on August 5, 2021, and August 23,

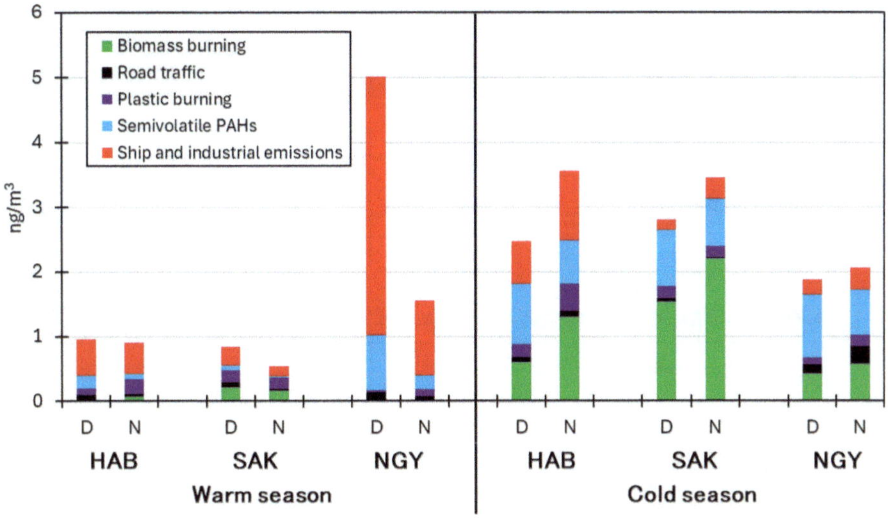

Fig. 6 Average estimated concentrations from identified sources for each season at the three sites (HAB: Habikino; SAK: Sakurai; NGY: Nagoya) (Fig. 6 in Ikemori et al. 2025)

2022, which corresponded to days when the concentrations of semi-volatile PAHs were also high. Together, these sources contributed more than 80% to the total PAH concentration. These results suggest that the port and industrial areas of Nagoya are sources of high PAH emissions. Distinguishing between ships and industries will require conducting observations at several locations near these areas and analyzing the relationship between tracer concentrations and the wind direction and speed.

3.3 *Risk Assessment*

For the risk assessment, the mean TEQ values were calculated for the identified sources in each season at the three sites (Fig. 8). In the warm season, the contribution of ship and industrial emissions was very high and exceeded 80% in Habikino, 50% in Sakurai, and 90% in Nagoya. To reduce the risk, the ship and industrial emissions need to be further separated to identify individual sources. In the cold season, biomass burning made a significant contribution in each city: It was the second-largest contributor in Habikino and Nagoya at about 30% and the largest contributor in Sakurai at over 60%. Biomass burning is recognized worldwide as a major source of $PM_{2.5}$. These results suggest that biomass burning may also be a very important source of toxic PAHs. Other studies have also reported on the high risks associated with biomass burning (Fushimi et al. 2021; Xue et al. 2024). Further studies are expected on the contribution of biomass burning to $PM_{2.5}$ and its toxicity.

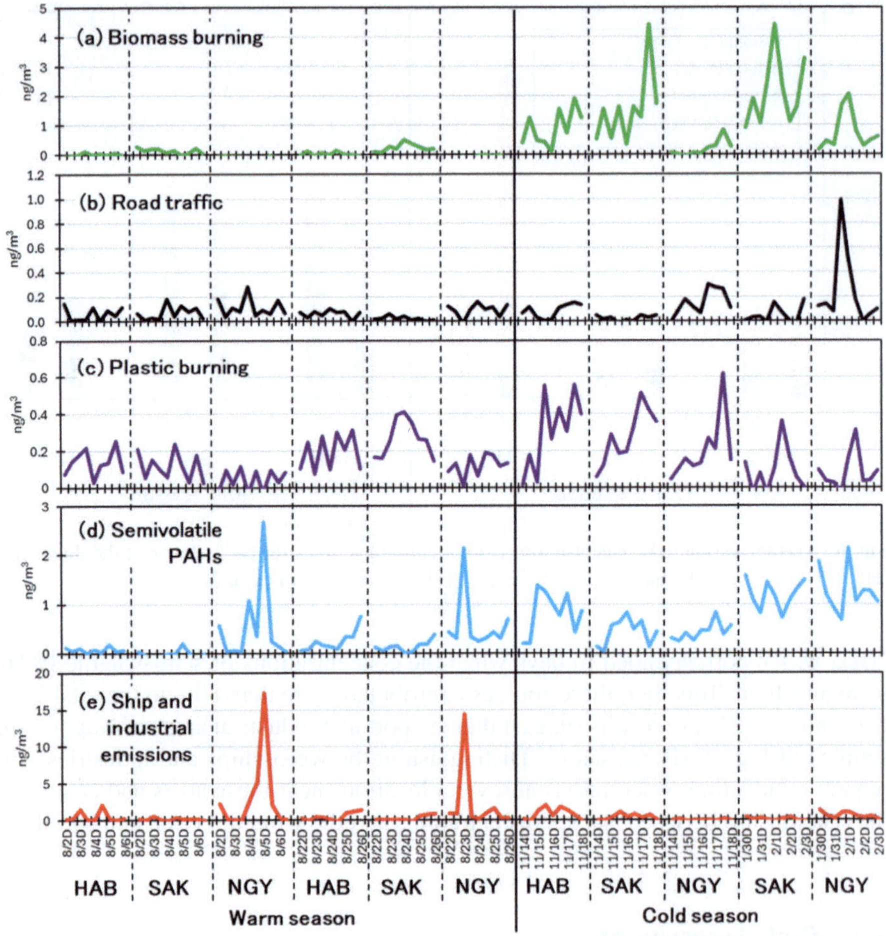

Fig. 7 Daytime variations of estimated concentrations of the identified sources for each season at the three sites (HAB: Habikino; SAK: Sakurai; NGY: Nagoya) (Fig. 7 in Ikemori et al. 2025)

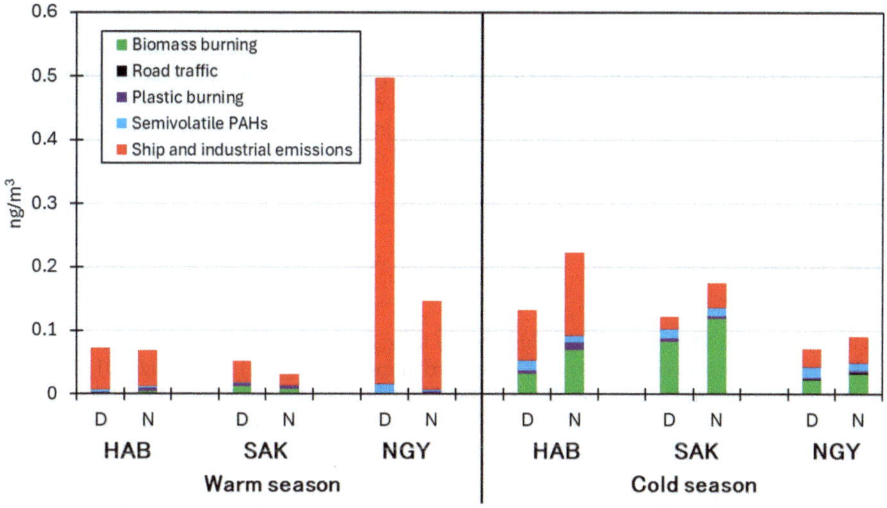

Fig. 8 Average estimated TEQ of the identified sources for each season at the three sites (HAB: Habikino; SAK: Sakurai; NGY: Nagoya)

4 Conclusion

A source contribution analysis was performed on $PM_{2.5}$ measurements from three Japanese cities to identify sources of PAHs such as biomass burning and their contributions. Five sources were identified: biomass burning, road traffic, plastic combustion, semi-volatile PAHs, and ship and industrial emissions. The contribution of biomass burning increased during the cold season at all sites and was particularly large at Sakurai at over 50%. The risk assessment demonstrated that PAHs from biomass burning made a significant contribution to the total TEQ of each city at about 30% for Habikino and Nagoya and over 60% for Sakurai. These results suggest that biomass burning may be a very important source of toxic PAHs. The source contribution analysis and risk assessment presented here are expected to help other countries and regions understand the sources and risks of PAHs and to accumulate essential data for their control.

Acknowledgements Author thanks Mr. Yuki Murakami of Nara Prefecture Landscape and Environment Center, Ms. Megumi Takabayashi of Waterworks Bureau of Nara prefecture, Ms. Rie Nishimura of Research Institute of Environment, Agriculture and Fisheries, Osaka Prefecture, Ms. Mami Hiramatsu of Department of Environment, Agriculture, Forestry and Fisheries, Osaka Prefecture, and Mr. Maku Ueda of Nagoya City Institute for Environmental Sciences for their sampling, analysis for chemical components and Dr. Ayako Yoshino, Dr. Satoru Chatani, Dr. Kei Sato, and Dr. Seiji Sugata of National Institute for Environmental Studies for discussion of the data.

Competing Interests This research was financially supported in part by a Grant in Aid for Scientific Research (No. 23K17047) from the Japan Society for the Promotion of Science, Japan, and

Science and Technology Research Partnership for Sustainable Development (SATREPS) in collaboration between Japan Science and Technology Agency (JST, JPMJSA2102) and Japan International Cooperation Agency (JICA), and a Type II joint research between the National Institute for Environmental Studies and the local environmental research institutes in Japan (1921AH001 and 2224AH001). This research was conducted as a Type II joint research between the National Institute for Environmental Studies and the local environmental research institutes in Japan.

References

Andreae, M. O., & Merlet, P. (2001). Emission of trace gases and aerosols from biomass burning. *Global Biogeochemical Cycles*, *15*(4), 955–966. https://doi.org/10.1029/2000gb001382

Bhattarai, H., Saikawa, E., Wan, X., Zhu, H., Ram, K., Gao, S., Kang, S., Zhang, Q., Zhang, Y., Wu, G., Wang, X., Kawamura, K., Fu, P., & Cong, Z. (2019). Levoglucosan as a tracer of biomass burning: Recent progress and perspectives. *Atmospheric Research*, *220*, 20–33. https://doi.org/10.1016/j.atmosres.2019.01.004

Bressi, M., Sciare, J., Ghersi, V., Mihalopoulos, N., Petit, J.-E., Nicolas, J. B., Moukhtar, S., Rosso, A., Féron, A., Bonnaire, N., Poulakis, E., & Theodosi, C. (2014). Sources and geographical origins of fine aerosols in Paris (France). *Atmospheric Chemistry and Physics*, *14*(16), 8813–8839. https://doi.org/10.5194/acp-14-8813-2014

Chanana, I., Sharma, A., Kumar, P., Kumar, L., Kulshreshtha, S., Kumar, S., & Patel, S. K. S. (2023). Combustion and stubble burning: A major concern for the environment and human health. *Fire*, *6*(2), 79. https://doi.org/10.3390/fire6020079

Chen, J., Li, C., Ristovski, Z., Milic, A., Gu, Y., Islam, M. S., Wang, S., Hao, J., Zhang, H., He, C., Guo, H., Fu, H., Miljevic, B., Morawska, L., Thai, P., LAM, Y. F., Pereira, G., Ding, A., Huang, X., & Dumka, U. C. (2017). A review of biomass burning: Emissions and impacts on air quality, health and climate in China. *Science of the Total Environment*, *579*, 1000–1034. https://doi.org/10.1016/j.scitotenv.2016.11.025

Chuang, M.-T., Fu, J. S., Lin, N.-H., Lee, C.-T., Gao, Y., Wang, S.-H., Sheu, G.-R., Hsiao, T.-C., Wang, J.-L., Yen, M.-C., Lin, T.-H., Thongboonchoo, N., & Chen, W.-C. (2015). Simulating the transport and chemical evolution of biomass burning pollutants originating from Southeast Asia during 7-SEAS/2010 Dongsha experiment. *Atmospheric Environment*, *112*, 294–305. https://doi.org/10.1016/j.atmosenv.2015.04.055

Eiguren-Fernandez, A., Miguel, A. H., Lu, R., Purvis, K., Grant, B., Mayo, P., Di Stefano, E., Cho, A. K., & Froines, J. (2008). Atmospheric formation of 9,10-phenanthraquinone in the Los Angeles air basin. *Atmospheric Environment*, *42*(10), 2312–2319. https://doi.org/10.1016/j.atmosenv.2007.12.029

Fushimi, A., Saitoh, K., Hayashi, K., Ono, K., Fujitani, Y., Villalobos, A. M., Shelton, B. R., Takami, A., Tanabe, K., & Schauer, J. J. (2017). Chemical characterization and oxidative potential of particles emitted from open burning of cereal straws and rice husk under flaming and smoldering conditions. *Atmospheric Environment*, *163*, 118–127. https://doi.org/10.1016/j.atmosenv.2017.05.037

Fushimi, A., Nakajima, D., Furuyama, A., Suzuki, G., Ito, T., Sato, K., Fujitani, Y., Kondo, Y., Yoshino, A., Ramasamy, S., Schauer, J. J., Fu, P., Takahashi, Y., Saitoh, K., Saito, S., & Takami, A. (2021). Source contributions to multiple toxic potentials of atmospheric organic aerosols. *Science of the Total Environment*, *773*, 145614. https://doi.org/10.1016/j.scitotenv.2021.145614

Hayakawa, K. (2024a). Polycyclic aromatic hydrocarbons from vegetation burning and health effects. *Global Environmental Research*, *27*(1), 13–20

Hayakawa, K. (2024b). Long-term and seasonal changes in emission sources of atmospheric particulate-bound pyrene and 1-nitropyrene in four selected cities in the Western Pacific. *Atmosphere*, *15*(6), 634. https://doi.org/10.3390/atmos15060634

Hayakawa, K., Tang, N., Nagato, E.G., Toriba, A., Sakai, S., Kano, F., Goto, S., Endo, O., Arashidani, K., & Kakimoto, H. (2018). Long term trends in atmospheric concentrations of polycyclic aromatic hydrocarbons and nitropolycyclic aromatic hydrocarbons: A study of Japanese cities from 1997 to 2014. *Environmental Pollution, 233*, 474–482. https://doi.org/10.1016/j.envpol.2017.10.038

Hayakawa, K., Tang, N., Toriba, A., & Nagato, E.G. (2020). Calculating sources of combustion-derived particulates using 1-nitropyrene and pyrene as markers. *Environmental Pollution, 265*, 114730. https://doi.org/10.1016/j.envpol.2020.114730

Hoshi, J., & Saito, S. (2020). Estimating the contribution of biomass burning to atmospheric organic particles at the central Tokyo metropolitan area using levoglucosan and radiocarbon. *Journal of Japan Society for Atmospheric Environment, 55*, 204–220 (In Japanese).

Huang, C., Chen, C. H., Li, L., Cheng, Z., Wang, H. L., Huang, H. Y., Streets, D. G., Wang, Y. J., Zhang, G. F., & Chen, Y. R. (2011). Emission inventory of anthropogenic air pollutants and VOC species in the Yangtze River Delta region, China. *Atmospheric Chemistry and Physics, 11*(9), 4105–4120. https://doi.org/10.5194/acp-11-4105-2011

Ichikawa, Y., & Naito, S. (2019). Analysis of organic compositions of primary $PM_{2.5}$ emitted from biomass burning sources. *Journal of Japan Society for Atmospheric Environment, 54*, 161–177 (in Japanese).

Ikemori, F., Uranishi, K., Asakawa, D., Nakatsubo, R., Makino, M., Kido, M., Mitamura, N., Asano, K., Nonaka, S., Nishimura, R., & Sugata, S. (2021a). Source apportionment in $PM_{2.5}$ in central Japan using positive matrix factorization focusing on small-scale local biomass burning. *Atmospheric Pollution Research, 12*(3), 162–172. https://doi.org/10.1016/j.apr.2021.01.006

Ikemori, F., Nakayama, T., & Hasegawa, H. (2019). Characterization and possible sources of nitrated mono- and di-aromatic hydrocarbons containing hydroxyl and/or carboxyl functional groups in ambient particles in Nagoya, Japan. *Atmospheric Environment, 211*, 91–102. https://doi.org/10.1016/j.atmosenv.2019.05.009

Ikemori, F., Yamagami, M., Hisatsune, K., & Osada, K. (2022). Enhanced $PM_{2.5}$ concentrations at urban areas causing by biomass burning-derived aerosol transported from north areas to south areas in the Nobi plain. *Earozoru Kenkyu, 37*, 45–56 (In Japanese).

Ikemori, F., Murakami, Y., Takabayashi, M., Nishimura, R., Hiramatsu, M., Ueda, M., Yoshino, A., Chatani, S., Sato, K., & Sugata, S. (2025). Source contribution analysis of polycyclic aromatic hydrocarbons in $PM_{2.5}$ at three Japanese cities using positive matrix factorization with organic tracers. *Atmosphere, 16(2)*, 175. https://doi.org/10.3390/atmos16020175

Ikemori, F., Nishimura, R., Saito, S., Akiyama, M., Yamamoto, S., Iijima, A., & Sugata, S. (2021b). Organic molecular tracers in $PM_{2.5}$ at urban sites during spring and summer in Japan: Impact of secondary organic aerosols on water-soluble organic carbon. *Atmosphere, 12*(5), 579. https://doi.org/10.3390/atmos12050579

Ikemori, F., Yamagami, M., & Sugata, S. (2018). Impact on adsorption of gaseous organic tracer compounds in $PM_{2.5}$ collected on quartz filter. *Journal of Japan Society for Atmospheric Environment, 53*, 70–78 (In Japanese).

Ikemori, F., Uranishi, K., Sato, T., Fujihara, M., Hasegawa, H., & Sugata, S. (2021c). Time-resolved characterization of organic compounds in $PM_{2.5}$ collected at Oki Island, Japan, affected by transboundary pollution of biomass and non-biomass burning from Northeast China. *Science of the Total Environment, 750*, 142183. https://doi.org/10.1016/j.scitotenv.2020.142183

Kawamura, K., Ikemori, F., Furuuchi, M., & Hata, M. (2024). Multiple organic tracer analyses of biomass changes affected by the open burning of rice residues. *Atmospheric Pollution Research, 15(8)*, 102153. https://doi.org/10.1016/j.apr.2024.102153

Kojima, Y., Inazu, K., Hisamatsu, Y., Okochi, H., Baba, T., & Nagoya, T. (2010). Changes in concentration levels of polycyclic aromatic compounds associated with airborne particulate matter in downtown Tokyo after introducing government diesel vehicle controls. *Asian Journal of Atmospheric Environment, 4*(1), 1–8. https://doi.org/10.1007/bf03654893

Křůmal, K., & Mikuška, P. (2020). Mass concentrations and lung cancer risk assessment of PAHs bound to PM1 aerosol in six industrial, urban and rural areas in the Czech Republic, Central

Europe. *Atmospheric Pollution Research, 11*(2), 401–408. https://doi.org/10.1016/j.apr.2019.11.012

Kumagai, K., Iijima, A., Shimoda, M., Saitoh, Y., Kozawa, K., Hagino, H., & Sakamoto, K. (2010). Determination of dicarboxylic acids and levoglucosan in fine particles in the Kanto Plain, Japan, for source apportionment of organic aerosols. *Aerosol and Air Quality Research, 10*(3), 282–291. https://doi.org/10.4209/aaqr.2009.11.0075

Ministry of the Environment Government of Japan (MOE). (2020). *Automobile NOx and PM low*. https://www.env.go.jp/air/car/noxpm.html

Nisbet, I. C. T., & LaGoy, P. K. (1992). Toxic equivalency factors (TEFs) for polycyclic aromatic hydrocarbons (PAHs). *Regulatory Toxicology and Pharmacology, 16*, 290–300. https://doi.org/10.1016/0273-2300(92)90009-X

Norris, G., Duvall, R., Brown, S., & Bai, S. (2014). *EPA positive matrix factorization (PMF) 5.0 fundamentals and user guide*. US Environmental Protection Agency

Pardo, M., Li, C., Zimmermann, R., & Rudich, Y. (2024). Health impacts of biomass burning aerosols: Relation to oxidative stress and inflammation. *Aerosol Science and Technology, 58*(10), 1093–1113. https://doi.org/10.1080/02786826.2024.2379551

Petetin, H., Sauvage, B., Parrington, M., Clark, H., Fontaine, A., Athier, G., Blot, R., Boulanger, D., Cousin, J.-M., Nédélec, P., & Thouret, V. (2018). The role of biomass burning as derived from the tropospheric CO vertical profiles measured by IAGOS aircraft in 2002–2017. *Atmospheric Chemistry and Physics, 18*(23), 17277–17306. https://doi.org/10.5194/acp-18-17277-2018

Pratt, G. C., Herbrandson, C., Krause, M. J., Schmitt, C., Lippert, C. J., McMahon, C. R., & Ellickson, K. M. (2018). Measurements of gas and particle polycyclic aromatic hydrocarbons (PAHs) in air at urban, rural and near-roadway sites. *Atmospheric Environment, 179*, 268–278. https://doi.org/10.1016/j.atmosenv.2018.02.035

Ravindra, K., Singh, T., Mor, S., Singh, V., Mandal, T. K., Bhatti, M. S., Gahlawat, S. K., Dhankhar, R., Mor, S., & Beig, G. (2019). Real-time monitoring of air pollutants in seven cities of North India during crop residue burning and their relationship with meteorology and transboundary movement of air. *Science of the Total Environment, 690*, 717–729. https://doi.org/10.1016/j.scitotenv.2019.06.216

Rogge, W. F., Hildemann, L. M., Mazurek, M. A., Cass, G. R., & Simoneit, B. R. T. (1993). Sources of fine organic aerosol. 2. Noncatalyst and catalyst-equipped automobiles and heavy-duty diesel trucks. *Environmental Science & Technology, 27*(4), 636–651. https://doi.org/10.1021/es00041a007

Simoneit, B. R. T., Schauer, J. J., Nolte, C. G., Oros, D. R., Elias, V. O., Fraser, M. P., Rogge, W. F., & Cass, G. R. (1999). Levoglucosan, a tracer for cellulose in biomass burning and atmospheric particles. *Atmospheric Environment, 33*(2), 173–182. https://doi.org/10.1016/s1352-2310(98)00145-9

Simoneit, B. R. T., Medeiros, P. M., & Didyk, B. M. (2005). Combustion products of plastics as indicators for refuse burning in the atmosphere. *Environmental Science & Technology, 39*(18), 6961–6970. https://doi.org/10.1021/es050767x

Tian, J., Chow, J. C., Cao, J., Han, Y., Ni, H., Chen, L.-W. A., Wang, X., Huang, R., Moosmüller, H., & Watson, J. G. (2015). A biomass combustion chamber: Design, evaluation, and a case study of wheat straw combustion emission tests. *Aerosol and Air Quality Research, 15*(5), 2104–2114. https://doi.org/10.4209/aaqr.2015.03.0167

Tobiszewski, M., & Namieśnik, J. (2012). PAH diagnostic ratios for the identification of pollution emission sources. *Environmental Pollution, 162*, 110–119. https://doi.org/10.1016/j.envpol.2011.10.025

Uranishi, K., Ikemori, F., Nakatsubo, R., Shimadera, H., Kondo, A., Kikutani, Y., Asano, K., & Sugata, S. (2017). Identification of biased sectors in emission data using a combination of chemical transport model and receptor model. *Atmospheric Environment, 166*, 166–181. https://doi.org/10.1016/j.atmosenv.2017.06.039

van der Werf, G. R., Randerson, J. T., Giglio, L., Collatz, G. J., Mu, M., Kasibhatla, P. S., Morton, D. C., DeFries, R. S., Jin, Y., & van Leeuwen, T. T. (2010). Global fire emissions and the contribution of deforestation, savanna, forest, agricultural, and peat fires (1997–2009). *Atmospheric Chemistry and Physics, 10*(23), 11707–11735. https://doi.org/10.5194/acp-10-11707-2010

Wan, X., Kawamura, K., Ram, K., Kang, S., Loewen, M., Gao, S., Wu, G., Fu, P., Zhang, Y., Bhattarai, H., & Cong, Z. (2019). Aromatic acids as biomass-burning tracers in atmospheric aerosols and ice cores: A review. *Environmental Pollution, 247*, 216–228. https://doi.org/10.1016/j.envpol.2019.01.028

Xue, Q. Q., Tian, Y. Z., Song, D. L., Huang, F. X., & Feng, Y. C. (2024). Variations of source-specific risks for inhalable particle-bound PAHs during long-term air pollution controls in a Chinese megacity: Impact of gas/particle partitioning. *Atmospheric Environment, 331*, 120565. https://doi.org/10.1016/j.atmosenv.2024.120565

Yamagami, M., Ikemori, F., Nakashima, H., Hisatsune, K., & Osada, K. (2019). Decreasing trend of elemental carbon concentration with changes in major sources at Mega city Nagoya, Central Japan. *Atmospheric Environment, 199*, 155–163. https://doi.org/10.1016/j.atmosenv.2018.11.014

Zhang, X., Hecobian, A., Zheng, M., Frank, N. H., & Weber, R. J. (2010). Biomass burning impact on $PM_{2.5}$ over the southeastern US during 2007: Integrating chemically speciated FRM filter measurements, MODIS fire counts and PMF analysis. *Atmospheric Chemistry and Physics, 10*(14), 6839–6853. https://doi.org/10.5194/acp-10-6839-2010

Plastics as Sorption Substrates for Polycyclic Aromatic Hydrocarbons

Edward Gou Nagato

Abstract PAHs are hydrophobic compounds and are thus prone to partitioning onto hydrophobic substrates such as plastics. For this reason, the two are extensively studied as sorbate -sorbent pairs, as it is thought that plastics are a vector for the transport of PAHs into environmental and biological systems. However, PAHs are more than just the mere 16 PAHs that are commonly studied and include a wide array of functionalized congeners, which will affect how sorption occurs on plastic surfaces. Further complicating the interaction is that plastics are liable to being degraded and thus changing as adsorption substrates. In this chapter, some of these aspects are discussed.

Keywords Plastic crystallinity · Plastic oxidation · Adsorption isotherms · Adsorption kinetics · PAH metabolism

1 Introduction

Plastics are a ubiquitous environmental contaminant, with one recent estimate of plastic waste emissions into aquatic ecosystems at 20–53 Mt annually (Borrelle et al. 2020). In general, the most commonly produced plastics are high and low density polyethylene (PE), polypropylene (PP), polystyrene (PS), polyvinylchloride (PVC), polyurethane (PU), polyethylene terephthalate (PET) and acrylic fibers (Costigan et al. 2022; Geyer et al. 2017; Gong et al. 2023). While much has been said about the toxicological effects of plastics, specifically the smaller particles such as microplastics and nanoplastics (Li et al. 2023; Zhao et al. 2023), it is not clear that the plastic itself is the main toxicological issue. It is more likely that it is the capacity for plastics to serve as carriers of other contaminants, in particular hydrophobic contaminants (Mato et al. 2001; Yu et al. 2020), that makes plastics a true toxicological threat. The structures of plastics are important as they determine the interactions formed with

E. G. Nagato (✉)
Faculty of Life and Environmental Science, Shimane University, Matsue, Japan
e-mail: nagato@life.shimane-u.ac.jp

K. Hayakawa and Y. Aoki (eds.), *A New Era of Polycyclic Aromatic Hydrocarbons*,
https://doi.org/10.1007/978-981-96-7695-8_6

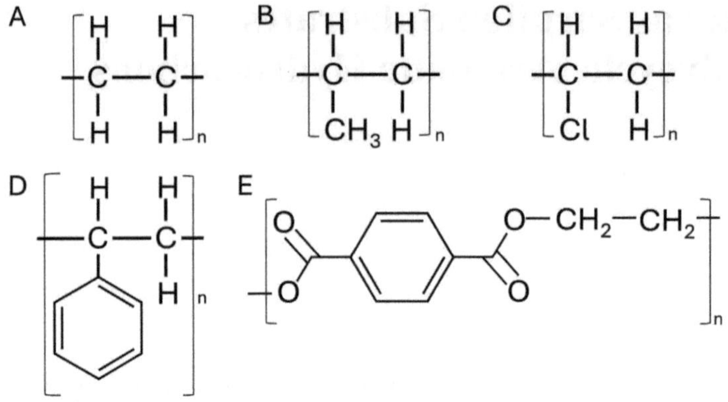

Fig. 1 The structures of some of the commonly used plastics in industry. **a** polyethylene, **b** polypropylene, **c** polyvinylchloride, **d** polystyrene, **e** polyethylene terephthalate

other contaminants, such as the PAHs. The structures of commonly used plastics is shown in Fig. 1.

In general, these interactions are determined by hydrophobic partitioning and electrostatic interactions with contributions from other non-covalent interactions (namely, Van der Waal forces, hydrogen bonding, π–π interactions), although this depends on the type of plastic, the degree of weathering of the plastic and the type of functionality present on the adsorbate molecule. For example, PE would be able to form Van der Waal interactions with other compounds, but polystyrene has the additional capacity for π–π interactions and this will affect the latter's capacity as an adsorption substrate (Hüffer and Hofmann 2016). This is in addition to the other aspects of plastic structure such as the porosity, crystallinity and crosslinking and the presence of functional groups; all of which will affect the capacity of plastics for as contaminant vectors.

2 PAHs: The EPA 16 and Their Derivatives

In general, plastics are hydrophobic and will adsorb other organic compounds with high low K_{ow} values (Costigan et al. 2022; Lee et al. 2014). One of the contaminants that may be adsorbed to the surfaces of plastics are the polycyclic aromatic hydrocarbons (PAHs) which are a ubiquitous environmental pollutant. PAHs are fused benzene rings that are the product of the incomplete combustion of organic matter (Reizer et al. 2022). In general, studies focus on the 16 priority PAHs, as outlined by the US EPA (See Chap. 12), though narrowly defining PAHs to these 16 congeners risks ignoring the many types of PAHs that actually exist; specifically, those containing reactive functional groups such as hydroxyl, carbonyl and nitro groups (Fig. 2). Thus, within PAH research there have been calls for there to be greater consideration for

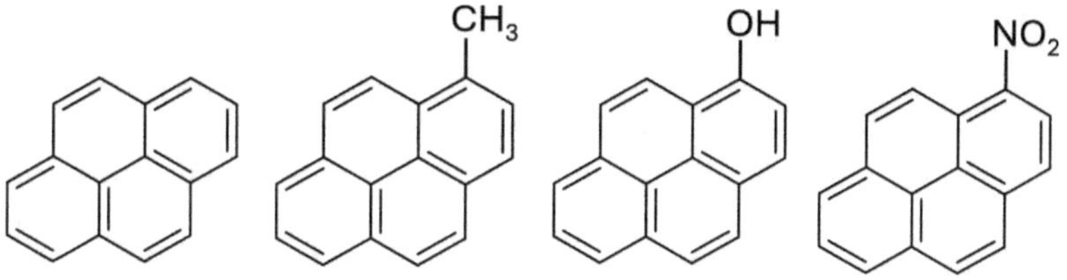

Fig. 2 The functionalization of PAHs will affect the partitioning behavior of PAHs. From left to right are pyrene, 1-methylpyrene, 1-hydroxypyrene and 1-nitropyrene

the PAHs beyond the narrowly defined 16 priority PAHs (An et al. 2022; Andersson and Achten 2015; Keith 2014).

Carbonyl and hydroxyl groups are one type of functional group known to affect the biological activity of the PAHs (Idowu et al. 2019; Park et al. 2006; Wenger et al. 2008). These oxidized PAHs (OPAHs) are the products of the abiotic interaction with radicals in the atmosphere (Lee et al. 2015; Vione et al. 2004) and through the metabolism of PAHs, for example via CYP enzymes (Elovaara et al. 2006; Shimada and Fujii-Kuriyama 2004). Benzo[a]pyrene is extensively studied, which is toxic after it has been metabolized to its oxidized congener, first as an epoxy adduct, then as a dihydroxyl, to form the DNA adduct that makes it carcinogenic (Shiizaki et al. 2017; Stowers and Anderson 1985). However, this type of oxidation is common in the environment as well and is the product of the microbial degradation of the parent PAHs, into these more reactive congeners (Tidwell et al. 2016; Wilcke et al. 2014). Related are the nitrated PAHs (NPAHs), which contain an -NO_2 group which makes the molecule highly reactive, and these are formed primarily through high temperature reactions, such as in diesel engines (Lara et al. 2023). The formation of these types of functional groups will allow PAHs to form interactions with solid substrates beyond hydrophobic interactions and π-π interactions (i.e., hydrogen bonding), but this may also inhibit their ability to adsorb to hydrophobic surfaces such as plastics.

Although these are known to exist at orders lower than the parent PAHs, they are the true toxic forms of PAHs and warrant more attention than they are given. The OPAHs and NPAHs are in addition to the various alkylated congeners (Qian et al. 2022) and halogenated PAHs (Goswami et al. 2023; Sun et al. 2021a) that also exist, which will have the effect of increasing the hydrophobicity of the PAH molecule. The outcome of functionalization, whether in the direction of more polar or less polar, is a change in the partitioning coefficients (viz. the log K_{ow}) and the adsorption of these PAHs to surfaces such as plastics. This then means that any discussion on PAH adsorption needs to differentiate the basic 16 that are commonly studied from the other functionalized PAH congeners.

3 PAH Adsorption to Plastics

The discussion of contaminant sorption is really a discussion on the process of the contaminant adhering to the surface of a solid (adsorption) and the movement into the bulk of the solid (absorption) and the extent to which this occurs depends on the type of plastic, how it was initially manufactured, the degree to which it has been weathered in the environment, as well as the type of PAH. With respect to the main PAHs studied (the EPA 16) the adsorption of PAHs onto plastics is thought to be driven primarily by hydrophobic interactions (Bao et al. 2021; Guo et al. 2012; Hou et al. 2023; O'Connor et al. 2016; Zhang et al. 2020), given that both plastics and PAHs are generally hydrophobic. However, this seems to not just apply to the main EPA 16 PAHs. Yu et al. (2020) have found that the naphthalene and naphthalene derivatives (i.e., those containing amine, hydroxyl and carboxylic groups), even if there are polar functional groups on the PAH molecule, adsorption to plastics was primarily determined by the hydrophobicity of the PAHs. This hydrophobicity is supported by the results of Kong et al. (2023) with NPAHs, where despite the polar nitro group, the adsorption is still determined by the hydrophobicity of the rest of the NPAH molecule (Kong et al. 2023), regardless of type of plastic adsorbent. Similar results are seen by others as well (Wang and Wang 2018; Zhao et al. 2020). Nevertheless, the addition of polar functional groups do weaken the adsorption affinity for plastics as shown by example by Zhang et al. (2020), where carboxylized phenanthrene had lower adsorption compared to phenanthrene and methylated phenanthrene. The addition of a hydroxyl group has similarly been shown to inhibit the adsorption of naphthalene and phenanthrene to PVC, although the adsorption to PS is still comparatively high, given the capacity of PS to for the π–π interactions with the rest of the molecule (Bao et al. 2021, 2023).

This adsorption also depends on the plastic type. For example, PAHs have been shown to adsorb onto plastics such as PP and PET more than a more oxidized plastic like polylactic acid. However, in a comparative sense, PP with its methyl group may be more likely to adsorb PAHs compared to PET because of the larger structure of the PET molecule (Lončarski et al. 2021), but also the greater hydrophobicity of PP. Similarly, Hou et al. (2023) found more PAH adsorption on PE, compared to PMMA. Biodegradable plastics (i.e., PLA) in general adsorb less PAHs (Karapanagioti and Klontza 2008; Lončarski et al. 2021; Wang and Wang 2018) because of the greater polarity of the polymer. In addition, there are other factors to consider with plastic adsorption; specifically the chain organization itself and whether the amorphous region of the plastics is in a rubbery or glassy state, with the latter having a more condensed structure and thus less diffusivity for adsorption to occur. However, there glassy polymers have nanoscale pores, where adsorption can take place (Teuten et al. 2009).

The ease of accessibility of plastic standards and the consistency with which they can be shaped (i.e., made into spherical nanoplastics and microplastics of various sizes) has made them a model for studying the adsorption kinetics and adsorption isotherms of PAHs to plastics. However, there is the obvious caveat that these perfect spheres may not be entirely environmentally relevant as they do not reflect the heterogeneous structures of actual plastics found in the environment. In general, the sorption experiment entails the examination of an analyte of interest from the liquid phase (or gas) onto the surface of the substrate particle and then the diffusion into the particle (Seidensticker et al. 2018). There is a sequence of events with the adsorption. First is the mass transfer from the surrounding matrix (either liquid or gas) to the external surface of the plastics. After contact with the surface, pore filling occurs and there is the intraparticle diffusion into the interior (Hüffer and Hofman 2016; Liu et al. 2023).

Many studies report the kinetics and the adsorption isotherms of PAHs to various plastic substrates (i.e., Lončarski et al. 2021; Sörensen et al. 2020; Zhao et al. 2020). The interaction of a sorbate (whether in liquid or gas) with a solid substrate is a process of sorption and desorption which will in time will reach equilibrium. From this, concentrations of the target sorbate in solution and on the particle surface are measured and these data points are fitted with various models to provide insight into the adsorption capacity and efficiency of the adsorption process. For example, these models can indicate whether there is uptake into the bulk polymer or whether there is a dominance of adsorption at the surface layer (Hüffer and Hofmann 2016). While there are many models that have been developed, the main models used are the Langmuir and Freundlich isotherms (Costigan et al. 2022; Hüffer and Hofmann 2016; Li et al. 2018; Lončarski et al. 2021; Zhang et al. 2024), with the former indicating that the adsorption occurs uniformly as a monolayer on the adsorption substrate, while the latter indicates that there is heterogeneity on the adsorption surface (Xu and Li 2010). However, other models such as the BET isotherm have been used in describing multilayer adsorption, as has been suggested with PAH derivative adsorption onto PU (Frescura et al. 2024). As PAHs become larger, there is a longer time to reach adsorption equilibrium and lower K_{ow} PAHs are more likely to desorb from the substrate as well (Chen et al. 2020; Hou et al. 2023; Kong et al. 2023). In addition to the sorption isotherms, adsorption kinetics provide insight into the rate limiting components of the adsorption processes and is generally limited by the extent to which Fickian diffusion occurs. In general, the descriptions are based on pseudo-first order and pseudo second order models, with the former indicating physisorption as the dominant process and the latter chemisorption (Bao et al. 2021; Emecheta et al. 2022; Frescura et al. 2024; Ho and McKay 1999; Kong et al. 2023; Lončarski et al. 2021; Udenby et al. 2022; Wang and Wang 2018; Yu et al. 2020; Zhao et al. 2020), although there is concern over the overreliance on these two particular models, in particular the pseudo-second order models (Revellame et al. 2020).

4 PAHs Adsorbed to Plastics in the Environment

It is in this adsorption capacity that plastics can increase the presence of PAHs found in the environment; in particular in the marine environment where they are substrates on which PAHs are found at higher concentrations than in the surrounding sediments and waters (Rios and Jones 2015; Schiferle et al. 2023; Teuten et al. 2007). This is particularly the case as particles become smaller and the total surface area increases (Lončarski et al. 2021), although there is indication that at the nano scale, aggregation of the particles may reduce the adsorption capacity as the colloid destabilizes (Wang et al. 2019). It has also been shown that seawater results in more PAHs adsorbing to plastics compared to freshwater (Frescura et al. 2024; Liu et al. 2023; Zhang et al. 2020), an effect akin to the salting out effect, making plastics in marine environments particularly good vectors for PAH transport.

Globally, PAHs are found adsorbed onto plastics in the environment and the concentration ranges reflect the heterogeneity of the PAH distributions in the global environment. Looking strictly at the 16 EPA PAHs: One study found that sum PAHs were between 11.2 and 7714 ng/g in China (Eastern Guongdong; Shi et al. 2020) and in 136.3–1586.9 ng/g on plastic pellets in the coastal areas of the Yellow sea and Bohai Sea (Zhang et al. 2015). Phenanthrene, chrysene and fluoranthene were dominant, though Benzo[a]pyrene was highest on pellets. Another study found 16.37–27.4 ng/g (León et al. 2019) in the Spanish Mediterranean coast. PAHs in the Indonesian coast had an average concentration of 552 ng/g on PE and PP, while plastics sampled in the North-Atlantic gyre had concentrations ranging from 11.2 to 255 mg/g (Bouhroum et al. 2019) and samples off the Taiwan coast had concentrations ranging between 104 and 3595 ng/g (Chen et al. 2020). It has been observed that sites closer to urban areas had higher PAH concentrations, likely the result of the proximity to anthropogenic sources (Hirai et al. 2011). However, from a larger perspective, there is doubt with regards to how relevant plastics are as a vector for PAH transport into marine environments in the first place, as one study found that while organic pollutants (such as PAHs) were bound to plastics, these concentrations were still orders lower than of those contained on suspended particulate matter, of which plastics constituted little to the total outflow (Mai et al. 2020). Therefore, some caution is warranted when considering how much these plastics are vectors for PAHs into the environment.

A number of studies have also examined the PAH content on beached plastics (Fisner et al. 2013a; Gorman et al. 2019; Hirai et al. 2011; Karapanagioti and Klontza 2008; Rochman et al. 2012; Rodriguez et al. 2020; Yeo et al. 2022). Rochman et al. (2012) found that PS contains comparatively high PAH concentrations (including alkylated PAHs) compared to other commonly used plastics and this may be due to the greater distance between chain segments that arises from having a bulky phenyl group, allowing for chemical diffusion to occur. This combined with the capacity for π–π interactions between the aromatic structures of PAHs and PS makes PS particularly apt for PAH adsorption (Hüffer and Hofmann 2016; Uber et al. 2019; Yu et al. 2020). This is also in addition to the planar structure of PAHs, which allow for a greater overlap of the PAH and PS π-clouds (Jonker and Koelmans 2002; Liu et al. 2016). A similar situation is also presented by PU, where π interactions resulted in more adsorption of both neutral and polar PAHs, compared to biodegradable plastic such as PBS and PCL (Zhao et al. 2020). Interestingly, despite having a lower log K_{ow} than Atrazine, 1-naphthylamine had greater adsorption to PS (whereas Atrazine could not be determined) and this was attributed to the π–π interactions as well (Zhao et al. 2020). The types of PAHs found on the surface tend to be the smaller 2–4 ring pahs (Gorman et al. 2019; Rochman et al. 2013; Rodriguez et al. 2020; Yeo et al. 2022).

With respect to where the PAHs originate from, it is unclear. However, in one study of marine plastic fragments, methylated PAHs, a sign of a petrogenic PAH origin, were comparatively high, relative to the parent PAHs, providing some clue as to the origins of the PAHs (Hirai et al. 2011). Fisner et al. (2013a) in their study of buried sediment, found that PAH composition varied by depth. However, more interesting is the inclusion of the methylated congeners of these priority 16 PAHs in the analysis. With a range of 386–1996 ng/g of total PAHs, the priority PAHs 198–1042 ng g of the total, indicating that despite the limited range of alkylated PAHs studied, they still constituted a major portion of the PAHs found at the sites and similar results were found in the authors' other study of beach pellets (Fisner et al. 2013b). This underscores the need to think beyond the EPA 16, as these types of studies can provide insight into the origins of the PAHs found on the plastics, especially given that these congeners, such as the nitrated and chlorinated congeners, have a much higher toxicity (Goswami et al. 2023; Hou et al. 2023; Kong et al. 2023). However, while plastics are adsorbing PAHs from the surrounding environment, it has been shown that PAHs are found in non-weathered, virgin plastics (Van et al. 2012), suggesting that these may have been included in the initial production of the plastics. This has also been shown for recycled plastics as well (Alassali et al. 2020).

5 Plastic Degradation

Like many pollutants, plastics are liable to being transformed once they are in the environment. This is made manifest at a visible level in the surface cracking, pitting and flaking that are easily identifiable (Deng et al. 2022; Maddison et al. 2023); all of which serve to increase the surface area of these plastics for further adsorption to occur. The rate of this degradation depends not only on plastic type, but also in what form it is produced; for example, whether as a film or a fiber (Chamas et al. 2020). At a surface level one estimate has the degradation rate of up to 469.73 um per year (Maddison et al. 2023) and this has an influence on the amount of microplastics that end up in the environment, but also to what extent these plastics will serve as vectors for the transport of other contaminants. It is little wonder then that many studies of contaminant adsorption onto plastic surfaces include aged plastics (Karapanagioti and Klontza 2008; Moncrieffe et al. 2023; Sankoda and Saito 2025; Udenby et al. 2022; Zhang et al. 2020), whether this chemically aging the plastics with H_2O_2 (viz. Fenten's reagent) or is done in *uv* aging chambers.

5.1 Polymer Oxidation and PAH Adsorption

At a molecular level, plastics undergo transformations that are broadly categorized as either increases in oxidation or changes in crystallinity. In the environment, the oxidation is a photo-oxidation process induced by *uv* radiation and is typically assessed using FTIR to calculate various indices such as the carbonyl and hydroxylation indices (Almond et al. 2020; Andrady et al. 2022; Phan et al. 2022). The added functionality from the carbonyl and hydroxyl group increases the capacity for hydrogen bonding to other contaminants (Guo et al. 2012). A plastic with polar functionality should result in a greater interaction with surrounding water molecules, which would reduce the adsorption of hydrophobic contaminants to the surface (Teuten et al. 2009). The oxidation of the plastic surfaces may be why the degradation of plastics has been shown to reduce the hydrophobicity of plastics and thus reduce the capacity to adsorb hydrophobic contaminants like the EPA 16 PAHs (Bouhroum et al. 2019; Emecheta et al. 2022; Fries and Zarfl 2012; Schiferle et al. 2023). However, the functionalization appears to increase the adsorption of functionalized PAHs, as seen by Yu et al. (2020), where derivatives of naphthalene (i.e., aminated) could adsorb more strongly to carboxylated microplastics, compared to the non-functionalized plastics.

5.2 Crystallization

In addition to oxidation, the arrangement of the polymer chain is also important. Broadly speaking, the polymer chain can be arranged in a highly packed, crystalline form, or can be irregular and loosely organized and thus amorphous (Mohanan et al. 2020), the latter which exists in either a rubbery or glassy state, depending on the glass transition temperature of the polymer. The polymer chain in the rubbery state is more mobile, so it is expected that there is greater adsorption occurring in these domains, compared to the crystallized regions (Emecheta et al. 2022; Endo et al. 2011; Guo et al. 2012; Zhang et al. 2024). In general, plastics in the environment, when exposed to *uv* radiation become more crystallized (Phan et al. 2022). The crystallized regions are tightly packed and are less likely to adsorb contaminants. In studies of environmental plastics, the increased crystallization resulted in the slower diffusion of phenanthrene into the eroded plastic pellets as well (Karapanagioti and Klontza 2008) and the importance of rubbery domains has also been found by Guo et al. (2012) in their examination of phenanthrene, naphthalene and 1-naphthol.

6 The Role of the Organic Matter Corona

The adsorption of PAHs onto plastics will also be affected by the presence of other environmental components. For example, the presence of organic matter such as humic acids can form coronas around the surface (Natarajan et al. 2020; Sun et al. 2023; Tang et al. 2021) and this has been extensively studied in the context of nanoplastic colloidal stability (Hakim and Kobayashi 2021; Liu et al. 2024; Singh et al. 2019). However, this will also affect the adsorption of contaminants onto the surface (Jin et al. 2020; Li et al. 2018). The presence of organic matter has been shown to reduce the adsorption to plastic surfaces, with Li et al. (2018) studying nitrogen, oxygen and sulfur containing PAH congeners in their study. Within the actual film itself, there is evidence of PAH accumulation (Bhagwat et al. 2021) with concentrations ranging from 18.3 to 176 ug/g and larger ring PAHs being dominant. However, microbes in these corona environments are capable of degrading PAHs, although it is not clear what types of harmful intermediaries are formed during this process. Jin et al. (2020) observed that in the summer, where there is more microbial activity, PAH concentrations declined suggesting their degradation by microbes, which contrasted with the winter where there would be steady accumulation of PAHs due to the decrease in microbial activity.

7 Plastics as a Vector for PAH Transport into Biological Systems

Contaminants adsorbed onto plastic surfaces are liable to desorb once inside an organism. The extent to which this desorption occurs depends on the surrounding environment, as has been demonstrated in gastric and small intestine conditions (Schiferle et al. 2023). Humans are susceptible to exposure, with for example one study showing that PAHs migrate from food grade plastics into foods (Kuźmicz and Ciemniak 2017; Loudon et al. 2025; Schweighuber et al. 2019). Desorption rates depend on the plastic structure as well (Kida and Koszelnik 2021; Wang and Wang 2018). More interesting, is the greater presence of alkylated PAH congeners, compared to the parent PAHs (Loudon et al. 2025), which underscores the importance of not restricting our analyses of PAHs to the basic 16 PAHs. Once inside the human body, the plastic associated PAH will be in a different environment, where desorption may take place. This has been demonstrated to be the case with simulated gastrointestinal fluid (Hou et al. 2023), where PAH desorption was greater, as the fluid would disrupt the interaction of the PAH and plastic surface.

Plastics as a vector for contaminant adsorption have been investigated in a range of organisms, including copepods (Sörensen et al. 2020), brine shrimp (Martinez-Alvarez et al. 2022), *Daphnia* spp. (Ma et al. 2016), zebrafish embryos (Martinez-Alvarez et al. 2022), zebra danio (Trevisan et al. 2020), bivalves (Pittura et al. 2018; Sun et al. 2021b) and even in the plant root rhizome (Abbasi et al. 2021). In general, the exposure to microplastics and PAHs concurrently results in greater exposure to the organism, confirming the vector effect of plastics and PAH (Ma et al. 2016; Martinez-Alvarez et al. 2022; Sun et al. 2021b), although this is not always the case, with for example microplastics having been observed to decrease the bioavailability of phenanthrene to marine medaka (Li et al. 2020) and Benzo[a]Pyrene to marine mussels (Wang et al. 2023), and to protect microorganisms from PAH exposure (Kleinteich et al. 2018), given that they can act as an adsorption substrate that prevents entry of the PAHs into the biological systems.

8 Conclusion

Studies of PAH adsorption to plastics are extensive and are the interactions between these sorbates-sorbents are a model system for understanding sorption behavior in laboratory environments. However, in the environment, the situation is complicated by factors such as *uv* light and microbial degradation, where both PAHs and plastics can be transformed into different adsorbates and adsorbents. As a result, it is still not entirely clear what the fate of PAHs are in the environment, when they are adsorbed to plastic surfaces. In addition, there is still the need to fully understand whether plastics and PAHs present an additive toxicological threat, or whether the plastic plays a protective role by sequestering PAHs from entering biological systems.

References

Abbasi, S., Moore, F., Keshavarzi, B. 2021. PET-microplastics as a vector for polycyclic aromatic hydrocarbons in a simulated plant rhizosphere zone. Environmental Technology & Innovation. 21: 101370

Alassali, A., Calmano, W., Gidarakos, E., Kuchta, K. 2020. The degree and source of plastic recyclates contamination with polycyclic aromatic hydrocarbons. RSC Advances. 10: 44989–44996

Almond, J., Sugumaar, P., Wenzel, M.N., Hill, G., Wallis, C. 2020. Determination of the carbonyl index of polyethylene and polypropylene using specified area under band methodology with ATR-FTIR spectroscopy. E-polymers. 20: 369–381

An, Z., Li, X., Yuan, Y., Duan, F., Jiang, J. 2022. Large contribution of non-priority PAHs in atmospheric fine particles: Insights from time-resolved measurement and nontarget analysis. Environment International. 163: 107193

Andersson, J.T., Achten, C. 2015. Time to say goodbye to the 16 EPA PAHs? Toward an up-to-date use of PACs for environmental purposes. Polycyclic Aromatic Compounds. 35: 330–354

Andrady, A.L., Barnes, P.W., Bornman, J.F., Gouin, T., Madronich, S., White, C.C., Zepp, R.G., Jansen, M.A.K. 2022. Oxidation and fragmentation of plastics in a changing environment; from UV-radiation to biological degradation. Science of the Total Environment. 851: 158022

Bao, Z.Z., Chen, Z.F., Zhong, Y., Wang, G., Qi, Z., Cai, Z. 2021. Adsorption of phenanthrene and its monohydroxy derivatives on polyvinyl chloride microplastics in aqueous solution: Model fitting and mechanism analysis. Science of the Total Environment. 764: 142889

Bao, Z.Z., Lu, S.Q., Wang, G., Cai, Z., Chen, Z.F. 2023. Adsorption of 2-hydroxynaphthalene, naphthalene, phenanthrene, and pyrene by polyvinyl chloride microplastics in water and their bioaccessibility under in vitro human gastrointestinal system. Science of The Total Environment. 871: 162157

Bhagwat, G., Carbery, M., Tran, T.K.A., Grainge, I., O'Connor, W., Palanisami, T. 2021. Fingerprinting plastic-associated inorganic and organic matter on plastic aged in the marine environment for a decade. Environmental Science & Technology. 55: 7407–7417

Borrelle, S.B., Ringma, J., Law, K.L., Monnahan, C.C., Lebreton, L., McGivern, A., Murphy, E., Jambeck, J., Leonard, G.H., Hilleary, M.A., Eriksen, M., Possingham, H.P., De Frond, H., Gerber, L.R., Polidoro, B., Tahir, A., Bernard, M., Mallos, N., Barnes, M., Rochman, C.M. 2020. Predicted growth in plastic waste exceeds efforts to mitigate plastic pollution. Science. 369: 1515–1518

Bouhroum, R. Boulkamh, A., Asia, I., Labarillier, S., Ter Halle, A., Syakti, A.D., Doumenq, P., Malleret, I., Wong-Wah-Chung, P. 2019. Concentrations and fingerprints of PAHs and PCBs adsorbed onto marine plastic debris from the Indonesian Cilacap coast and the North Atlantic gyre. Regional Studies in Marine Science. 29: 100611

Chamas, A., Moon, H., Zheng, J., Qiu, Y., Tabassum, T., Jang, J.H., Abu-Omar, M., Scott, S.L., Suh, S. 2020. Degradation rates of plastics in the environment. ACS Sustainable Chemistry & Engineering. 8: 3494–3511

Chen, C.F., Ju, Y.R., Lim, Y.C., Hsu, N.H., Lu, K.T., Hsieh, S.L., Dong, C.D., Chen, C.W. 2020. Microplastics and their affiliated PAHs in the sea surface connected to the southwest coast of Taiwan. Chemosphere. 254: 126818

Costigan, E., Collins, A., Hatinoglu, M.D., Bhagat, K., MacRae, J., Perreault, F., Apul, O. 2022. Adsorption of organic pollutants by microplastics: Overview of a dissonant literature. Journal of Hazardous Materials Advances. 6: 100091

Deng, H., Su, L., Zheng, Y., Du, F., Liu, Q.X., Zheng, J., Zhou, Z., Shi, H. 2022. Crack Patterns of Environmental Plastic Fragments. Environmental Science & Technology. 56: 63–6414

Elovaara, E., Mikkola, J., Stockman-Juvala, H., Luukkanen, L., Keski-Hynnilä, H., Kostiainen, R., Pasanen, M., Pelkonen, O., Vainio, H. 2006. Polycyclic aromatic hydrocarbon (PAH) metabolizing enzyme activities in human lung, and their inducibility by exposure to naphthalene, phenanthrene, pyrene, chrysene, and benzo(a)pyrene as shown in the rat lung and liver. Archives of Toxicology. 81: 169–182

Emecheta, E.E., Borda, D.B., Pfohl, P.M., Wohlleben, W., Hutzler, C., Haase, A., Roloff, A. 2022. A comparative investigation of the sorption of polycyclic aromatic hydrocarbons to various polydisperse micro- and nanoplastics using a novel third-phase partition method. Microplastics and Nanoplastics. 2:29

Endo, S., Hale, S.E., Goss, K.U., Arp, H.P.H. 2011. Equilibrium partition coefficients of diverse polar and nonpolar organic compounds to polyoxymethylene (POM) passive sampling devices. Environmental Science & Technology. 45: 10124–10132

Fisner, M., Taniguchi, S., Moreira, F., Bicego, M.C., Turra, A. 2013a. Polycyclic aromatic hydrocarbons (PAHs) in plastic pellets: Variability in the concentration and composition at different sediment depths in a sandy beach. Marine Pollution Bulletin. 70: 219–226

Fisner, M., Taniguchi, S., Majer, A.P., Bicego, M.C., Turra, A. 2013b. Concentration and composition of polycyclic aromatic hydrocarbons (PAHs) in plastic pellets: Implications for small-scale diagnostic and environmental monitoring. Marine Pollution Bulletin. 76: 349–354

Frescura, L.M., Brummelhaus de Menezes, B., Funari Jr., R.A., Mortari, S.R., de Moraes Bastos, A.F., Barcellos da Rosa, M. 2024. Polycyclic aromatic hydrocarbon derivatives onto polar microplastics of polyurethane: equilibrium, thermodynamics, and kinetics of monolayer-multilayer adsorption. Environmental Science and Pollution Research. 31: 55158–55168

Fries, E., Zarfl, C. 2012. Sorption of polycyclic aromatic hydrocarbons (PAHs) to low and high density polyethylene. Environmental Science and Pollution Research. 19: 1296–1304

Geyer, R., Jambeck, J.R., Law, K.L. 2017. Production, use, and fate of all plastics ever made. Science Advances. 19: e1700782

Gong, H., Li, R., Li, F., Guo, X., Xu, L., Gan, L., Yan, M., Wang, J. 2023. Toxicity of nanoplastics to aquatic organisms: Genotoxicity, cytotoxicity, individual level and beyond individual level. 443: 130266

Gorman, D., Moreira, F.T., Turra, A., Fontenelle, F.R., Combi, T., Bícego, M.C., Martins, de Castro Martins, C. 2019. Organic contamination of beached plastic pellets in the South Atlantic: Risk assessments can benefit by considering spatial gradients. Chemosphere. 223: 608–615

Goswami, P., Ohura, T., Subasinghe, S., Wickrama-Arachchige, A.U.K., Takeuchi, S., Imaki, M., Miizuma, Y., Watanabe, M., Guruge, K.S. 2023. Voyaging of halogenated polycyclic aromatic hydrocarbons, an emerging group of pollutants, on micro-mesoplastics in the marine environment. Journal of Hazardous Materials. 460: 132502

Guo, X., Wang, X., Zhou, X., Kong, X., Tao, S., Xing, B. 2012. Sorption of four hydrophobic organic compounds by three chemically distinct polymers: Role of chemical and physical composition. Environmental Science & Technology. 46: 7252–7259

Hakim, A., Kobayashi, M. 2021. Aggregation and aggregate strength of microscale plastic particles in the presence of natural organic matter: Effects of ionic valence. Journal of Polymers and the Environment. 29: 1921–1929

Hirai, H., Takada, H., Ogata, Y., Yamashita, R., Mizukawa, K., Saha, M., Kwan, C., Moore, C., Gray, H., Laursen, D., Zettler, E.R., Farrington, J.W., Reddy, C.M., Peacock, E.E., Ward, M.W. 2011. Organic micropollutants in marine plastics debris from the open ocean and remote and urban beaches. Marine Pollution Bulletin. 62: 1683–1692

Ho, Y.S., McKay, G. 1999. Pseudo-second order model for sorption processes. Process Biochemistry. 34: 451–465

Hou, G., Zhao, X., Zhao, T., Wu, X., Pu, S., Tang, Z., Wu, F. 2023. The adsorption of PAHs on microplastics and desorption in the simulated human digestive system. Chemical Engineering Journal. 473: 145157

Hüffer, T., Hofmann, T. 2016. Sorption of non-polar organic compounds by micro-sized plastic particles in aqueous solution. Environmental Pollution. 214: 194–201

Idowu, O., Sempler, K.T., Ramadass, K., O'Connor, W., Hansbro, P., Thavamani, P. 2019. Beyond the obvious: Environmental health implications of polar polycyclic aromatic hydrocarbons. Environment International. 123: 543–557

Jin, M., Yu, X., Yao, Z., Tao, P., Yu, X., Zhao, J.L., Peng, J. 2020. How biofilms affect the uptake and fate of hydrophobic organic compounds (HOCs) in microplastic: Insights from an In situ study of Xiangshan Bay, China. Water Research. 184: 116118

Jonker, M.T.O., Koelmans, A.A. 2002. Sorption of polycyclic aromatic hydrocarbons and polychlorinated biphenyls to soot and soot-like materials in the aqueous environment: Mechanistic considerations. Environmental Science & Technology. 36: 3725–3734

Karapanagioti, H.K., Klontza, I. 2008. Testing phenanthrene distribution properties of virgin plastic pellets and plastic eroded pellets found on Lesvos island beaches (Greece). Marine Environmental Research. 65: 283–290

Keith, L.H. 2014. The source of U.S. EPA's sixteen PAH priority pollutants. Polycyclic Aromatic Compounds. 35: 147–16

Kida, M., Koszelnik, P. 2021. Investigation of the presence and possible migration from microplastics of phthalic acid esters and polycyclic aromatic hydrocarbons. Journal of Polymers and the Environmennt. 29: 599–611

Kleinteich, J., Seidensticker, S., Marggrander, N., Zarfi, C. 2018. Microplastics reduce short-term effects of environmental contaminants. Part II: Polyethylene particles decrease the effect of polycyclic aromatic hydrocarbons on microorganisms. International Journal of Environment Research and Public Health. 15: 287

Kong, X., Zhang, J., Ji, Q., Li, C., Chen, X., Cao, X., Zhu, F., Yang, S., Li, S., He, H. 2023. Insights into adsorption mechanisms of nitro polylcyclic aromatic hydrocarbons on common microplastic particles: Experimental studies and modeling. Chemosphere. 320: 138050

Kuźmicz, K., Ciemniak, A. 2017. Assessing contamination of smoked sprats (Sprattus sprattus) with polycyclic aromatic hydrocarbons (PAHs) and changes in its level during storage in various types of packaging. Journal of Environmental Science and Health, Part B. 53: 1–11

Lara, S., Villanueva, F., Cabañas, B., Sagrario, S., Aranda, I., Soriano, J.A., Martin, P. 2023. Determination of polycyclic aromatic compounds (PAH, nitro-PAH and oxy-PAH) in soot collected from a diesel engine operating with different fuels. Science of the Total Environment. 900: 165755

Lee, H., Shim, W. J., Kwon, J.H. 2014. Sorption capacity of plastic debris for hydrophobic organic chemicals. Science of the Total Environment. 470-471: 1545–1552

Lee, J.Y., Lane, D.A., Kim, Y.P. 2015. Formation of polyaromatic hydrocarbon (PAH)-quinones during the gas phase reactions of PAHs with the OH radical in the atmosphere. Environmental Chemistry. 12: 307–315

León, V.M., García-Agüera, I., Moltó, V., Fernández-González, V., Llorca-Pérez, L., Andrade, J.M., Muniategui-Lorenzo, S., Campillo, J.A. 2019. PAHs, pesticides, personal care products and plastic additives in plastic debris from Spanish Mediterranean beaches. Science of The Total Environment. 670: 672–684

Li, R., Tan, H., Zhang, L., Wang, S., Wang, Y., Yu, K. 2018. The implications of water extractable organic matter (WEOM) on the sorption of typical parent, alkyl, and N/O/S-containing polycyclic aromatic hydrocarbons (PAHs) by microplastics. Ecotoxicology and Environmental Safety. 156: 176–182

Li, Y., Wang, J., Yang, G., Lu, L., Zheng, Y., Zhang, Q., Zhang, X., Tian, H., Wang, W., Ru, S. 2020. Low level of polystyrene microplastics decreases early developmental toxicity of phenanthrene on marine medaka (*Oryzias melastigma*). Journal of Hazardous Materials. 385: 121586

Li, Y., Tao, L., Wang, Q., Wang, F., Li, G., Song, M. 2023. Potential health impact of microplastics; A review of environmental distribution, human exposure, and toxic effects. Environment & Health. 1: 249–257

Liu, L., Fokkink, R., Koelmans, A.A. 2016. Sorption of polycyclic aromatic hydrocarbons to polystyrene nanoplastic. Environmental Toxicology and Chemistry. 35: 1650–1655

Liu, W., Tang, H., Yang, B., Li, C., Chen, Y., Huang, T. 2023. Molecular level insight into the different interaction intensity between microplastics and aromatic hydrocarbon in pure water and seawater. Science of the Total Environment. 862: 160786

Liu, T., Zhang, Y., Gutierrez, L., Zheng, X., Benedetti, M., Croué, J.P. 2024. Impact of dissolved organic matter chacateristics and inorganic species on the stability and removal by coagulation of nanoplastics in aqueous media. Chemosphere. 366: 143482

Lončarski, M., Gvoić, V., Prica, M., Cveticanin, L., Agbaba, J., Tubić, A. 2021. Sorption behavior of polycyclic aromatic hydrocarbons on biodegradable polylactic acid and various nondegradable microplastics: Model fitting and mechanisms analysis. Science of the Total Environment. 785: 147289

Loudon, K.B., Tomy, T.M.Z., Liebzeit, E.C., Halldorson, T., Xia, Z., Sambanthan, S., Hoang, D.L., Vitharan, N., Tomy, G.T. 2025. Estimating dietary exposure to polycyclic aromatic compounds from food grade plastics. Environmental Science Advances. 4: 292–305

Maddison, C., Sathish, C.I., Lakshmi, D., Wayne, O., Palanisami, T. 2023. An advanced analytical approach to assess the long-term degradation of microplastics in the marine environment. NPJ Materials Degradation. 7: 59

Ma, Y., Huang, A., Cao, S., Sun, F., Wang, L., Guo, H., Ji, R. 2016. Effects of nanoplastics and microplastics on toxicity, bioaccumulation, and environmental fate of phenanthrene in fresh water. Environmental Pollution. 219: 166–173

Mai, L., He, H., Bao, L.J., Liu, L.Y., Zeng, E.Y. 2020. Plastics are an insignificant carrier of riverine organic pollutants to the coastal oceans. Environmental Science & Technology. 54: 15852–15860

Martinez-Alvarez, I., Le Menach, K., Devier, M.H., Cajaraville, M.P., Budzinski, H., Orbea, A. 2022. Screening of the toxicity of polystyrene nano- and microplastics alone and in combination with benzo(a)pyrene in brine shrimpe larvae and zebrasifh embryos. Nanomaterials. 12: 941

Mato, Y., Isobe, T., Takada, H., Kanehiro, H., Ohtake, C., Kaminuma, T. 2001. Plastic resin pellets as a transport medium for toxic chemicals in the marine environment. Environmental Science & Technology. 35: 318–324

Rios, L.M., Jones, P.R. 2015. Characterisation of microplastics and toxic chemicals extracted from microplastic samples from the North Pacific Gyre. Environmental Chemistry. 12: 611–617

Mohanan, N., Montazer, Z., Sharma, P.K., Levin, D.B. 2020. Microbial and enzymatic degradation of synthetic plastics. Frontiers in Microbiology. 11: 580709

Moncrieffe, R., Masry, M., Cai, B., Rossignol, S., Kamari, A., Poirier, L., Bertrand, S., Wong-Wah-Chung, P., Zalouk-Vergnoux, A. 2023. Study of the ageing and the sorption of polyaromatic hydrocarbons as influencing factors on the effects of microplastics on blue mussel. Aquatic Toxicology. 262: 106669

Natarajan, L., Omer, S., Jetly, N., Jenifer, M.A., Chandrasekaran, N., Suraishkumar, G.K., Mukherjee, A. 2020. Eco-corona formation lessens the toxic effects of polystyrene nanoplastics towards marine microalgae *Chlorella sp*. Environmental Research. 188: 109842

O'Connor, I.A., Golsteijn, L., Hendriks, A.J. 2016. Review of the partitioning of chemicals into different plastics: Consequences for the risk assessment of marine plastic debris. Marine Pollution Bulletin. 113: 17–24

Park, J.H., Troxel, A.B., Harvey, R.G., Penning, T.M. 2006. Polycyclic aromatic hydrocarbon (PAH) o-quinones produced by the aldo-keto-reductases (AKRs) generate abasic sites, oxidized pyrimidines, and 8-oxo-dguo via reactive oxygen species. Chemical Research in Toxicology. 19: 719–728

Phan, S., Padilla-Gamiño, J.L., Luscombe, C.K. 2022. The effect of weathering environments on microplastic chemical identification with Raman and IT spectroscopy: Part 1. Polyethylene and polypropylene. Polymer Testing. 116: 107752

Pittura, L., Avio, C.G., Giuliani, M.E., D'Errico, G., Keiter, S.H., Cormier, B., Gorbi, S., Regoli, F. 2018. Microplastics as vehicles of environmental PAHs to marine organisms: Combined chemical and physical hazard to the Mediterranean Mussels, *Mytilus galloprovincialis*. Frontiers in Marine Science. 5: 103

Qian, Y., Xu, Z., Hong, X., Luo, Z., Gao, X., Tie, C., Liang, H. 2022. Alkylated polycyclic aromatic hydrocarbons are the largest contributor to polycyclic aromatic compound concentrations in the topsoil of Huaibei Coalfield, China. International Journal of Environmental Research and Public Health. 19: 12733

Reizer, E., Viskolcz, B., Fiser, B. 2022. Formation and growth mechanisms of polycyclic aromatic hydrocarbons: A mini-review. Chemosphere. 291: 132793

Revellame, E.D., Fortela, D.L., Sharp, W., Hernandez, R., Zappi, M.E. 2020. Adsorption kinetic modeling using pseudo-first order and pseudo-second order rate laws: A review. Cleaner Engineering and Technology. 1: 100032

Rochman, C.M., Manzano, C., Hentschel, B.T., Simonich, S.L.M., Hoh, E. 2012. Polystyrene plastic; A source and sink for polycyclic aromatic hydrocarbons in the marine environment. Environmental Science & Technology. 47: 13976–13984

Rochman, C., Hoh, E., Hentschel, B.T., Kaye, S. 2013. Long-term field measurement of sorption of organic contaminants to five types of plastic pellets: Implications for plastic marine debris. Environmental Science & Technology. 47: 1646–1654

Rodriguez, C., Fossatti, M., Carrizo, D., Sanchez-Garcia, L., Teixeira de Mello, F., Weinstein, F., Lozoya, J.P. 2020. Mesoplastics and large microplastics along a use gradient on the Uruguay Atlantic coast: Types, sources, fates, and chemical loads. Science of the Total Environment. 721: 13773

Sankoda, K., Saito, K. 2025. Effect of sunlight aging on physicochemical properties and sorption capacities of environmental microplastics: implications for contamination by PAHs. Environmental Science and Pollution Research

Schiferle, E.B., Ge, W., Reinhard, B.M. 2023. Nanoplastics weathering and polycyclic aromatic hydrocarbon mobilization. ACS Nano. 17: 5773–5784

Schweighuber, A., Himmelsbach, M., Buchberger, W., Klampfl, C.W. 2019. Analysis of polycyclic aromatic hydrocarbons migrating from polystyrene/divinylbenzene-based food contact materials. Monatshefte für Chemie. 150: 901–906

Seidensticker, S., Grathwohl, P., Lamprecht, J., Zarfl, C. 2018. A combined experimental and modeling study to evaluate pH-dependent sorption of polar and non-polar compounds to polyethylene and polystyrene microplastics. Environmental Sciences Europe. 30:30

Shi, J., Sanganyado, E., Wang, L., Li, P., Li, X., Liu, W. 2020. Organic pollutants in sedimentary microplastics from eastern Guangdong: Spatial distribution and source identification. Ecotoxicology and Environmental Safety. 193: 110356

Shiizaki, K., Kawanishi, M., Yagi, T. 2017. Modulation of benzo[a]pyrene-DNA adduct formation by CYP1 inducer and inhibitor. Genes and Environment. 39: 14

Shimada, T., Fujii-Kuriyama, Y. 2004. Metabolic activation of polycyclic aromatic hydrocarbons to carcinogens by cytochromes P450 1A1 and1B1. Cancer Science. 19: 1–6

Singh, N., Tiwari, E., Khandelwal, N., Darbha, G.K. 2019. Understanding the stability of nanoplastics in aqueous environments: effect of ionic strength, temperature, dissolved organic matter, clay, and heavy metals. Environmental Science: Nano. 6: 2968–2976

Sörensen, L., Rogers, E., Altin, D., Salaberria, I., Booth, A.M. 2020. Sorption of PAHs to microplastic and their bioavailability and toxicity to marine copepods under co-exposure conditions. Environmental Pollution. 258: 113844.

Stowers, S.J., Anderson, M.W. 1985. Formation and persistence of benzo(a)pyrene metabolite-DNA adducts. Environmental Health Perspectives. 62: 31–39

Sun, Q., Lu, F.H., Ni, H.G. 2021a. Microplastic-water partitioning of two states halogenated PAHs: Solutes and sol. Environmental Research. 202: 111634

Sun, S., Shi, W., Tang, Y., Han, Y., Du, X., Zhou, W., Zhang, W., Sun, C., Liu, G. 2021b. The toxic impacts of microplastics (MPs) and polycyclic aromatic hydrocarbons (PAHs) on haematic parameters in a marine bivalve species and their potential mechanisms of action. Science of the Total Environment. 783: 147003

Sun, H., Jiao, R., Yu, J., Wang, D. 2023. Combined effects of particle size and humic acid corona on the aggregation kinetics of nanoplastics in aquatic environments. Science of the Total Environment. 901: 165987

Tang, S., Lin, L., Wang, X., Sun, X., Yu, A. 2021. Adsorption of fulvic acid onto polyamide 6 microplastics: Influencing factors, kinetics modeling, site energy distribution and interaction mechanisms. Chemosphere. 272: 129638

Teuten, E.L., Rowland, S.J., Galloway, T.S., Thompson, R.C. 2007. Potential for plastics to transport hydrophobic contaminants. Environmental Science & Technology. 41: 7759–7764

Teuten, E.L., Saquing, J.M., Knappe, D.R.U., Barlaz, M.A., Jonsson, S., Björn, A., Rowland, S.J., Thompson, R.C., Galloway, T.S., Yamashita, R., Ochi, D., Watanuki, Y., Moore, C., Viet, P.H., Tana, T.S., Prudente, M., Boonyatumanond, R., Zakaria, M.P., Akkhavong, K., Ogata, Y., Hirai, H., Iwasa, S., Mizukawa, K., Hagino, Y., Imamura, A., Saha, M., Takada, H. 2009. Transport and release of chemicals from plastics to the environment and to wildlife. Philosophical Transactions of the Royal Society B. 364: 2027–2045

Tidwell, L.G, Allan, S.E., O'Connell, S.G., Hobbie, K.A., Smith, B.W., Anderson, K.A. 2016. PAH and OPAH flux during the Deepwater Horizon accident. Environmental Science & Technology. 50: 7489–7497

Trevisan, R., Uzochukwu, D., Di Giulio, R.T. 2020. PAH sorption to nanoplastics and the Trojan horse effect as drivers of mitochondrial toxicity and PAH localization in zebrafish. Frontiers in Environmental Science. 8:78

Uber, T.H., Hüffer, T., Planitz, S., Schmidt, T.C. 2019. Sorption of non-ionic organic compounds by polystyrene in water. Science of the Total Environment. 682: 348–355

Udenby, F.A.O., Almuhtaram, H., McKie, M.J., Andrews, R.C. 2022. Adsorption of fluoranthene and phenanthrene by virgin and weathered polyethylene microplastics in freshwaters. Chemical Engineering Journal. 473: 145157

Van, A., Rochman, C.M., Flores, E.M., Hill, K.L., Vargas, E., Vargas, S.A., Hoh, E. 2012. Persistent organic pollutants in plastic marine debris found on beaches in San Diego, California. Chemosphere. 86: 258–263

Vione, D., Barra, S., de Gennaro, G., de Rienzo, M., Gilardoni, S., Perrone, M.G., Pozzoli, L. 2004. Polycyclic Aromatic Hydrocarbons in the Atmosphere: Monitoring, Sources, Sinks and Fate. II: Sinks and Fate. Annali di Chimica. 94: 257–268

Wang, W., Wang, J. 2018. Different partition of polycyclic aromatic hydrocarbon on environmental particulates in freshwater: Microplastics in comparison to natural sediment. Ecotoxicology and Environmental Safety. 147: 648–655

Wang, J., Liu, X., Liu, G., Zhang, Z., Wu, H., Cui, B., Bai, J., Zhang, W. 2019. Size effect of polystyrene microplastics on sorption of phenanthrene and nitrobenzene. Ecotoxicological and Environmental Safety. 173: 331–338

Wang, Y., Zhang, M., Ding, G., Shi, H., Cong, Y., Li, Z., Wang, J. 2023. Polystyrene microplastics alleviate adverse effects of benzo[a]pyrene on tissues and cells of the marine mussel, *Mytlius galloprovincialis.* Aquatic Toxicology. 256: 106430

Wenger, D., Gerecke, A.C., Heeb, N.V., Schmid, P., Hueglin, C., Naegeli, H., Zenobi, R. 2008. In vitro estrogenicity of ambient particulate matter: contribution of hydroxylated polycyclic aromatic hydrocarbons. Journal of Applied Toxicology. 29: 223–232

Wilcke, W., Kiesewetter, M., Bandowe, B.A.M. 2014. Microbial formation and degradation of oxygen-containing polycyclic aromatic hydrocarbons (OPAHs) in soil during short-term incubation. Environmental Pollution. 184: 385–390

Xu, X.R., Li, X.Y. 2010. Sorption and desorption of antibiotic tetracycline on marine sediments. Chemosphere. 78: 430–436

Yeo, B.G., Mizukawa, K., Takada, H., Tait, H., Frechou, J. 2022. Polycyclic aromatic hydrocarbons (PAHs) in new unexposed and beached expanded polystyrene foams. Environmental Monitoring and Contaminants Research. 2: 14–21

Yu, H., Yang, B., Waigi, M.G., Peng, F., Li, Z., Hu, X. 2020. The effects of functional groups on the sorption of naphthalene on microplastics. Chemosphere. 261: 127592

Zhang, W., Ma, X., Zhang, Z., Wang, Y., Wang, J., Wang, J., Ma, D. 2015. Persistent organic pollutants carried on plastic resin pellets from two beaches in China. Marine Pollution Bulletin. 99: 28–34

Zhang, P., Huang, P., Sun, H., Ma, J., Li, B. 2020. The structure of agricultural microplastics (PY, PU and UF) and their sorption capacities for PAHs and PHE derivatives under various salinity and oxidation treatments. Environmental Pollution. 257: 113525

Zhang, D., Zhang, Z., Liu, H., Zou, J., Yin, L., Liu, X., Zhang, Y., Qu, J., Peijnenburg, W.J.G.M. 2024. Insights into the effect of crystallinity on the sorption of organic pollutants to microplastics. Environmental Science and Pollution Research. 31: 42202–42211

Zhao, L., Rong, L., Xu, J., Lian, J., Wang, L., Sun, H. 2020. Sorption of five organic compounds by polar and nonpolar microplastics. Chemosphere. 257: 127206

Zhao, B., Rehati, P., Yang, Z., Cai, Z., Guo, C., Li, Y. 2023. The potential toxicity of microplastics on human health. Science of the Total Environment. 912: 168946

Toxicity and Adverse Health Effects

Carcinogenic Risk Assessment of Polycyclic Aromatic Hydrocarbons in the Ambient Air Using In Vivo Mutagenicity Data

Yasunobu Aoki

Abstract Polycyclic aromatic hydrocarbons (PAHs) and their related compounds absorbed on airborne particles are suspected of causing lung cancer, and it is necessary to develop the approaches for assessing the carcinogenic risk of PAHs. In the 1990s, we observed that the level of DNA adducts was dramatically increased in the lungs of rats after 4-week inhalation of ambient air from the Tokyo metropolitan area. This observation led us to evaluate the in vivo mutagenicity of airborne particles and their components by means of transgenic mouse gene mutation assay (*gpt* delta mouse), and we found elevated mutation frequency in the lungs of mice exposed to diesel exhaust or diesel exhaust particles. We then conducted a real-world study to evaluate the in vivo mutagenicity of airborne particles containing a mixture of PAHs. Extracts from airborne particles collected in 1980, 1990, and 2010 from the Tokyo metropolitan area were instilled to *gpt* delta mouse lungs, and the 1990 extract showed the most potent in vivo mutagenicity. In addition, G-to-T transversion, a landmark mutation of oxidative stress, was markedly induced in the extract-instilled lungs. Our findings indicate that the carcinogenic risk of PAHs absorbed on airborne particles can be quantitatively assessed from in vivo mutagenicity.

Keywords Ambient air · In vivo mutagenicity · Lung cancer · Risk assessment · Transgenic rodent assay

1 Introduction

The polycyclic aromatic hydrocarbons (PAHs) are a class of organic compounds in which the member chemicals possess at least two (generally 2–6) six-atom, aromatic fused rings. PAHs are generated by incomplete combustion of fossil fuels and biomass, and consequently these compounds are emitted to the environment, mainly the ambient air as reviewed in the references (WHO 2000, 2010, 2021; European

---

Y. Aoki (✉)
National Institute for Environmental Studies, Tsukuba, Ibaraki, Japan
e-mail: aoki.yasunobu@nies.go.jp

K. Hayakawa and Y. Aoki (eds.), *A New Era of Polycyclic Aromatic Hydrocarbons*,
https://doi.org/10.1007/978-981-96-7695-8_7

Communities 2001a; Kelly et al. 2021; CAREX Canada). Major sources of PAHs in ambient air are the exhausts from automobiles, steel mills, and power plants. In the workplace, the main source of PAHs is the burning of fossil fuels for manufacturing.

PAHs incorporated into the body can have potent mutagenicity and carcinogenicity, and exposure to PAHs in ambient air is strongly suspected to cause a variety of cancers including lung and breast (WHO 2021; Waddingham et al. 2024). Indeed, several PAHs are currently classified as IARC Group 1 (human carcinogens) and Group 2A (probable human carcinogens) carcinogens by the International Agency for Research on Cancer under the World Health Organization (WHO 2010). Recently, other adverse effects posed by PAHs have been recognized, including cardiotoxicity, developmental toxicity, and reproductive toxicity (WHO 2021; Mallah et al. 2022; Venkatraman et al. 2024), prompting the development of comprehensive approaches for evaluation of the health risks of PAHs (Hsieh et al. 2021; de Oliveira Galvão et al. 2024).

This chapter begins by presenting a short overview of approaches suitable for assessing the carcinogenic risk of PAHs in ambient air. Then, the discussion moves on to our group's recent studies using transgenic rodent-based assays (Thybaud et al. 2003; Lambert et al. 2005) to examine the in vivo mutagenicity of ambient air containing mixtures of PAHs. To conclude, the application of in vivo mutagenicity data for assessing the carcinogenic risk of ambient PAHs is discussed.

2 Approaches for Assessing the Carcinogenic Risk of PAHs

PAHs in the air are mainly found absorbed on particulate matter and are inhaled as a mixture of various congeners; therefore, risk assessments of the carcinogenicity of PAHs must consider simultaneous exposure to a mixture of PAHs and their related compounds (WHO 2000; European Communities 2001a, b). To this end, two approaches are commonly used: the whole-mixture approach or the component-based approach (U.S. EPA 2000, 2007; OECD 2018).

Whole-mixture approach This approach is most often used for assessing the risk of complex mixtures that contain unknown chemicals, under the assumption that the mixture is a single chemical (OECD 2018). However, it can also be used to assess environmental mixtures for which dose–response data are available (U.S. EPA 2000) or for mixtures for which toxicity and exposure data are available. The advantage of whole-mixture approaches is that the cumulative risk of a mixture of chemicals can be assessed, even if some components of the mixture cannot be identified.

To evaluate the risk associated with exposure to a mixture, a reference dose/concentration (RfD/RfC) value for a mixture is estimated. When the NOAEL (no observed adverse effect level), LOAEL (lowest observed adverse effect level), or BMDL (95% confidence lower limit of benchmark dose at benchmark ratio) values can be derived from dose–response (toxicity) data obtained using epidemiological data or animal experimental data of a mixture, the RfD/RfC of a mixture is calculated using formula (1). The uncertainty factor is set to avoid underestimation of the risk due to uncertainties about the differences in the sensitivity to the chemical between species/individuals, and it is usually set at over 100. The procedure used for the whole mixture is the same as that for a single chemical assessment, when the composition of the test mixture is functionally the same as what is found in the environment (OECD 2018; U.S. EPA 2000).

$$\mathrm{RfD/RfC}\left(\mathrm{mg/kg/day/mg/m^3}\right) = \mathrm{NOAEL,\ LOAEL,\ or\ BMDL/uncertainty\ factor} \tag{1}$$

To assess the carcinogenic risk of a mixture, the slope factor ($(\mathrm{mg/kg/day})^{-1}$) or unit risk ($(\mathrm{mg/m^3})^{-1}$) of the mixture is calculated using a benchmark ratio (usually excess cancer risk = 0.1) and the BMDL as follows:

$$\mathrm{Slope\ factor\ or\ unit\ risk} = \mathrm{Benchmark\ ratio/BMDL} \tag{2}$$

An example of the whole-mixture approach is carcinogenic risk assessment of coke oven emissions containing PAHs. The unit risk of lung cancer caused by PAHs in ambient air is estimated to be 6.2×10^{-4} (μg/m^3 benzene extract from the workplace)$^{-1}$ by the WHO based on dose–response data for the relationship between the concentration of PAHs in the benzene extract of airborne particles sampled from workplaces and the incidence of lung cancer among the workers; the dose–response data were derived from an epidemiological study of workers exposed to coke oven emissions in the USA (WHO 2000). The unit risk of lung cancer was estimated as 8.7×10^{-5} (ng/m^3 B[*a*]P)$^{-1}$, assuming that the concentration of benzo[*a*]pyrene (B[*a*]P), the most carcinogenic PAH, as a surrogate of the PAHs as a whole, in benzene extract was 0.71%. Based on this unit risk, the 1 in 100,000 excess lung cancer risk is estimated to be 0.1 ng/m^3 B[*a*]P (WHO 2000).

Component-based approach This approach is used to estimate the risk of mixtures of chemicals by using exposure and hazard data for each component of the mixture, and this approach is most appropriate when all of the components of a mixture are defined (OECD 2018). There are two types of component-based approach. The first is the Hazard Index (HI) approach. HI is defined as the sum of the hazard quotients (HQ, exposure level per RfD or RfC of an individual component) for each component of the mixture and is usually calculated as follows:

$$\mathrm{HI} = \Sigma \mathrm{Ei}/(\mathrm{RfDi\ or\ RfCi}) = \Sigma \mathrm{HQi} \tag{3}$$

where Ei, RfDi, RfCi, HQi are the exposure level (dose/concentration), RfD, RfC, and HQ, respectively, of component i. If the HI for a mixture is < 1, it is generally assumed that the risk associated with exposure to the mixture is acceptable, whereas if it is $\geqq 1$, the risk is of potential concern.

The second component-based approach is the relative potency factor (RPF) approach. RPF is defined as the relative toxicity of an individual chemical (component) compared to an index chemical, and it is calculated using either general formula

(4) or formula (5) when carcinogenicity specifically is being examined.

$$\text{RPFi} = \text{toxicity of chemical i/toxicity of index chemical} \tag{4}$$

$$\begin{aligned}\text{RPFi} = {} & \text{slope factor of chemical i/slope factor of index chemical} \\ & \text{or unit risk of chemical i/unit risk of index chemical}\end{aligned} \tag{5}$$

The potency of the toxicity of a mixture is assessed by reporting the index chemical equivalent dose (ICED) calculated by using on the dose–response data for an index chemical:

$$\text{ICED} = \Sigma \text{Ei} \times \text{RPFi} \tag{6}$$

To assess the carcinogenicity of PAHs by the RPF approach, B[*a*]P is usually used as the index chemical. The RPFs of various PAHs, derived using in vivo carcinogenicity or mutagenicity data, have been reported by several agencies and researchers (European Communities 2001b; U.S. EPA 2010). However, some RPF values derived using in vitro approaches, such as an antibody-forming-cell response assay (Zaccaria and McClure 2013) and activation of DNA damage-signaling protein checkpoint kinase 1 (de Oliveira Galvão et al. 2024), have also been reported.

3 Assessment of Carcinogenic Risk Associated with Inhalation of Airborne Particles Containing Mixtures of PAHs

PAHs in the ambient air are mainly absorbed on airborne particles. Over the past three decades, the Japanese Ministry of the Environment has monitored the concentration of B[*a*]P on airborne particles in urban areas (Ministry of the Environment 2024). Although the average concentration of B[*a*]P in the urban ambient air has gradually decreased from 0.61 ng/m^3 (range, 0.050–3.7 ng/m^3) in 1998 to 0.16 ng/m^3 (0.0092–3.2 ng/m^3) in 2022, the most recent values indicate that the carcinogenic risk of PAHs still cannot be ignored. Therefore, the assessment for the carcinogenic risk of mixtures of PAHs in ambient air is needed.

3.1 In Situ Exposure Studies to Detect DNA Adducts Generated by Exposure to Ambient Air

Inhalation of a mixture of mutagens (e.g., PAHs) is suspected to cause lung cancer by inducing DNA damage, such as the formation of DNA adducts. In vivo mutagenicity data are informative for understanding how a mixture of pollutants in ambient air has the potential to induce lung cancer (Lodovici et al. 1998; Vineis and Husgafvel-Pursiainen 2005). Our group therefore conducted an in situ exposure study to examine DNA adduct formation in the lungs of rats that had inhaled ambient urban air containing airborne particles (Sato et al. 2003). In our study, rats were maintained in a small animal facility built beside an intersection with heavy traffic in the Tokyo metropolitan area and directly exposed to airborne particles (suspended particulate matter with a diameter of < 11 μm at an average monthly concentration of 76.5 $\mu g/m^3$) for 4 to 60 weeks in 1996 and 1997. P^{32}-postlabeling assay was then used to assess the degree of formation of bulky DNA adducts in the lungs. After exposure to the outdoor air for 4 weeks, bulky DNA adduct formation was significantly increased by 17.1-fold compared to that in the non-exposed control rats, although the levels, while remaining somewhat elevated, were much lower at subsequent timepoints (Fig. 1) (Sato et al. 2003).

This amount of bulky DNA adducts was comparable to that observed in the lungs of smokers (11.0 ± 6.9 DNA adducts/10^8 nucleotides), which itself was significantly higher than that in non-smokers (4.7 ± 3.0 DNA adducts/10^8 nucleotides) (Gyorffy et al. 2004). Furthermore, Gyorffy et al. reported that the PAH-DNA adduct level in the smokers and non-smokers was 6.2 ± 5.1 DNA adducts/10^8 nucleotides and 5.1 ± 2.7 DNA adducts/10^8 nucleotides, respectively, as determined by competitive benzo[*a*]pyrene diol epoxide-DNA chemiluminescence immunoassay (Gyorffy et al.

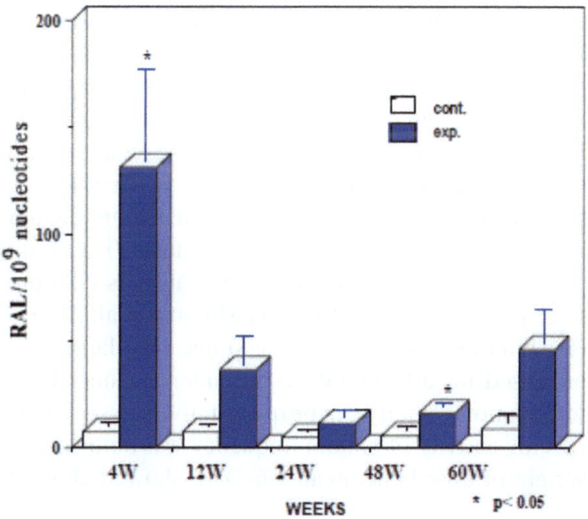

Fig. 1 Relative adduct levels (RALs) in the lungs of rats after exposure to ambient air. RAL values in the lungs of control (cont.) and exposed (exp.) rats were determined by P^{32}-postlabeling assay after exposure to ambient air for 4, 12, 24, 48, and 60 weeks (4W, 12W, 24W, 48W, and 60W, respectively). $n = 3$ in all groups. Bars indicate SE. Asterisks indicate statistical difference from control ($p < 0.05$). Figure reproduced from Sato et al. (2003)

2004), suggesting that the bulky DNA adducts detected by P^{32}-postlabeling assay were mainly PAH-DNA adducts. In the real world, the amounts of bulky DNA adducts in white blood cells have been used for biomonitoring of PAH exposure in residents and workers, and the DNA adduct levels in people living or working in heavily air-polluted areas have been shown to be significantly higher than in those residing in non-polluted areas (Merlo et al. 1997; Peluso et al. 1998; Sato and Aoki 2002; Gyorffy et al. 2004; Ayi-Fanou et al. 2011; Hadrup et al. 2019; Munnia et al. 2023). Together, these results show that PAHs in outdoor air, especially those absorbed on airborne particles, have the ability to induce DNA adduct formation.

3.2 Analysis of In Vivo Mutagenicity of Model Air Pollutants in Transgenic Rodents

Although the studies discussed in Sect. 3.1 clearly indicated that PAHs in outdoor air, especially those absorbed on airborne particles, have the ability to induce DNA adduct formation, it remained unclear whether these DNA adducts produced by PAHs and other related air pollutants are fixed as mutations in the lungs. Therefore, our group conducted a model study in which we evaluated the in vivo mutagenicity of diesel exhaust and its components, as model air pollutants, in transgenic rodents. Transgenic rodents produced using, for example, the Muta, Big Blue, and *gpt* delta systems are useful tools for examining the in vivo mutations induced by exposure to mixtures of pollutants. In these systems, a target gene for detecting mutations (lacZ encoding *Escherichia coli* β-galactosidase, lacI encoding the *E. coli* lactose operon repressor, or *gpt* encoding *E. coli* guanine phosphoribosyltransferase) carried on a λ phage shuttle vector is integrated into the rodents' genomic DNA (for review, see Lambert et al. 2005). To detect in vivo mutations in *gpt* delta rodents, the shuttle vector on the genomic DNA extracted from the organs is rescued to host *E. coli* after exposure to a mutagen, and the mutated target gene is detected as a phenotypic transformation in the host *E. coli*, which in this case is the appearance of 6-thioguanine-resistant colonies (Nohmi et al. 1996). Mutant frequency is defined as the number of mutated *gpt* genes per the number of *gpt* genes rescued to the host *E. coli*. Since the mutations in the *gpt* genes are easily detected by positive selection of the resistant colonies, our group has conducted several studies using *gpt* delta mice to detect in vivo mutations induced by exposure to various pollutants. The studies are summarized below.

As representative potent PAH mutagens contained in diesel exhaust, *gpt* delta mice were exposed to B[*a*]P (Hashimoto et al. 2005) or 1,6-dinitropyrene (1,6-DNP) (Hashimoto et al. 2006) via a single instillation into the lungs. Mutant frequency increased linearly in a dose-dependent manner for both chemicals, indicating that in vivo mutagenicity can be quantitatively assayed by using *gpt* delta mice. When the specific activity of mutant frequency was defined as mutant frequency per milligram weight of dosed chemical, 1,6-DNP showed about 20 times higher specific mutant

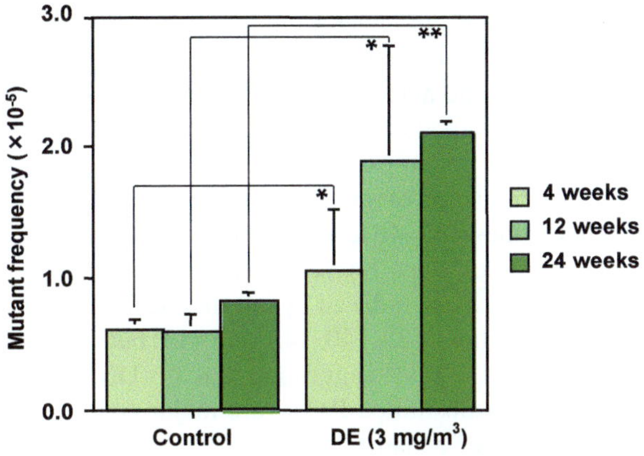

Fig. 2 Mutant frequency in the lungs of *gpt* delta mice after inhalation of diesel exhaust (DE). The mice were subjected to inhalation of 3 mg/m^3 diesel exhaust particles for 12 h per day for 4, 12, or 24 weeks, and then the mutant frequency in the lungs of non-exposed (control) and DE-exposed mice was determined. Original data were originally published in Table 1 in Hashimoto et al. (2007). $^*p < 0.05$, $^{**}p < 0.01$. Bars indicate SD

frequency activity than B[*a*]P. This higher mutagenicity of 1,6-DNA compared to B[*a*]P is consistent with the findings of Ames tests (Kubo et al. 2002).

As a model air pollutant, *gpt* delta mice were subjected to inhalation of diesel exhaust containing diesel exhaust particles (DEP) at a concentration of 3 mg/m^3 for 4 to 24 weeks (Fig. 2) (Hashimoto et al. 2007). Mutant frequency increased depending on the duration of inhalation of diesel exhaust. Our earlier P^{32}-postlabeling assay data had already shown that bulky DNA adduct formation was accelerated in the lungs of mice after only 4 weeks exposure to diesel exhaust at a concentration of 3 mg/m^3 (Aoki et al. 2001). Together, these findings indicated that DNA adducts induced by inhalation of diesel exhaust are indeed fixed as mutations in the lung.

To confirm the in vivo mutagenicity of diesel exhaust, DEP and its extract (DEP extract) were administered to the lungs of *gpt* delta mice by single instillation. The treatment resulted in a significant dose-dependent linear increase in mutant frequency. After treatment with 0.5 mg DEP or 0.2 mg DEP extract, the mutant frequency in the lung was 3.0- and 2.7-fold higher than that in the control mice, respectively. The mutagenic potency (mutant frequency mg^{-1}) of the DEP extract (5.6×10^{-5}) was double that of DEP (2.7×10^{-5}), suggesting that the in vivo mutagenicity of diesel exhaust was induced primarily by compounds present in the extract (Hashimoto et al. 2007). The mutated *gpt* genes were then sequenced to analyze the mutation spectra and to examine the distribution of mutation hotspots (mutation loci identified in three or more mice). G-to-A transition and G-to-T transversion (a landmark mutation induced by oxidative stress) were the single-base substitutions most frequently induced by these treatments. G-to-T transversion was the most frequent substitution induced by B[*a*]P, whereas G-to-A transition was the most frequent substitution induced by 1,6-DNP and diesel exhaust (Hashimoto et al. 2005, 2006, 2007).

Sequence analysis of mutated *gpt* genes from the diesel exhaust–exposed lungs of *gpt* delta mice revealed mutation hotspots on the *gpt* gene at nucleotides 64, 110, and 115 for G-to-A transitions, and nucleotides 185, 401, 402, 406, and 418 for G-to-A transitions and G-to-T transversions (Aoki 2017). Since the mutations at nucleotides 64 and 110 were also found in non-exposed mice, the components in diesel exhaust possibly enhanced the frequency of spontaneously induced mutations. In addition, the mutations at nucleotides 185, 401, 402, 406, and 418 may be a unique signature for exposure to diesel exhaust. Interestingly, these hotspots induced by diesel exhaust were different from those induced by intratracheally instilled B[*a*]P (i.e., G-to-T transversions at nucleotides 125, 140, 143, and 413) but similar to those induced by intratracheally instilled 1,6-DNP, suggesting that 1,6-DNP and related PAHs, as well as B[*a*]P and other unsubstituted PAHs, are major contributors to the in vivo mutagenicity of inhaled diesel exhaust.

3.3 Real-World Study: In Vivo Mutagenicity Induced by Airborne Particle Extracts

To examine the impact of mutagenic compounds in the real environment, our group evaluated the in vivo mutagenicity of airborne particles by *gpt* assay of extract obtained from particles collected in the Tokyo area over a period of 31 years (Aoki et al. 2018). From 1980 to 2010, a high-volume air sampler equipped with quartz-fiber filters was used to collect total suspended particles from the ambient air at two different but nearby districts in the Tokyo area (Shirokanedai District, 1980–2001; Kagurazaka District, 2000–2010). The sampling revealed that the concentration of SPM in the air had gradually decreased from 0.1 mg/m^3 in 1980 to 0.04 mg/m^3 in 2010. After the sampling, the quartz-fiber filters were stored at − 80 °C, and dichloromethane extracts were produced from the total suspended particles collected in 1980, 1990, and 2010. The extracts were administered to the lungs of *gpt* delta mice via a single instillation, and two indices of in vivo mutagenicity, mutant frequency (Fig. 3) and mutation spectrum, were determined in samples of lung tissue. A significant increase of total mutation frequency, defined as the frequency at which independent mutants arose in the same animal, compared with control was observed at certain doses of the 1990 and 2010 extracts. The frequency of G-to-T transversion, a landmark mutation of oxidative stress, was also significantly increased compared with control by both extracts at certain doses, implying that the extracts contained compounds that generate reactive oxygen species that then induce mutations. The intratracheal administration of the 1990 extract at a dose of 0.3 mg was calculated to be comparable to accumulative dose of the 1990 extract in the lung if mice were exposed to the urban air of 1990 for 700 days as previously reported (Aoki et al. 2018). Furthermore, the in vivo mutagenic potency, defined as mutant frequency per milligram-extract, of the 1990 extract was higher than that of the 1980 and 2010 extracts, but the Ames activity of the 1980 extract was higher than that of the 1990 extract (Table 1). The components in the 1990 extract that caused the higher in vivo mutation frequency need to be identified.

When we examined the nucleotides in the *gpt* gene at which mutations were induced in the extract-treated lungs, we identified nucleotides 64, 110, and 115 as mutation hotspots (mutation foci identified in three or more mice) for G-to-A transition and nucleotide 406 as a mutation hotspot for G-to-T transversion. These mutation hotspots were also observed to be induced by diesel exhaust, a model air pollutant. The G-to-A transitions at nucleotides 64 and 110 were possibly spontaneously induced, and 1,6-DNP or related compounds may contribute to induction of these G-to-A transitions (Aoki 2017). The mutation at nucleotide 406 was G-to-T transversion within a TGAA sequence, which has since been identified as a mutation hotspot induced by oxidative stress in the small intestine of *gpt* delta mice after

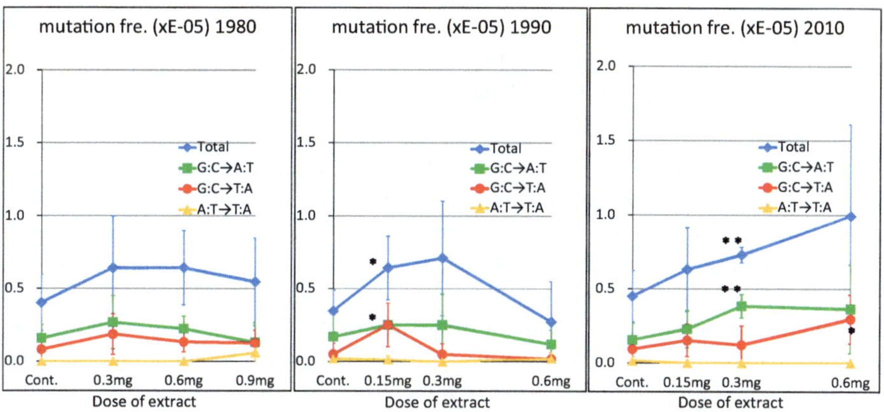

Fig. 3 Mutation frequencies in the lungs of *gpt* delta mice administered extracts obtained from particles collected in the Tokyo area in 1980, 1990, and 2010. The data used to estimate the total mutation frequencies and the number of animals in each group were originally published in Table S1 in Aoki et al. (2018). The frequencies of base substitutions (G-to-A transition, G-to-T transversion, and A-to-T transversion) were calculated as described in Aoki et al. (2018). Panels show the mutation frequency (fre.) in mice administered the 1980, 1990, and 2010 extract. Symbols and bars indicate average values and SD. *$p < 0.05$, **$p < 0.01$. Figures reproduced from Aoki et al. (2018)

Table 1 Mutagenicity of extracts produced form ambient air collected in the Tokyo metropolitan area in 1980, 1990, and 2010

Sampling year	In vivo potency*	Concentration of the extract (mg/m^3)	In vitro (Ames test)**
1980	0.8×10^{-5}	0.095	1302
1990	1.9×10^{-5}	0.080	870
2010	0.93×10^{-5}	0.036	

**gpt* average mutation frequency per extract
**His^+ revertants per milligram-extract in TA98 without S9 (representative Ames test results)
Tables reproduced from Aoki et al. (2018)

oral administration of potassium bromate, a well-characterized inducer of oxidative stress (Aoki et al. 2020). The implication, therefore, is that this mutation may be a unique molecular signature of oxidative stress, which, if true, would indicate that particles collected from urban air, as well as diesel exhaust, a model air pollutant, contain components that induce oxidative stress.

Together, the experiments discussed in this section indicate that the in vivo mutations observed in the lungs of the extract-treated mice are the result of both the generation of reactive oxygen species and the induction of bulky DNA adduct formation. However, exactly which components of the extract, or which subset of PAHs, are responsible remains unknown. Further studies are needed to identify these active components, quantify their atmospheric presence in airborne particles, and elucidate the mechanisms through which they contribute to the incidence of cancer.

4 Assessment of the Carcinogenic Risk of Airborne Particles Using In Vivo Mutagenicity Data

Our group has reported that the TD_{50} (defined as the dose at which cancers were induced in half of the experimental animals) of lung cancer can be predicted by using in vivo mutagenicity data (Aoki 2017). A linear correlation was found between log[T/I] and the log TD_{50} of lung cancer obtained from an animal study, where T is the total dose of chemical and I is the induced mutant frequency (the mutant frequency of the treatment group minus the mutant frequency of the control group) in transgenic mice estimated by in vivo mutation assay $\times$ 10^5 (Fig. 4). The equation describing this linear correlation for lung cancer in mice was

$$\log[T/I] = 1.6719 \log TD_{50} - 0.023 \tag{7}$$

We estimated the T/I of the 1990 extract to be 14.7 mg/kg based on our previous publication (Aoki (2017). Next, TD_{50} was predicted to be 5.2 mg/kg/day from the value of T/I using formula (7), and this predicted TD_{50} value was converted to TD_{50} for a daily inhalation; 3.9 mg/m^3 of extract for a mouse, assuming a body weight and inhalation volume of 0.03 kg and 0.04 m^3, respectively (Aoki 2017). This estimated TD_{50} for a daily inhalation is 370-fold higher than the estimated average concentration of the components of the extract in ambient air in 1990 (Aoki et al. 2018). When this TD_{50} for a daily inhalation (3900 μg/m^3) is assumed to be the benchmark dose at the 50% benchmark ratio of lung cancer, the unit risk for the 1990 extract was estimated to be 1.3×10^{-4} (μg-extract/m^3)$^{-1}$ (benchmark ratio/benchmark dose at 50% benchmark ratio = 0.5/3900), although this is an unusual estimation of unit risk because a unit risk is usually estimated from a benchmark dose at 10% benchmark ratio. This unit risk was compared with the unit risk of human lung cancer estimated from epidemiological study.

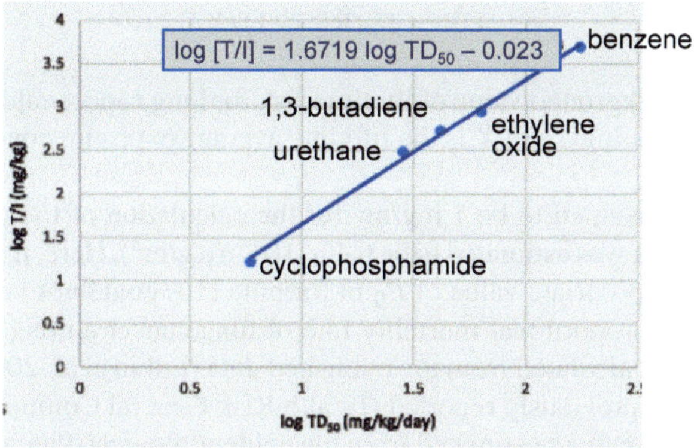

Fig. 4 Plot of log[T/I] versus log TD_{50}. T is the total dose of the indicated chemical, and I is the induced mutant frequency in the lungs of transgenic mice, estimated by in vivo mutation assay $\times 10^5$. TD_{50} was defined as the dose at which cancers were induced in half of the experimental animals. The plots indicate the harmonic mean of the TD_{50} and log[T/I] of each chemical that was calculated as described in (Aoki 2017). This figure is slightly modified by adding the formula and the names of chemicals from that presented originally as Fig. 1 in Aoki (2017)

Several epidemiological analyses examining the relationship between level of exposure to mixtures of PAHs and incidence of lung cancer have been reported (Armstrong et al. 2004; Petit et al. 2019). Among these analyses, Armstrong et al. (1994) only estimated the linear relationship between the risk ratio (RR) for cumulative exposure to benzene-soluble extract containing PAHs obtained from ambient particulate matter collected in working place at an aluminum foundry (WP extract) and epidemiological data for lung cancer incidence among the foundry workers. The RR was calculated as provided by Armstrong et al. (1994) as follows:

$$\mathrm{RR} = 1 + 0.031 \times a \tag{8}$$

where a is the cumulative exposure to the WP extract calculated as the lifetime average exposure concentration (mg/m^3 of WP extract) $\times$ lifetime duration of exposure (years). This RR can be converted to RR of environmental exposure by adjusting for duration of exposure. When exposure-days in a year and exposure-hours in a day are assumed to be 240 days/year and 8 h/day for occupational exposure and 365 days/year and 24 h/day for environmental exposure, respectively, the formula becomes

$$\mathrm{RR} = 1 + 0.031 \times (365/240) \times (24/8) \times a \tag{9}$$

When the lifetime average exposure concentration was assumed to be 1 mg/m^3 of WP extract, the RR of environmental exposure for a lifetime environmental exposure of 70 years was calculated as 10.9 by using formula (9).

The unit risk (UR) can then be derived using an average relative risk model (Cummings 2009):

$$\text{UR} = P_0(\text{RR} - 1)/X \quad (10)$$

where P_0 is the background value of lifetime risk for lung cancer calculated by using a life-table methodology, and X is the lifetime average exposure concentration (mg/m^3).

When X was assumed to be 1 mg/m^3 for the calculation of the unit risk of WP extract, the unit risk was estimated to be 6.1×10^{-4}/(μg/m^3). Here, it should be noted that because an appropriate value of P_0 in formula (10) could not be obtained from the literature, the proportional mortality rate of lung cancer among white males in the USA (0.0616) calculated from the published data (Lubin et al. 2000) was used as the value of P_0, as previously reported (Health Risk General Committee 2010). The unit risk of the WP extract estimated from the epidemiological data in the workplace was close to that predicted using the in vivo mutagenicity data (i.e., 1.3×10^{-4}/(μg/m^3)), indicating that analysis of in vivo mutagenicity may be a useful means of predicting the carcinogenicity of airborne particles.

Finally, to examine whether B[*a*]P, the most carcinogenic PAH, is a major driver of the carcinogenic activity of the 1990 extract, we compared the carcinogenic activity of B[*a*]P to that of the other PAHs in the extract and then compared these findings to those for two large cities in the UK (London and Middlesbrough) and to the aluminum foundry used in the earlier epidemiological evaluation of the risk of lung cancer from exposure to workplace ambient air. The contribution was estimated using a component-based approach by dividing the relative concentration of B[*a*]P (i.e., 1) by the sum of the relative concentrations of B[*a*]P and six other carcinogenic PAHs (benz[*a*]anthracene, dibenzo[*ah*]anthracene, benzo[*b*]fluoranthene, benzo[*k*]fluoranthene, indeno[123-*cd*]pyrene, and chrysene; Table 2a) to B[*a*]P multiplied by the relative potency factor (RPF) of carcinogenicity of each PAH to B[*a*]P. Previously, Long et al. (2017) demonstrated that dose-additive predictions of in vivo mutagenicity based on concentration and in vivo mutagenicity-based RPF values of the PAHs listed in Table 2 and benzo(*ghi*)perylene result in values that are close to those determined experimentally, supporting that the component-based approach is applicable to risk assessment of PAH mixtures containing several major carcinogenic PAHs. The contribution of B[*a*]P to the carcinogenicity of the ambient air collected in the two cities was 37.5 and 44.6%, which was similar to the value obtained at the aluminum foundry (49.3%) (European Communities 2001b). The contribution of B[*a*]P to the carcinogenicity of the 1990 extract, which was calculated using composition data from Aoki et al. (2018), was also 55.2% (Table 2b), indicating that B[*a*]P was a major contributor to the carcinogenicity of both the 1990 extract and the extract from the aluminum foundry. These findings demonstrate not only that B[*a*]P is the main driver of the carcinogenicity of these PAH mixtures, but also that epidemiological studies in the workplace can be used to assess the risk of lung cancer caused by mixtures of PAHs in the ambient air.

Table 2 Contribution of benzo[*a*]pyrene (B[*a*]P) to the carcinogenic activity of PAHs

(a)

Site type	London			Middlesbrough			Aluminum Smelter		
	Average 1991–1995			Average 1993–1995					
PAH[a]	RC	RP	RC × RP	RC	RP	RC × RP	RC	RP	RC × RP
Benzo[a]pyrene	1	1	1	1	1	1	1	1	1
Benzo[a]anthracene	1.2	0.1	0.1	1.3	0.1	0.1	1.8	0.1	0.2
Dibenzo[ah]anthracene	0.3	1.91	0.6	0.5	1.91	1	0.1	1.91	0.2
Benzo[b]fluoranthene	2.4	0.11	0.3	2.4	0.11	0.3	3.8	0.11	0.4
Benzo[k]fluoranthene	2.1	0.03	0.1	1.9	0.03	0.1	1.2	0.03	0.1
Indeno[123cd]pyrene	1.8	0.08	0.1	1.5	0.08	0.1	0.7	0.08	0.1
Chrysene	2	0.03	0.1	2.3	0.03	0.1	3.5	0.03	* 0.1
% Contribution of B[a]P to total carcinogenic activity	44.6			37.5			49.3		

Cited from Table 17 in European Communities (2001b)
*This value was corrected from the original table

(b)

PAH	RC	RP	RC × RP
Benzo[a]pyrene	1	1	1
Benz[a]anthracene	1.1	0.1	0.11
Dibenzo[ah]anthracene	0.13	1.91	0.25
Benzo[b]fluoranthene	2.1	0.11	0.23
Benzo[k]fluoranthene	0.62	0.03	0.02
Indeno[123-cd]pyrene	1.7	0.08	0.14
Chrysene	2.0	0.03	0.06
Total			1.81
% contribution of B[a]P to total carcinogenic activity of extract			55.2

RC in this table was calculated using composition data from Aoki et al. (2018)
RC = Relative concentration to B[a]P
RP = Relative potency factor
% contribution of B[a]P to total carcinogenic activity of extract = (RC × RP) of B[a]P/$\sum$(RC × RP) × 100

5 Conclusion

The studies presented in this chapter indicate that the lung mutagenesis induced by airborne particle extracts containing mixtures of PAHs can be used to predict the risk of lung cancer associated with exposure to particulate matter in ambient air.

Acknowledgements I would like to thank Yukako Ishitsuka and Keiko Wada (JAPAN NUS Co., Ltd) and Dr. Michi Matsumoto (National Institute for Environmental Studies, Japan) for their helpful comments to the manuscript.

Competing Interests The author has no conflicts of interest to declare that are relevant to the content of this chapter. This chapter does not reflect the official policy of the National Institute for Environmental Studies and Ministry of the Environment of Japan.

References

Aoki, Y. (2017). "Evaluation of in vivo mutagenesis for assessing the health risk of air pollutants." Genes Environ **39**: 16.

Aoki, Y., D. Nakajima, M. Matsumoto, M. Yagishita, M. Matsumoto, R. Yanagisawa, S. Goto, K. Masumura and T. Nohmi (2018). "Change over time of the mutagenicity in the lungs of gpt delta transgenic mice by extract of airborne particles collected from ambient air in the Tokyo metropolitan area." Genes Environ **40**: 25.

Aoki, Y., H. Sato, N. Nishimura, S. Takahashi, K. Itoh and M. Yamamoto (2001). "Accelerated DNA adduct formation in the lung of the Nrf2 knockout mouse exposed to diesel exhaust." Toxicol Appl Pharmacol **173**(3): 154–160.

Aoki, Y., Y. Taniguchi, M. Matsumoto, M. Matsumoto, M. Ohno, K. Masumura, S. Sasaki, T. Tsuzuki, M. Yamamoto and T. Nohmi (2020). "Oxidative-stress-driven mutagenesis in the small intestine of the gpt delta mouse induced by oral administration of potassium bromate." Mutat Res Genet Toxicol Environ Mutagen **850–851**: 503136.

Armstrong, B., E. Hutchinson, J. Unwin and T. Fletcher (2004). "Lung cancer risk after exposure to polycyclic aromatic hydrocarbons: a review and meta-analysis." Environ Health Perspect **112**(9): 970–978.

Armstrong, B., C. Tremblay, D. Baris and G. Thériault (1994). "Lung cancer mortality and polynuclear aromatic hydrocarbons: a case-cohort study of aluminum production workers in Arvida, Quebec, Canada." Am J Epidemiol **139**(3): 250–262.

Ayi-Fanou, L., P. H. Avogbe, B. Fayomi, G. Keith, C. Hountondji, E. E. Creppy, H. Autrup, B. H. Rihn and A. Sanni (2011). "DNA-adducts in subjects exposed to urban air pollution by benzene and polycyclic aromatic hydrocarbons (PAHs) in Cotonou, Benin." Environ Toxicol **26**(1): 93–102.

CAREX Canada, PAHs Profile, https://www.carexcanada.ca/profile/polycyclic_aromatic_hydrocarbons/.

Cummings, P. (2009). "The relative merits of risk ratios and odds ratios." Arch Pediatr Adolesc Med **163**(5): 438–445.

de Oliveira Galvão, M. F., C. Scaramboni, B. Ünlü Endirlik, A. Vieira Silva, M. Öberg, S. A. Pozza, T. Watanabe, P. C. de Oliveira Rodrigues, P. de Castro Vasconcellos, I. Sadiktsis and K. Dreij (2024). "Application of an in vitro new approach methodology to determine relative cancer potency factors of air pollutants based on whole mixtures." Environ Int **190**: 108942.

European Communities (2001a) Ambient Air Pollution by Polycyclic Aromatic Hydrocarbons (PAH). Position Paper, (Office for Official Publications of the European Communities), https://ec.europa.eu/environment/archives/air/pdf/pp_pah.pdf.

European Communities (2001b) Ambient Air Pollution by Polycyclic Aromatic Hydrocarbons (PAH) Position Paper, Annexes, (Office for Official Publications of the European Communities), https://ec.europa.eu/environment/archives/air/pdf/annex_pah.pdf.

Gyorffy, E., L. Anna, Z. Gyori, J. Segesdi, J. Minárovits, I. Soltész, S. Kostic, A. Csekeo, M. C. Poirier and B. Schoket (2004). "DNA adducts in tumour, normal peripheral lung and bronchus, and peripheral blood lymphocytes from smoking and non-smoking lung cancer patients: correlations between tissues and detection by 32P-postlabelling and immunoassay." Carcinogenesis **25**(7): 1201–1209.

Hadrup, N., D. Mielżyńska-Švach, A. Kozłowska, M. Campisi, S. Pavanello and U. Vogel (2019). "Association between a urinary biomarker for exposure to PAH and blood level of the acute phase protein serum amyloid A in coke oven workers." Environ Health **18**(1): 81.

Hashimoto, A. H., K. Amanuma, K. Hiyoshi, Y. Sugawara, S. Goto, R. Yanagisawa, H. Takano, K. Masumura, T. Nohmi and Y. Aoki (2007). "Mutations in the lungs of gpt delta transgenic mice following inhalation of diesel exhaust." Environ Mol Mutagen **48**(8): 682–693.

Hashimoto, A. H., K. Amanuma, K. Hiyoshi, H. Takano, K. Masumura, T. Nohmi and Y. Aoki (2005). "In vivo mutagenesis induced by benzo[a]pyrene instilled into the lung of gpt delta transgenic mice." Environ Mol Mutagen **45**(4): 365–373.

Hashimoto, A. H., K. Amanuma, K. Hiyoshi, H. Takano, K. Masumura, T. Nohmi and Y. Aoki (2006). "In vivo mutagenesis in the lungs of gpt-delta transgenic mice treated intratracheally with 1,6-dinitropyrene." Environ Mol Mutagen **47**(4): 277–283.

Health Risk General Committee (2010) Health risk assessment on arsenic and its compounds. In: The Way to Countermeasure Hazardous Air Pollutants in the Future. The 9th Report, Ambient Air Division, The Central Environment Council (in Japanese). https://www.env.go.jp/press/files/jp/16392.pdf.

Hsieh, J. H., A. Sedykh, E. Mutlu, D. R. Germolec, S. S. Auerbach and C. V. Rider (2021). "Harnessing In Silico, In Vitro, and In Vivo Data to Understand the Toxicity Landscape of Polycyclic Aromatic Compounds (PACs)." Chem Res Toxicol **34**(2): 268–285.

Kelly, J. M., P. D. Ivatt, M. J. Evans, J. H. Kroll, A. I. H. Hrdina, I. N. Kohale, F. M. White, B. P. Engelward and N. E. Selin (2021). "Global Cancer Risk From Unregulated Polycyclic Aromatic Hydrocarbons." Geohealth **5**(9): e2021GH000401.

Kubo, T., K. Urano and H. Utsumi (2002). "Mutagenicity Characteristics of 255 Environmental Chemicals." Journal of Health Science **48**(6): 545–554.

Lambert, I. B., T. M. Singer, S. E. Boucher and G. R. Douglas (2005). "Detailed review of transgenic rodent mutation assays." Mutat Res **590**(1-3): 1–280.

Lodovici, M., V. Akpan, L. Giovannini, F. Migliani and P. Dolara (1998). "Benzo[a]pyrene diol-epoxide DNA adducts and levels of polycyclic aromatic hydrocarbons in autoptic samples from human lungs." Chem Biol Interact **116**(3): 199–212.

Long, A. S., C. L. Lemieux, R. Gagné, I. B. Lambert and P. A. White (2017). "Genetic Toxicity of Complex Mixtures of Polycyclic Aromatic Hydrocarbons: Evaluating Dose-Additivity in a Transgenic Mouse Model." Environ Sci Technol **51**(14): 8138–8148.

Lubin, J. H., L. M. Pottern, B. J. Stone and J. F. Fraumeni, Jr. (2000). "Respiratory cancer in a cohort of copper smelter workers: results from more than 50 years of follow-up." Am J Epidemiol **151**(6): 554–565.

Mallah, M. A., L. Changxing, M. A. Mallah, S. Noreen, Y. Liu, M. Saeed, H. Xi, B. Ahmed, F. Feng, A. A. Mirjat, W. Wang, A. Jabar, M. Naveed, J. H. Li and Q. Zhang (2022). "Polycyclic aromatic hydrocarbon and its effects on human health: An overeview." Chemosphere **296**: 133948.

Merlo, F., C. Bolognesi, M. Peluso, F. Valerio, A. Abbondandolo and R. Puntoni (1997). "Airborne levels of polycyclic aromatic hydrocarbons: 32P-postlabeling DNA adducts and micronuclei in white blood cells from traffic police workers and urban residents." J Environ Pathol Toxicol Oncol **16**(2-3): 157–162.

Ministry of the Environment, Japan Government (2024) https://www.env.go.jp/content/000258205.pdf.

Munnia, A., V. Bollati, V. Russo, L. Ferrari, M. Ceppi, M. Bruzzone, S. Dugheri, G. Arcangeli, F. Merlo and M. Peluso (2023). "Traffic-Related Air Pollution and Ground-Level Ozone Associated Global DNA Hypomethylation and Bulky DNA Adduct Formation." Int J Mol Sci **24**(3).

Nohmi, T., M. Katoh, H. Suzuki, M. Matsui, M. Yamada, M. Watanabe, M. Suzuki, N. Horiya, O. Ueda, T. Shibuya, H. Ikeda and T. Sofuni (1996). "A new transgenic mouse mutagenesis test system using Spi- and 6-thioguanine selections." Environ Mol Mutagen **28**(4): 465–470.

OECD (2018) Considerations for Assessing the Risks of Combined Exposure to Multiple Chemicals, Series on Testing and Assessment No. 296, Environment, Health and Safety Division, Environment Directorate, https://www.oecd.org/chemicalsafety/riskassessment/considerations-for-assessing-therisks-of-combined-exposure-to-multiplechemicals.pdf.

Peluso, M., F. Merlo, A. Munnia, F. Valerio, A. Perrotta, R. Puntoni and S. Parodi (1998). "32P-postlabeling detection of aromatic adducts in the white blood cell DNA of nonsmoking police officers." Cancer Epidemiol Biomarkers Prev **7**(1): 3–11.

Petit, P., A. Maître, R. Persoons and D. J. Bicout (2019). "Lung cancer risk assessment for workers exposed to polycyclic aromatic hydrocarbons in various industries." Environ Int **124**: 109–120.

Sato, H. and Y. Aoki (2002). "Mutagenesis by environmental pollutants and bio-monitoring of environmental mutagens." Curr Drug Metab **3**(3): 311–319.

Sato, H., K. T. Suzuki, H. Sone, Y. Yamano, J. Kagawa and Y. Aoki (2003). "DNA-adduct formation in lungs, nasal mucosa, and livers of rats exposed to urban roadside air in Kawasaki City, Japan." Environ Res **93**(1): 36–44.

Thybaud, V., S. Dean, T. Nohmi, J. de Boer, G. R. Douglas, B. W. Glickman, N. J. Gorelick, J. A. Heddle, R. H. Heflich, I. Lambert, H. J. Martus, J. C. Mirsalis, T. Suzuki and N. Yajima (2003). "In vivo transgenic mutation assays." Mutat Res **540**(2): 141–151

U.S. EPA (2000) Supplementary Guidance for Conducting Health Risk of Assessment of Chemical Mixtures, Risk Assessment Forum. U. S. Environmental Protection Agency, Risk Assessment Forum, Washington, DC, EPA/630/R-00/002, https://cfpub.epa.gov/neea/risk/recordisplay.cfm?deid=20533.

U.S. EPA (2007) Concepts, Methods and Data Sources for Cumulative Health Risk Assessment of Multiple Chemicals, Exposures and Effects: A Resource Document, U. S. Environmental Protection Agency, National Center for Environmental Assessment, Cincinnati, OH. EPA/600/R-06/013F, https://cfpub.epa.gov/si/si_public_record_report.cfm?Lab=NCEA&dirEntryId=190187.

U.S. EPA, (2010) Development of a Relative Potency Factor (Rpf) Approach for Polycyclic Aromatic Hydrocarbon (PAH) Mixtures (External Review Draft, Suspended) https://cfpub.epa.gov/ncea/iris_drafts/recordisplay.cfm?deid=194584#tab-3.

Venkatraman, G., N. Giribabu, P. S. Mohan, B. Muttiah, V. K. Govindarajan, M. Alagiri, P. S. Abdul Rahman and S. A. Karsani (2024). "Environmental impact and human health effects of polycyclic aromatic hydrocarbons and remedial strategies: A detailed review." Chemosphere **351**: 141227.

Vineis, P. and K. Husgafvel-Pursiainen (2005). "Air pollution and cancer: biomarker studies in human populations." Carcinogenesis **26**(11): 1846–1855.

Waddingham, C. M., P. Hinton, P. J. Villeneuve, J. R. Brook, E. Lavigne, K. Larsen, W. D. King, D. Wen, J. Meng, J. Zhang, E. Galarneau and S. A. Harris (2024). "Exposure to ambient polycyclic aromatic hydrocarbons and early-onset female breast cancer in a case-control study in Ontario, Canada." Environ Epidemiol **8**(5): e333.

WHO (2000) Polycyclic aromatic hydrocarbons, In, Air quality guidelines for Europe, 2nd ed., World Health Organization. Regional Office for Europe, 92, https://apps.who.int/iris/handle/10665/107335.

WHO (2010) IARC Working Group on the Evaluation of Carcinogenic Risks to Humans. Some Non-heterocyclic Polycyclic Aromatic Hydrocarbons and Some Related Exposures. Lyon (FR): International Agency for Research on Cancer. (IARC Monographs on the Evaluation of Carcinogenic Risks to Humans, No. 92.). https://www.ncbi.nlm.nih.gov/books/NBK321712/.

WHO (2021) Human health effects of polycyclic aromatic hydrocarbons as ambient air pollutants: report of the Working Group on Polycyclic Aromatic Hydrocarbons of the Joint Task Force on the Health Aspects of Air Pollution. Copenhagen: WHO Regional Office for Europe.

Zaccaria, K. J. and P. R. McClure (2013). "Using immunotoxicity information to improve cancer risk assessment for polycyclic aromatic hydrocarbon mixtures." Int J Toxicol **32**(4): 236–250.

Effects and Reactions of Polycyclic Aromatic Hydrocarbons and Their Derivatives in Respiratory System

Akiko Honda, Satsuki Takai, Takayuki Kameda, Natsuko Miyasaka, Megumi Nagao, Yasuto Matsui, and Hirohisa Takano

Abstract Polycyclic aromatic hydrocarbons (PAHs), originating from fossil fuel combustion, constitute one of the major components of fine particulate matter with diameter $\leq$ 2.5 μm. In addition to their suspected carcinogenic and mutagenic properties, these PAHs have been shown to aggravate inflammatory respiratory diseases, such as bronchial asthma. The quinone form of PAHs has a high reactive oxygen species (ROS) generation capacity. Thus, there are concerns regarding its health effects. Additionally, the hydroxy form of PAH has been detected in air and its presence in the atmosphere is due to the secondary reactions of PAHs. 1-Hydroxypyrene (1-OHPy) is also a metabolite of living organisms and its presence in urine is indicative of PAH exposure. However, its toxicity and health effects largely remain unknown. In this review, we outline the respiratory toxicity of the quinone and hydroxy forms of PAHs as well as the associated mechanisms and identify avenues for future research.

Keywords PM2.5 · Quinone · 1-hydroxypyrene · Metabolism · ROS · Respiratory disease

---

A. Honda (✉) · N. Miyasaka · M. Nagao
Department of Environmental Engineering, Graduate School of Engineering, Kyoto University, Kyoto, Japan
e-mail: akko@health.env.kyoto-u.ac.jp

S. Takai · Y. Matsui
Agency for Health, Safety and Environment, Kyoto University, Kyoto, Japan

T. Kameda
Graduate School of Energy Science, Kyoto University, Kyoto, Japan

N. Miyasaka · M. Nagao
Research & Collaboration Center, Kyoto University of Advanced Science, Kyoto, Japan

H. Takano
Institute for International Academic Research, Kyoto University of Advanced Science, Kyoto, Japan

Research Institute for Coexistence and Health Science, Kyoto University of Advanced Science, Kyoto, Japan

K. Hayakawa and Y. Aoki (eds.), *A New Era of Polycyclic Aromatic Hydrocarbons*,
https://doi.org/10.1007/978-981-96-7695-8_8

1 Introduction

Atmospheric air pollutants are known to adversely affect the human body, especially the respiratory system, and atmospheric particulate matter (PM), fine particulate matter with diameter $\leq$ 2.5 μm (PM2.5), can penetrate deeply into the respiratory tract, relative to larger particles, reaching the bronchioles and alveoli. Further, PM2.5, originating from anthropogenic (combustion of fossil fuels, industrial production, etc.) as well as natural sources (sand dust, volcanic activity, etc.), contains polycyclic aromatic hydrocarbons (PAHs), metals, metal oxides, incomplete combustion components, and ions (Nang et al. 2023), and owing to its toxic constituents, it has been shown to exert harmful effects on the respiratory, immune, and circulatory systems. However, the components contributing to these health effects have not been identified.

Airway epithelial cells, consisting of bronchial and alveolar epithelial cells, are constantly exposed to environmental contaminants owing to respiration and constitute the first physicochemical contact point between foreign substances and the respiratory system. When these airway epithelial cells are damaged or inflamed after exposure to PM2.5 or its components, these foreign substances can easily penetrate into the airways, triggering both innate and adaptive immune responses. Specifically, the activation of phagocytic cells, such as macrophages and neutrophils in innate immunity and various immune cells, such as lymphocytes, and eosinophils in acquired immunity can exacerbate respiratory diseases (Honda and Sagawa 2023). Therefore, maintaining the barrier function of the airway epithelium is important to avoid excessive induction of subsequent immune responses.

2 Toxicity of the Quinone Form of PAHs

PAHs are suspected to be carcinogenic and mutagenic and have been also shown to aggravate inflammatory respiratory diseases, such as bronchial asthma. Specifically, an animal study has shown that exposure to benzo[a]pyrene, a representative carcinogen and mutagen, leads to an increase in the levels of interleukin (IL)-5, IL-13, and serum ovalbumin (OVA)-specific IgG_1 (Yanagisawa et al. 2016). In an in vitro study, benzo[a]pyrene has increased CD86 expression which regulates T cell activation and immune responses, on antigen presenting cells (Chowdhury et al. 2017).

Quinones are compounds in which two hydrogen atoms in the benzene ring have been replaced by oxygen atoms, and owing to the presence of oxygen, quinones are cytotoxic at extremely low concentrations relative to common aromatic hydrocarbons. Their formation results from: (1) fuel combustion, which releases air pollutants, such as exhaust gases and PM2.5, into the atmosphere; (2) the oxidation of aromatic hydrocarbons in the presence of light, ozone, etc., and (3) the metabolism of aromatic hydrocarbons in living organisms.

Naphthoquinone has been shown to aggravate allergic airway inflammation characterized by increased peri-airway eosinophil and goblet cell counts in the bronchial epithelium and enhance IL-4, IL-5, IL-13 production and mucin 5AC expression (Inoue et al. 2007a, b). Exposure to phenanthrene quinone also increases eosinophil count in bronchoalveolar lavage fluid (BALF) and upregulates serum OVA-specific IgG_1 and IgE levels (Hiyoshi et al. 2005). Further, the quinone derivatives of phenanthrene, naphthalene, and pyrene cause damage to airway epithelial cells (Koike et al. 2014; Honda et al. 2022). Additionally, quinones with oxygen atoms at different positions exert different oxidative stress and cytotoxicity effects on airway epithelial cells (Motoyama et al. 2009), and it has also been suggested that the cytotoxicity of quinones may be due to the overproduction of H_2O_2 by ortho-quinoid PAHs in a redox cycle following the consumption of thiol group. Compared to the symmetrical and nonpolar C=C double bond, C=O double bonds are progressively more polarized, with carbon tending to be electron deficient. In para-quinones, the carbonyl is polar but symmetrical. In ortho-quinones, on the other hand, symmetry is low. This is why ortho-quinones tend to be more reactive than para-quinones (Monks et al. 1992). Further, quinones, which are electrophiles can modify the nucleophilic function of cysteine residues in proteins resulting in the formation of adducts. During this process, they disrupt signaling pathways, such as those involving kinases and transcription factors, thereby exerting adverse biological effects (Kumagai and Abiko 2017). Therefore, quinone compounds are attracting attention as PAHs that suppress the function of airway epithelial cells. The disruption of the barrier function of airway epithelial cells by quinone results in the passage of antigens and biological, chemical, physical substances into the lungs, potentially exacerbating respiratory diseases, such as bronchial asthma.

In a previous study, PM2.5 and Asian sand dust particles (ASD) were each heated to 360 °C for 30 min to inactivate attached chemical and biological substances. Subsequent exposure of cells to these heated particles in combination with quinone showed a higher oxidative stress induction capacity for heated-PM2.5 particles than for heated-ASD (Honda et al. 2023). Similarly, another study showed that combined exposure of airway epithelial cells to metals and quinone induces inflammatory responses and antioxidant metallothionein gene expression (Honda et al. 2017). These observations implied that the toxicity of quinones is enhanced when they act in combination with certain particulate components, and possibly, the interaction between particulate components and quinones is responsible for the development and aggravation of respiratory diseases.

Recently, N-(1,3-dimethylbutyl)-N′-phenyl-p-phenylenediamine (6-PPD), a chemical that is added to tires during manufacture to prevent their oxidative degradation, has attracted considerable attention. Reportedly, 6-PPD in tire wear dust reacts with ozone resulting in the release of 6-PPD quinone, which is associated with lethal toxicity in fish, into the environment (Tian et al. 2021). In the aquatic environment, the combined action of microplastics and 6-PPD quinone results in toxicity to fish. Specifically, zebrafish exposed to 6-PPD quinone show significant hyperlocomotion, and this effect is even stronger following co-exposure to 6-PPD quinone (50 μg/L) and polystyrene nanoparticles (3 mg/L). Transcriptomic analysis has shown that co-exposure to 6-PPD quinone and plastics influences the cellular processes associated with sterol biosynthesis, cholesterol metabolism, and muscle tissue development (Varshney et al. 2024). Another study in mice showed that oral exposure of mice to 6-PPD and 6-PPD quinone induces hepatotoxicity (Fang et al. 2023). Additionally, after oral exposure, 6-PPD quinone in mice undergoes biotransformation, primarily via hydroxylation and subsequent glucuronidation (Deng et al. 2024). Intraperitoneal exposure to 6-PPD quinone has also been shown to induce lung fibrosis (He et al. 2024). The toxicity of 6-PPD quinone to human hepatocytes has also been shown to be higher than that of 6-PPD (Qi et al. 2024). Conversely, Chen et al. (2024) showed that 6-PPD is more toxic to airway epithelial cells than 6-PPD quinone, contradicting the abovementioned findings. Therefore, 6-PPD metabolism in the lungs may differ from that in other organs, which may determine the sensitivity to 6-PPD quinone. Further studies are needed to clarify the reasons.

3 Toxicity of the Hydroxy Form of PAHs

Hydroxy derivatives of PAHs have also been detected in the air. Hayakawa (2023) showed that the concentrations of pyrene derivatives in air varies in the order: pyrene > quinoid pyrenes > hydroxy pyrenes > nitropyrenes, implying that hydroxy pyrenes rank second in abundance after quinoid pyrenes. Reportedly, hydroxy pyrenes are produced via secondary PAH reactions and metabolism in vivo, and urinary 1-OHPy level has been used as a biomarker of PAHs inhalation exposure (Huang et al. 2023). In a study involving human volunteers, 1-OHPy, owing inhalation exposure to pyrene, was detected in urine samples from the subjects (Brzeźnicki et al. 1997). However, the toxicity of 1-OHPy and its effects on the human body largely remain unclear.

In one of our previous studies, we exposed airway epithelial cells (BEAS-2B) to pyrene and pyrene derivatives (Fig. 1; 4,5-pyrenequinone (4, 5-PQ), 1,8-PyQ + 1,6-PyQ, and 1-OHPy) and thereafter, investigated the effects of these chemicals on airway epithelial cells.

The results obtained showed that pyrene had little or no effect on the cellular activity of BEAS-2B, whereas 4,5-PyQ and 1,8-PyQ + 1,6-PyQ decreased cell

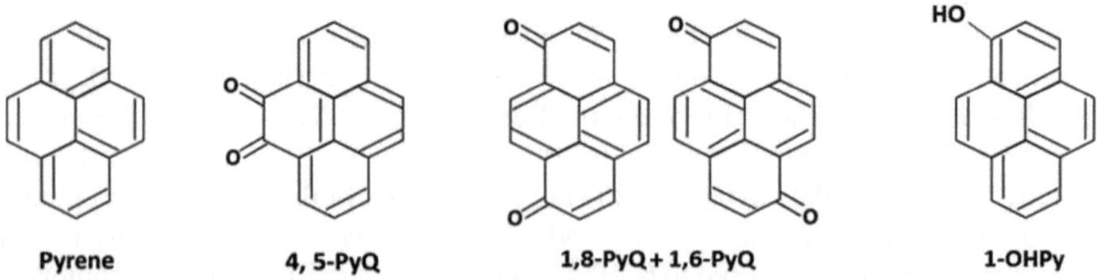

Fig. 1 Chemical structural formula of pyrene and its derivatives

activity by more than 50% at concentrations of 3 μM and 10 μM or higher, respectively, and 1-OHPy decreased BEAS-2B cellular activity by more than 50% at concentrations of 30 μM. 4,5-PyQ induced a marked decrease in cellular activity even at low concentrations and showed greater toxicity than 1,8-PyQ + 1,6-PyQ (Fig. 2).

These quinones also induced intracellular ROS production (Fig. 3), consistent with the results of an oxidative capacity study using dithiothreitol assay reported by Okubo et al. (2021). 4,5-PyQ -induced excessive ROS production can induce premature cell death. The toxicity of the hydroxy form 1-OHPy was not as strong as those of 4,5-PyQ and 1,8-PyQ + 1,6-PyQ. However, it showed a toxicity at high concentrations, but 1-OHPy, unlike 4,5-PyQ and 1,8-PyQ + 1,6-PyQ, did not induce intracellular ROS production. Therefore, it is possible that the death of BEAS-2B owing to 1-OHPy exposure, unlike was the case with pyrene quinone-induced cell death, did not involve ROS production but another mechanism.

From the results by Sagawa et al. (2025) and the present study, Raman, fluorescence microscopy and high performance liquid chromatography (HPLC) analysis

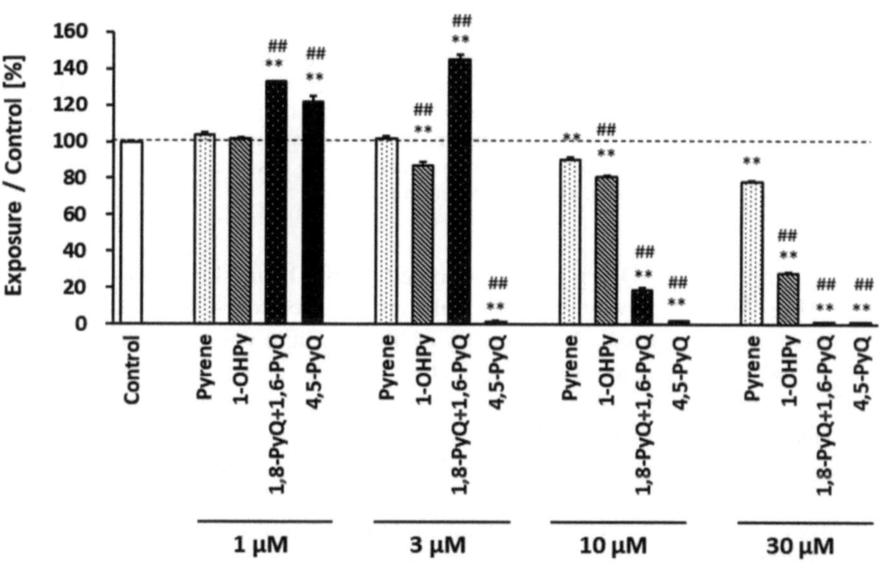

Fig. 2 Effects of pyrene and its derivatives on the viability of human airway epithelial cells. Pyrene and its derivatives were dissolved in dimethyl sulfoxide (DMSO). After BEAS-2B cells reached semi-confluency in LHC-9 medium (fetal bovine serum (FBS) free), they were exposed to pyrene, 4,5-PyQ, 1,8-PyQ + 1,6-PyQ and 1-OHPy at 0, 1, 3, 10 or 30 μM for 24 h. A water-soluble tetrazolium salt (WST)-1 assay was performed to measure cell viability using the Premix WST-1 Cell Proliferation Assay System (TaKaRa Bio, Kusatsu, Shiga, Japan). After incubating BEAS-2B cells with WST-1 reagent, absorbance was measured at a wavelength of 450 nm and a reference wavelength of 630 nm. Cell viability results were expressed as the percentage of viable cells compared to untreated cells (control, 0.1% DMSO). $^{**}p < 0.01$ versus control, $^{\#\#}p < 0.01$ versus pyrene at the same concentration

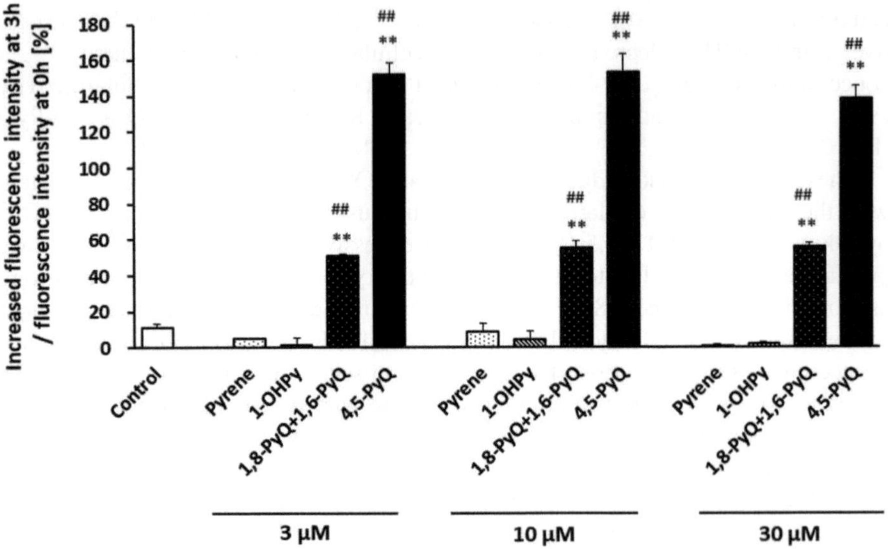

Fig. 3 Effects of pyrene and its derivatives on the ROS production in human airway epithelial cells. A fluorescent probe, 5-(and-6)-chloromethyl-2′,7′-dichlorodihydrofluorescein diacetate acetyl ester (CM-H_2DCFDA) was used to measure intracellular ROS production. BEAS-2B were incubated with 5 μM CM-H_2DCFDA for 30 min, allowing the dye to enter the cells. The cells were then washed to remove extracellular dye and exposed to pyrene and its derivatives. Increased fluorescence intensity after 3 h exposure (excitation at 485 nm and emission at 530 nm) was measured. $^{**}p < 0.01$ versus control, $^{\#\#}p < 0.01$ versus pyrene at the same concentration

have shown that the 1-OHPy invades into cell (Table 1) and localizes in the cytoplasm (Fig. 4) when airway epithelial cells are exposed to 1-OHPy. Additionally, Leu293 on the aryl hydrocarbon receptor (AhR) potentially forms hydrogen bonds with 1-OHPy (Wang et al. 2024). Miao et al. (2022), via oral exposure of mice to 1-OHPy and experiments involving an immortalized proximal tubule epithelial cell line (HK-2), revealed that 1-OHPy binds to AhR which is a transcription factor that activates the drug-metabolizing enzymes CYPA1A, CYP1A2, and CYP1B1, thereby inducing renal fibrosis. There is possibility that extracellular 1-OHPy invades in the cytoplasm, and some 1-OHPy may bind to AhR and translocate into the nucleus. Then, 1-OHPy is metabolized by drug-metabolizing enzymes to compounds such as 1,8-, and 1,6-dihydroxypyrene (Shimada et al. 2016), and might exert toxicity in the process. Further studies are needed to elucidate the mechanism of 1-OHPy toxicity.

Interestingly, exposure of BEAS-2B to pyrene and intracellular HPLC analysis revealed the presence of 1-OHPy, indicating that 1-OHPy is produced in airway epithelial cells as a pyrene metabolite, and the concentration was 0.001–0.003 fmol/cell (Table 1). Reportedly, pyrene can be converted to 1-OHPy, 1,8-, and 1,6-dihydroxypyrene, or even 1,8- and 1,6-PyQ by microorganisms (Wunder et al. 1994). However, in our study, neither 1,8-PyQ nor 1,6-PyQ was detected in the BEAS-2B exposed to pyrene and 1-OHPy (data not shown). Further studies are

Table 1 Quantification of pyrene and 1-OHPy after exposure to pyrene and 1-OHPy in human airway epithelial cells

Sample	Pyrene (fmol/cell)	1-OHPy (fmol/cell)	1-OHPy/Pyrene (%)
Control	N.D	N.D	–
10 μM-Pyrene	1.9 ± 0.12	0.0018 ± 0.0003	0.096
30 μM-Pyrene	7.6 ± 1.82	0.0021 ± 0.0003	0.029
10 μM-1-OHPy	–	3.1 ± 0.5538	–

N.D.: not detected. (–): not available

BEAS-2B were exposed to pyrene and 1-OHPy at 0, 10 or 30 μM for 24 h, and washed twice with phosphate-buffered saline (PBS). Trypsin- ethylenediaminetetraacetic acid (EDTA) without phenol red was added on cells and the cell suspension was collected through a 40 μm cell strainer. Sample were centrifuged at 400 g × 5 min (4 °C), and the supernatant was aspirated and suspended in PBS. After cell counting, ultrapure water was added, frozen and thawed, and sonicated for 1 min. Cell membranes were disrupted in three steps: osmotic shock by addition of ultrapure water, freeze–thaw, and sonication. Liquid–liquid partitioning was performed between dichloromethane and water using a separating funnel. The dichloromethane layer was evaporated to dryness, followed by solvent replacement with 0.5 mL of methanol. The pretreated sample was analyzed by High Performance Liquid Chromatography (HPLC) with fluorescence detection. The details of the analysis conditions were as follows: An Inertsil ODS-3 column (3 μm particle size, 3 mmφ inner diameter × 150 mm length, GL science) was used at a temperature of 40 °C. The mobile phase consisted of methanol/ water in a 7/3 ratio (v/v) for Py and a 1/1 ratio (v/v) for 1-OHPy. The flow rate was set at 0.5 mL/ min, and detection was carried out at wavelengths of 240 nm for excitation and 387 nm for emission

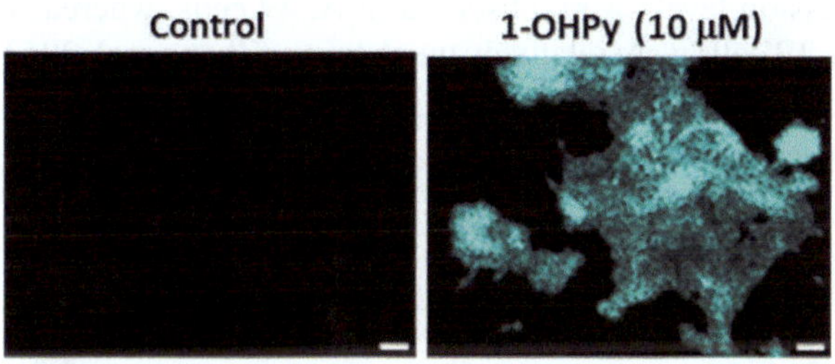

Fig. 4 Intracellular visualization of 1-OHPy (light blue) in human airway epithelial cells using fluorescence microscopy. BEAS-2B were exposed to 1-OHPy at 0 and 10 μM for 24 h. Cells were fixed in 4% paraformaldehyde and observed under a fluorescence microscope (excitation at 365 nm and emission at 420 nm). Scale bar = 50 μm

needed to determine whether the same metabolism can occur in respiratory cells as in microorganisms.

4 Metabolism of PAHs

The body employs foreign-body metabolism (drug metabolism) to protect itself from foreign substances. Foreign-body metabolic reactions are classified as phase I reactions, such as oxidation, reduction, and hydrolysis, and phase II reactions, such as conjugation. Following exposure via the airway, PAHs may be locally metabolized in the lungs or enter systemic circulation from the lungs and become metabolized in the liver (Gerde et al. 2001).

According to Castell et al. (2005), in human lung tissue and airway epithelial cells (A549), *CYP1B1* is the most expressed gene among *CYP1A1*, *CYP1A2*, and *CYP1B1*. In our study using BEAS-2B and A549 cells, 1-OHPy exposure upregulated *CYP1B1* gene expression 1.88- and 3.58-fold, respectively, indicating the possible existence of a 1-OHPy metabolic pathway in airway epithelial cells mediated by *CYP1B1* (Fig. 5). However, metabolism can either cause the foreign substance to lose its toxicity or show increased toxicity. Additionally, PAH detoxification requires rapid metabolism by phase II enzymes following phase I reactions.

Phase II reaction enzymes, such as uridine diphosphate glucuronosyl-transferases (UGTs), are classified under two major subfamilies, UGT1 and UGT2, which are involved in the glucuronidation of various endogenous compounds and foreign biological substances. In a study focused on *UGT1A1, UGT1A3, UGT1A7, UGT1A8, UGT1A9, UGT1A10,* and *UGT2B7* gene expression in multiple lung cell types, including BEAS-2B, A549, H358, and HBEC-KT cells, the highest *UGT1A3* and *UGT2B7* expression levels were observed in A549 cells, whereas their expression levels in BEAS-2B cells were relatively much lower (Zhang et al. 2013). The results of

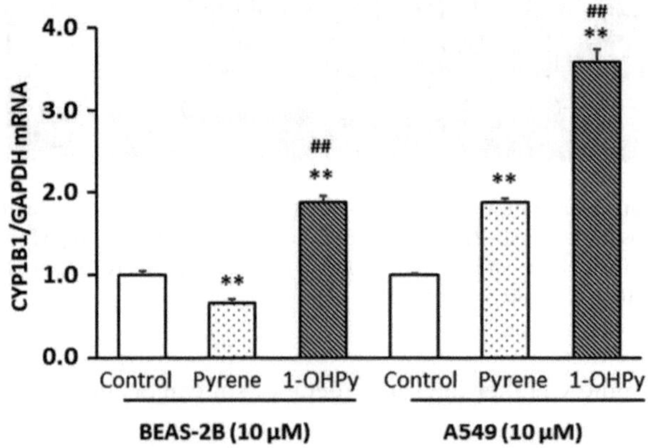

Fig. 5 Effects of pyrene and its derivatives on *CYP1B1* mRNA expression in human airway epithelial cells. BEAS-2B and A549 were exposed to pyrene and 1-OHPy at 0 and 10 μM for 24 h without FBS condition. Total RNA was extracted and reverse transcribed to cDNA. PCR amplification was performed with TaqMan probe and pair primers (Life technologies). The relative intensity was normalized to glyceraldehyde-3-phosphate dehydrogenase (GAPDH) as an endogenous control gene. **$p < 0.01$ versus control, ##$p < 0.01$ versus pyrene at the same concentration

our previous study showed greater sensitivity to 1-OHPy and quinone for BEAS-2B than for A549 (Fig. 6), and differences in the *UGT1A3* expression may be related to cytotoxicity although quinone is observed in cell death due to excessive ROS production (Fig. 7). Glucuronide conjugates of 1-OHPy have been detected in human urine (Strickland and Kang 1999; Kakimoto et al. 2008). In a study involving A549 cells, the monoglucuronide of 9,10-dihydroxyphenanthrene was identified as a metabolite of 9,10-phenanthrenequinone, generated through 9,10-dihydroxyphenanthrene, which terminates its redox cycling and is transported to extracellular space (Taguchi et al. 2008). In future, it will be necessary to conduct further studies to clarify the metabolites and toxicity of PAHs and their derivatives. At least for pyrene and its derivatives, the following metabolic pathway is speculated from previous reports and our research results. After pyrene is taken up into airway epithelial cells, 1-OHPy is partially generated. Pyrene or 1-OHPy that transferred to the systemic circulation may be further metabolized and excreted in urine. On the other hand, 1-OHPy that remains in airway epithelial cells may activate drug-metabolizing enzymes by stimulating AhR in the cells. 1-OHPy may then be further metabolized by CYP1B1 and other enzymes to generate 1,8- or 1,6-di-OHPy as intermediate metabolites of 1,8- or 1,6-PyQ, which may then form glucuronide and be excreted outside the cells. On the other hand, quinone compounds that may be generated by reactions in the atmosphere or metabolism in the body may induce cell death from inside and outside the cells via the production of ROS (Fig. 8). 1-OHPy exerts cytotoxicity, although weaker than its quinone form. The underlying mechanism has not been clarified, and further research is needed, considering into the metabolism of PAHs. We believe that in future, research on the metabolism of PAHs during inhalation exposure will progress, leading to further elucidation of aggravating respiratory disease by PAHs.

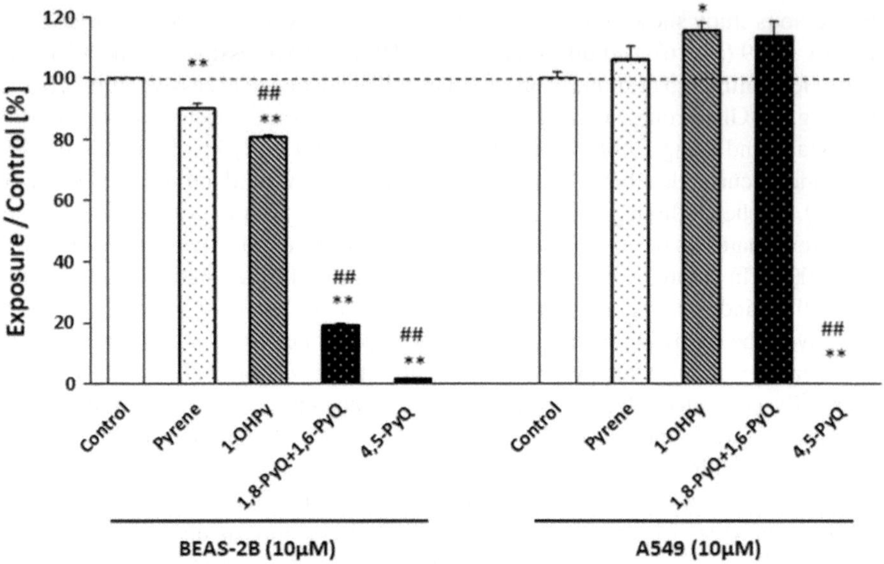

Fig. 6 Effects of pyrene and its derivatives on the viability of human airway epithelial cells. BEAS-2B and A549 were exposed to pyrene, 4,5-PyQ, 1,8-PyQ + 1,6-PyQ and 1-OHPy at 0 and 10 μM for 24 h without FBS condition. A water-soluble tetrazolium salt (WST)-1 assay was performed to measure cell viability using the Premix WST-1 Cell Proliferation Assay System (TaKaRa Bio, Kusatsu, Shiga, Japan). After incubating cells with WST-1 reagent, absorbance was measured at a wavelength of 450 nm and a reference wavelength of 630 nm. Cell viability results were expressed as the percentage of viable cells compared to untreated cells (control, 0.1% DMSO). $^{**}p < 0.01$, $^{*}p < 0.05$ versus control, $^{\#\#}p < 0.01$ versus pyrene at the same concentration

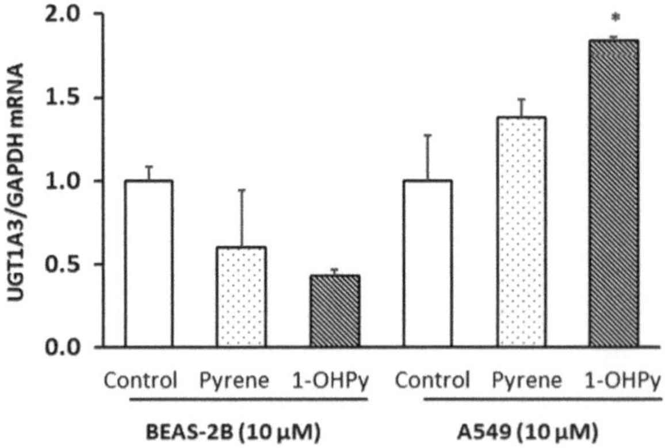

Fig. 7 Effects of pyrene and its derivatives on *UGT1A3* mRNA expression in human airway epithelial cells. BEAS-2B and A549 were exposed to pyrene and 1-OHPy at 0 and 10 μM for 24 h without FBS condition. Total RNA was extracted and reverse transcribed to cDNA. PCR amplification was performed with TaqMan probe and pair primers (Life technologies). The relative intensity was normalized to glyceraldehyde-3-phosphate dehydrogenase (GAPDH) as an endogenous control gene. $^{*}p < 0.05$ versus control

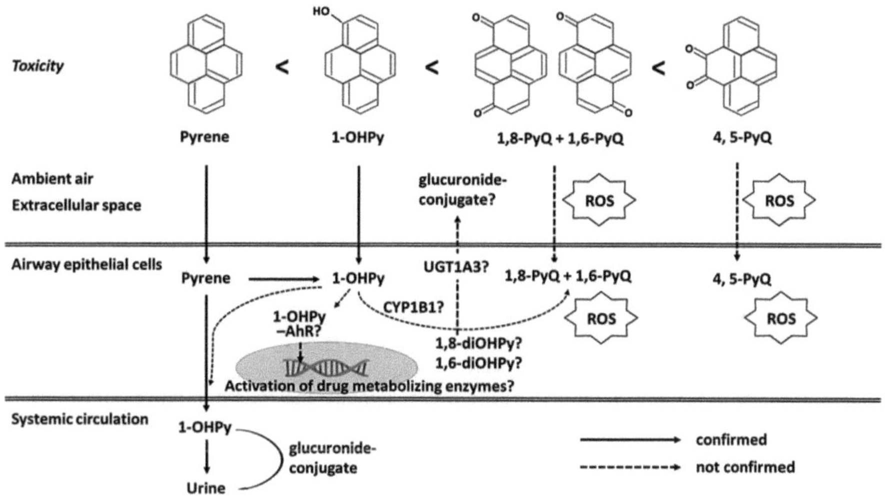

Fig. 8 Hypothesis of reactions of pyrene and their derivatives in respiratory system

Acknowledgements This work was partly supported by the Core Research for Evolutional Science and Technology (CREST) program of the Japan Science and Technology Agency (JST), Japan [grant number JPMJCR19H3].

References

Brzeźnicki, S., Jakubowski, M., & Czerski, B. (1997). Elimination of 1-hydroxypyrene after human volunteer exposure to polycyclic aromatic hydrocarbons. International Archives of Occupational and Environmental Health, 70, 257–260. https://doi.org/10.1007/s004200050216

Castell, J. V., Donato, M. T., & Gómez-Lechón, M. J. (2005). Metabolism and bioactivation of toxicants in the lung. The in vitro cellular approach. Experimental and Toxicologic Pathology: Official Journal of the Gesellschaft fur Toxikologische Pathologie, 57, 189–204. https://doi.org/10.1016/j.etp.2005.05.008

Chen, H., Xie, M., Li, W., Tan, L., Cai, X., Shen, M., & Li, R. (2024). Detection of 6-PPD and 6-PPDQ in airborne particulates and assessment of their toxicity in lung cells. Chemosphere, 364, 143205. https://doi.org/10.1016/j.chemosphere.2024.143205

Chowdhury PH, Kitamura G, Honda A, Sawahara T, Hayashi T, Fukushima W, Kudo H, Ito S, Yoshida S, Ichinose T, Ueda K, Takano H. Synergistic effect of carbon nuclei and polyaromatic hydrocarbons on respiratory and immune responses. Environ Toxicol. 2017 Sep;32(9):2172–2181.

Deng, M., Ji, X., Peng, B., & Fang, M. (2024). *In Vitro* and *In Vivo* biotransformation profiling of 6PPD-Quinone toward their detection in human urine. Environmental Science & Technology, 58, 9113–9124. https://doi.org/10.1021/acs.est.4c01106

Fang, L., Fang, C., Di, S., Yu, Y., Wang, C., Wang, X., & Jin, Y. (2023). Oral exposure to tire rubber-derived contaminant 6PPD and 6PPD-quinone induce hepatotoxicity in mice. Science of the Total Environment, 869, 161836. https://doi.org/10.1016/j.scitotenv.2023.161836

Gerde, P., Muggenburg, B. A., Lundborg, M., & Dahl, A. R. (2001). The rapid alveolar absorption of diesel soot-adsorbed benzo[a]pyrene: bioavailability, metabolism and dosimetry of an inhaled particle-borne carcinogen. Carcinogenesis, 22, 741–749. https://doi.org/10.1093/carcin/22.5.741

Hayakawa, K. (2023). Polycyclic aromatic hydrocarbons from vegetation burning and health Effects. Global Environmental Research, 27, 13–20.

He, W., Chao, J., Gu, A., & Wang, D. (2024). Evaluation of 6-PPD quinone toxicity on lung of male BALB/c mice by quantitative proteomics. Science of the Total Environment, 922, 171220. https://doi.org/10.1016/j.scitotenv.2024.171220

Hiyoshi, K., Takano, H., Inoue, K. I., Ichinose, T., Yanagisawa, R., Tomura, S., & Kumagai, Y. (2005). Effects of phenanthraquinone on allergic airway inflammation in mice. Clinical and Experimental Allergy: Journal of the British Society for Allergy and Clinical Immunology, 35, 1243–1248. https://doi.org/10.1111/j.1365-2222.2005.02297.x

Honda, A., Chowdhury, P. H., Ito, S., Okano, H., Onishi, T., Kawaryu, Y., Ueda, K., & Takano, H. (2017). Synergic effects of 9,10-phenanthrenequinone and cadmium on pro-inflammatory responses in airway epithelial cells. Environmental Toxicology and Pharmacology, 52, 276–279. https://doi.org/10.1016/j.etap.2017.04.019

Honda, A., Inoue, K. I., Takai, S., Kameda, T., Ueda, K., & Takano, H. (2022). Effects of oxidized pyrenes on the biological responses in the human bronchial epithelial cells. Applied Sciences, 12, 9664. https://doi.org/10.3390/app12199664

Honda, A., Inoue, K. I., Higashihara, M., Ichinose, T., Ueda, K., & Takano, H. (2023). Differential pattern of cell death and ROS production in human airway epithelial cells exposed to quinones combined with heated-PM2.5 and/or Asian sand dust. International Journal of Molecular Sciences, 24, 10544. https://doi.org/10.3390/ijms241310544

Honda, A., & Sagawa, T. (2023). Biological mechanisms of respiratory effects of particulate matter and its interaction with COVID-19. Global Environmental Research, 27, 21–26.

Huang, X., Li, Z., Zhang, T., Zhu, J., Wang, X., Nie, M., Harada, K., Zhang, J., & Zou, X. (2023). Research progress in human biological monitoring of aromatic hydrocarbon with emphasis on the analytical technology of biomarkers. Ecotoxicology and Environmental Safety, 257, 114917. https://doi.org/10.1016/j.ecoenv.2023.114917

Inoue, K., Takano, H., Hiyoshi, K., Ichinose, T., Sadakane, K., Yanagisawa, R., Tomura, S., & Kumagai, Y. (2007a). Naphthoquinone enhances antigen-related airway inflammation in mice. The European Respiratory Journal, 29, 259–267. https://doi.org/10.1183/09031936.00033106

Inoue, K., Takano, H., Ichinose, T., Tomura, S., Yanagisawa, R., Sakurai, M., Sumi, D., Cho, A. K., Hiyoshi, K., & Kumagai, Y. (2007b). Effects of naphthoquinone on airway responsiveness in the presence or absence of antigen in mice. Archives of Toxicology, 81, 575–581. https://doi.org/10.1007/s00204-007-0186-5

Kakimoto, K., Toriba, A., Ohno, T., Ueno, M., Kameda, T., Tang, N., & Hayakawa, K. (2008). Direct measurement of the glucuronide conjugate of 1-hydroxypyrene in human urine by using liquid chromatography with tandem mass spectrometry. Journal of Chromatography. B, Analytical Technologies in the Biomedical and Life sciences, 867, 259–263. https://doi.org/10.1016/j.jchromb.2008.04.015

Koike, E., Yanagisawa, R., & Takano, H. (2014). Toxicological effects of polycyclic aromatic hydrocarbons and their derivatives on respiratory cells. Atmospheric Environment, 97, 529–536. https://doi.org/10.1016/j.atmosenv.2014.04.003

Kumagai, Y., & Abiko, Y. (2017). Environmental electrophiles: Protein adducts, modulation of redox signaling, and interaction with persulfides/plysulfides. Chemical Research in Toxicology, 30, 203–219. https://doi.org/10.1021/acs.chemrestox.6b00326

Miao, H., Wu, X. Q., Wang, Y. N., Chen, D. Q., Chen, L., Vaziri, N. D., Zhuang, S., Guo, Y., Su, W., Ma, S. X., Zhang, H. Q., Shang, Y. Q., Yu, X. Y., Zhao, Y. L., Mao, J. R., Gao, M., Zhang, J. H., Zhao, J., Zhang, Y., Zhang, L., … Cao, G. (2022). 1-Hydroxypyrene mediates renal fibrosis through aryl hydrocarbon receptor signalling pathway. British Journal of Pharmacology, 179, 103–124. https://doi.org/10.1111/bph.15705

Monks, T. J., Hanzlik, R. P., Cohen, G. M., Ross, D., & Graham, D. G. (1992). Quinone chemistry and toxicity. *Toxicology and applied pharmacology*, *112*(1), 2–16. https://doi.org/10.1016/0041-008x(92)90273-u

Motoyama, Y., Bekki, K., Chung, S. W., Tang, N., Kameda, T., Toriba, A., Taguchi, K., & Hayakawa, K. (2009). Oxidative stress more strongly induced by *ortho*- than *para*-quinoid polycyclic aromatic hydrocarbons in A549 cells. Journal of Health Science, 55, 845–850. https://doi.org/10.1248/jhs.55.845

Nang, N., Yan, Z., Zhang, Y., Chen, R., Qin, G., & Sang, N. (2023). Overview of $PM_{2.5}$ and health outcomes: Focusing on components, sources, and pollutant mixture co-exposure. Chemosphere, 323, 138181. https://doi.org/10.1016/j.chemosphere.2023.138181

Qi, Y., Qiu, A., Wei, X., Huang, Y., Huang, Q., & Huang, W. (2024). Effects of 6PPD-quinone on human liver cell lines as revealed with cell viability assay and metabolomics analysis. Toxics, 12, 389. https://doi.org/10.3390/toxics12060389

Okubo, R., Kameda, T., & Tohno, S. (2021). Evaluation of oxidative potential of pyrenequinone isomers by the dithiothreitol (DTT) Assay. Polycyclic Aromatic Compounds, 42, 5152–5159. https://doi.org/10.1080/10406638.2021.1925711

Sagawa, T., Ishikawa, R., Akaji, S., Honda, A., Kameda, T., & Takano, H. (2025) Relationship between bio-distribution of environmental particles and induction of biological and immune response in the respiratory system. pp 141–153. In: Baba, Y., Hanayama, R., Akita, H., Yasui, T. (eds) Extracellular Fine Particles. Springer, Singapore. https://doi.org/10.1007/978-981-97-7067-0_11

Shimada, T., Takenaka, S., Murayama, N., Kramlinger, V. M., Kim, J. H., Kim, D., Liu, J., Foroozesh, M. K., Yamazaki, H., Guengerich, F. P., & Komori, M. (2016). Oxidation of pyrene, 1-hydroxypyrene, 1-nitropyrene and 1-acetylpyrene by human cytochrome P450 2A13. Xenobiotica; The Fate of Foreign Compounds in Biological Systems, 46, 211–224. https://doi.org/10.3109/00498254.2015.1069419

Strickland, P., & Kang, D. (1999). Urinary 1-hydroxypyrene and other PAH metabolites as biomarkers of exposure to environmental PAH in air particulate matter. Toxicology Letters, 108, 191–199. https://doi.org/10.1016/s0378-4274(99)00089-2

Taguchi, K., Shimada, M., Fujii, S., Sumi, D., Pan, X., Yamano, S., Nishiyama, T., Hiratsuka, A., Yamamoto, M., Cho, A. K., Froines, J. R., & Kumagai, Y. (2008). Redox cycling of 9,10-phenanthraquinone to cause oxidative stress is terminated through its monoglucuronide conjugation in human pulmonary epithelial A549 cells. Free Radical Biology & Medicine, 44, 1645–1655. https://doi.org/10.1016/j.freeradbiomed.2008.01.024

Tian, Z., Zhao, H., Peter, K. T., Gonzalez, M., Wetzel, J., Wu, C., Hu, X., Prat, J., Mudrock, E., Hettinger, R., Cortina, A. E., Biswas, R. G., Kock, F. V. C., Soong, R., Jenne, A., Du, B., Hou, F., He, H., Lundeen, R., Gilbreath, A., ... Kolodziej, E. P. (2021). A ubiquitous tire rubber-derived chemical induces acute mortality in coho salmon. Science (New York, N.Y.), 371, 185–189. https://doi.org/10.1126/science.abd6951

Varshney, S., O'Connor, O. L., Gora, A. H., Rehman, S., Kiron, V., Siriyappagouder, P., Dahle, D., Kögel, T., Ørnsrud, R., & Olsvik, P. A. (2024). Mixture toxicity of 6PPD-quinone and polystyrene nanoplastics in zebrafish. Environmental Pollution (Barking, Essex: 1987), 348, 123835. https://doi.org/10.1016/j.envpol.2024.123835

Wang, M., Luo, N., Gao, Y., Li, G., & An, T. (2024). Pyrene and its derivatives increase lung adverse effects by activating aryl hydrocarbon receptor transcription. Science of the Total Environment, 916, 170030. https://doi.org/10.1016/j.scitotenv.2024.170030

Wunder, T., Kremer, S., Sterner, O., & Anke, H. (1994). Metabolism of the polycyclic aromatic hydrocarbon pyrene by *Aspergillus niger* SK 9317. Applied Microbiology and Biotechnology, 42, 636–641. https://doi-org.kyoto-u.idm.oclc.org/https://doi.org/10.1007/BF00173932

Yanagisawa, R., Koike, E., Win-Shwe, T. T., Ichinose, T., & Takano, H. (2016). Low-dose benzo[a]pyrene aggravates allergic airway inflammation in mice. Journal of Applied Toxicology, 36, 1496–1504. https://doi.org/10.1002/jat.3308

Zhang, L., Huang, M., Blair, I. A., & Penning, T. M. (2013). Interception of benzo[a]pyrene-7,8-dione by UDP glucuronosyltransferases (UGTs) in human lung cells. Chemical Research in Toxicology, 26, 1570–1578. https://doi.org/10.1021/tx400268q

Exposure to PAHs and Asthma/Cough Diseases

Akinori Hara and Hiroyuki Nakamura

Abstract Polycyclic aromatic hydrocarbons (PAHs) are common air pollutants that are associated with respiratory diseases, including asthma, particularly in children. PAHs are present in both particulate matter and gas phase, with exposure routes, including inhalation, skin contact, and contaminated food ingestion. Epidemiological evidence suggests that PAH exposure contributes to decreased lung function, asthma exacerbation, and obstructive lung diseases, with potential variations based on individual characteristics. Although several PAHs have carcinogenic potential, this chapter highlights their role in nonmalignant respiratory diseases. Mechanistic studies underscore causal pathways; however, to clarify exposure timing, specific PAH effects, and nonmalignant health impacts, further research is warranted.

Keywords Atmosphere · Benzo[a]anthracene · Respiratory disease · Airway · Lungs · Biomarker

1 Introduction

Air pollution is the primary contributor to morbidity and premature mortality worldwide. The Global Burden of Disease study estimated that airborne particulate matter (PM) pollution would be the leading contributor to the global disease burden in 2021, accounting for 8.0% (95% uncertainty interval, 6.7–9.4) of the total disability-adjusted life years (DALYs) (Brauer et al. 2024). Although global all-age DALY counts declined by 42.6% (26.5–54.6) for household air pollution from solid fuels, they increased by 40.6% (21.5–62.9) for ambient PM pollution from 2000 to 2021 (Brauer et al. 2024). Without proactive measures, the death rates could double by 2050, with the greatest increases occurring in South and East Asia (Vohra et al. 2021).

PM is the most intensively investigated component of air pollutants and is strongly linked to several health outcomes. Some of the strongest associations have been

A. Hara · H. Nakamura (✉)
Department of Hygiene and Public Health, Faculty of Medicine, Institute of Medical, Pharmaceutical and Health Sciences, Kanazawa University, Kanazawa, Ishikawa, Japan
e-mail: hnakamu@staff.kanazawa-u.ac.jp

K. Hayakawa and Y. Aoki (eds.), *A New Era of Polycyclic Aromatic Hydrocarbons*,
https://doi.org/10.1007/978-981-96-7695-8_9

reported between fine particles (< 2.5 μm [$PM_{2.5}$]) and the development or exacerbation of adverse respiratory diseases, including asthma, chronic obstructive pulmonary disease (COPD), respiratory infections, and lung cancer (Låg et al. 2020). Moreover, a causal relationship has been established between PM and several diseases with evidence from experimental studies involving human, animal, and cellular models (Låg et al. 2020). Although recent findings have indicated that the chemical composition of PM is significant for several biological and toxicological effects of PM, the precise mechanisms involved in respiratory diseases may vary with biological endpoints.

Among the organic chemicals in PM, some of the polycyclic aromatic hydrocarbons (PAHs) are believed to cause detrimental effects on health. PAHs are a large group of organic compounds containing two or more fused aromatic (benzene) rings (WHO Regional Office for Europe 2021). Anthropogenic sources of these chemicals primarily result from incomplete combustion or pyrolysis of organic materials, including fossil fuels, biofuels, automobile exhaust, and tobacco smoke (WHO Regional Office for Europe 2021).

PAHs are absorbed into the body through inhalation, skin contact, and contaminated food ingestion (WHO Regional Office for Europe 2021). High-molecular-weight PAHs with five or more aromatic rings are mainly detected bound to PM, whereas smaller PAHs containing four or less aromatic rings are predominantly noted in the gas phase (Låg et al. 2020). However, low-molecular-weight PAHs also tend to predominate PAHs bound to PM owing to their higher total airborne concentrations (Låg et al. 2020). The relative amount of PAHs adhering to the PM depends on the source but, to a large degree, also influenced by temperature (Låg et al. 2020). Therefore, the levels of PAHs attaching to the PM during winter may be higher than those during summer when highly volatile PAHs predominantly evaporate and exist in the gaseous state (Låg et al. 2020). Generally, PAHs in the ambient gas state exist for < 1 day, whereas particle-associated PAHs may persist for weeks and undergo long-range atmospheric transport (WHO Regional Office for Europe 2021). Although accumulating evidence has shown that several of the PAHs in ambient air have not only carcinogenic potential, which may contribute to malignant disease development, but also affect the development and exacerbation of nonmalignant respiratory diseases, including asthma and COPD.

This chapter reviews epidemiological findings on the association between ambient PAHs and nonmalignant respiratory diseases or processes related to the diseases, experimental studies showing the essential aspects of PAH toxicity, and evidence-based strategies for disease prevention.

2 Epidemiology

Several epidemiological studies have demonstrated that airborne PAHs are associated with the development and exacerbation of different respiratory symptoms and diseases, including asthma, impaired respiratory functions, and COPD. Inhaled

PAHs can more significantly affect respiratory health outcomes than those from other sources (WHO Regional Office for Europe 2021). A systematic review has provided evidence of the overall association between PAH exposure and respiratory diseases in humans; however, some studies have reported no such associations (Nwaozuzu et al. 2021).

The following sections are based on recent evidence on the effects of ambient PAHs on respiratory diseases and symptoms.

Childhood asthma

Studies have reported the association between PAH exposure and an increased risk of the development and exacerbation of childhood asthma and its related symptoms (Karimi et al. 2015; Låg et al. 2020; WHO Regional Office for Europe 2021). A case–control study of 195 children showed that serum PAH levels were strongly associated with the biomarkers of childhood asthma (Al-Daghri et al. 2013; Låg et al. 2020). The levels of PAHs measured in the study were correlated with biomarkers, including immunoglobulin E (IgE), resistin, granulocyte–macrophage colony-stimulating factor, and interferon-γ as well as asthma-related cytokines (Al-Daghri et al. 2013). Another case–control study in children showed that exposure to PAHs measured in urine 1-hydroxypyrene was associated with asthma (Wang et al. 2017). The latter study suggested that oxidative stress mediated the effects of PAH exposure on asthma, based on the findings that the exposure was also associated with the oxidative stress marker, 8-hydroxy-2′-deoxyguanosine (8-OHdG), in the urine.

Most of the studies reported that the associations were not limited to B[a]P but also included low-molecular-weight PAHs, including naphthalene, 1,2-benzanthracene, chrysene, phenanthrene, and pyrene (Låg et al. 2020).

Respiratory function and symptoms in children

Although exposure to air pollution is known to impair respiratory functions, studies investigating the relationship between ambient PAHs and pulmonary functions in children remain limited (Låg et al. 2020; WHO Regional Office for Europe 2021). No association has been reported between PAH exposure and respiratory symptoms in children with asthma living near an industrial complex and petroleum refineries in Canada (Smargiassi et al. 2014) nor in those living in an inner-city cohort in New York City (NY, USA) (Miller et al. 2010). A cross-sectional study in the US has demonstrated an association between annual PAHs and pulmonary function in children without asthma; however, no such association has been observed in those with asthma (Padula et al. 2015). Moreover, a cohort study in the US has reported that adverse associations between 1/9-hydroxyphenanthrene and persistent wheeze were observed among girls (Sherris et al. 2024). Collectively, these findings suggest that the impact of atmospheric PAHs on pediatric respiratory function is influenced by the PAH type and clinical characteristics. To investigate the relationship between ambient PAHs and pulmonary functions in children, considering their background, further studies are warranted.

Prenatal and postnatal exposure: effects on respiratory symptoms and morbidity

Several studies have indicated that in utero PAH exposure of the fetus dysregulates lung development and causes respiratory symptoms soon after birth (WHO Regional Office for Europe 2021). In another study, the methylation of 5′-CpG island(s) of the acyl-CoA synthetase long-chain family member 3 in the umbilical cord white blood cell DNA was significantly associated with maternal airborne PAH exposure and asthma symptoms in children (Perera et al. 2009). These modifications are suggested to have profound implications for asthma development, particularly following prenatal and infant PAH exposure. Conversely, a cohort study in the US has reported that prenatal PAH exposure is not associated with asthma at 8–9 years old (Sherris et al. 2024).

Postnatal PAH exposure has been suggested to influence pediatric respiratory function, with the likelihood of separate and combined effects of transplacental and postnatal PAH exposure (WHO Regional Office for Europe 2021). A longitudinal birth cohort study in New York City (NY, USA) has shown that prenatal and postnatal pyrene exposure are associated with asthma-related conditions (asthma medication use and emergency department visits for asthma) in non-atopic children (Jung et al. 2012). Thus, the effects of early PAH exposure over time on respiratory consequences may differ according to asthma phenotype.

Adult asthma

Epidemiological studies have suggested that exposure to air pollution also influences asthma in adults; however, few studies have investigated the impact of PAH exposure on adult asthma (Låg et al. 2020). A case–control study has reported that the total urinary PAH metabolites measured are associated with an increased risk of adult asthma (odds ratio. 1.34 [95% confidential interval, 1.03–1.74]) (Huang et al. 2018). Furthermore, this study has shown that the relative risks for asthma are higher for some of the low-molecular-weight PAH metabolites (2-hydroxyfluorene, 4-hydroxyphenanthrene [OHPHE], 1-OHPHE, 2-OHPHE, and 1-hydroxypyrene), indicating that low-molecular-weight PAHs are associated with asthma in adults.

Respiratory function and symptoms in adults

In the general population and older adults, alterations in urinary PAH metabolites appear to be associated with significant decrements in lung function (Låg et al. 2020; WHO Regional Office for Europe 2021). Additionally, in occupational settings, associations between decreased lung function parameters and increased urinary PAH metabolite levels have been observed in coke oven workers or diesel engine testing workers (Låg et al. 2020; WHO Regional Office for Europe 2021).

PAHs have been associated with cough and respiratory symptoms, particularly in individuals with chronic cough conditions. Previous studies have shown that ambient PAH exposure is associated with cough in adult patients with chronic cough (Anyenda et al. 2016a, b). In addition, this relationship has been reported to be notably strong in patients without asthma. Furthermore, certain PAH species have been identified as potentially more challenging. Of the ambient PAHs measured, benzo[a]anthracene

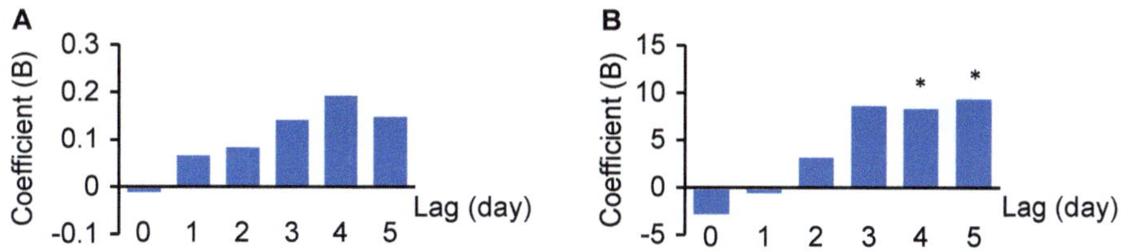

Fig. 1 Adjusted estimates of cough symptoms by each PAH exposure. **a** Total PAH, **b** BaA. Generalized estimating equations are used for investigating the relationship between cough occurrence and PAH exposure ($n = 98$). *p value < 0.05. Adjusted for monitoring sites, sex, age, body mass ranking, asthma, and daily $PM_{2.5}$, SO_2, and NO_2 concentrations. PAH, polycyclic aromatic hydrocarbon; BaA, Benz[a]anthracene; $PM_{2.5}$, particulate matter with $a \leq 2.5$-mm diameter; SO_2, sulfur dioxide; NO_2, nitrogen dioxide

(BaA) has been specifically linked to cough symptoms in individuals with chronic cough (Hara et al. 2022) (Fig. 1). Pyrene and benzo[ghi]perylene showed similar trends to BaA in their association with cough (Hara et al. 2022). Of note, regarding the timing of effects, BaA levels were positively associated with cough in the lag 4 and 5 models, suggesting a delayed effect (Hara et al. 2022) (Fig. 1). Additionally, a cross-sectional study in the US has provided epidemiological evidence linking PAH exposure to increased chronic cough risk (Jiang and Zhao 2024).

Although these findings suggest a clear association between PAH exposure and cough, particularly in individuals with pre-existing respiratory conditions, of note, the acute effects frequently attributed to PAH exposure (e.g., headache, nausea, and immediate respiratory irritation) may actually be caused by other agents present in the environment (Agency for Toxic Substances and Disease Registry (ATSDR) 2009).

3 Risk Modifiers

The risk modifiers of chemical exposures, including PAHs, on adverse health outcomes are presented in Table 1 (Woodruff 2024). A previous review has described the factors modifying the effects of air pollutants on asthma (Guarnieri and Balmes 2014). Young children with asthma have been considered a vulnerable population who are prone to air pollution-related adverse effects owing to their developing lungs, immature metabolic pathways, high ventilation rates per body weight, and increased time exercising outdoors. Even exposure in utero has been suggested to affect the postnatal risk of asthma and its exacerbations. Low birth weight, which may be associated with narrow airways during early childhood, is a risk factor for air pollution-related asthma symptomatology.

Young boys exhibit more prevalent and severe asthma exacerbations than young girls, and this disparity has been attributed, at least in part, to the relatively narrow airways in early life (Guarnieri and Balmes 2014). Boys and girls have shown no

Table 1 Risk modifiers of PAH exposure on adverse respiratory outcomes

EXTRINSIC ↑
Socioeconomic status
Discrimination
Racism
Access to health care
Food insecurity
Nutritional status
Smoking and alcohol use
Psychosocial stress
Underlying disease
Pregnancy
Age
Genetics
Sex
↓ INTRINSIC

consistent differences in the effects of air pollution on asthma; however, some studies have demonstrated sex-specific differences in responses. In adults, active asthma is more prevalent in females than that in males; however, no consistent sex-specific difference in risk has been reported for air pollution-related exacerbations. Recently, although no direct relationship is observed between airborne PAHs and asthma, an epidemiological study of middle-aged and older adult individuals in Japan has shown an association between airborne $PM_{2.5}$ levels and asthma in females but not in males (Hara et al. 2024) (Fig. 2). Older adults with asthma are likely to develop an increased risk of air pollution-related adverse effect; however, little evidence is available to support this notion.

Evidence supporting differences in ethnicity-specific susceptibility to air pollution-related asthma is scarce; however, the incidence of asthma has been associated with air pollution exposure in African–American and Latino children enrolled from several US cities and Puerto Rico. Low socioeconomic status-related influences may complicate the effects of ethnicity (Guarnieri and Balmes 2014). Children with asthma from such families may have greater exposure to air pollution and greater susceptibility to the effects of pollutants than children from families of higher socioeconomic status. Neighborhood characteristics (e.g., increased levels of crime, less green space, or poor food access), stress, and diet are the factors that may contribute to this susceptibility (Guarnieri and Balmes 2014).

Dietary factors can contribute to the susceptibility to pollutant-induced health effects. The body of evidence on the protective effects of a fruit- and vegetable-rich

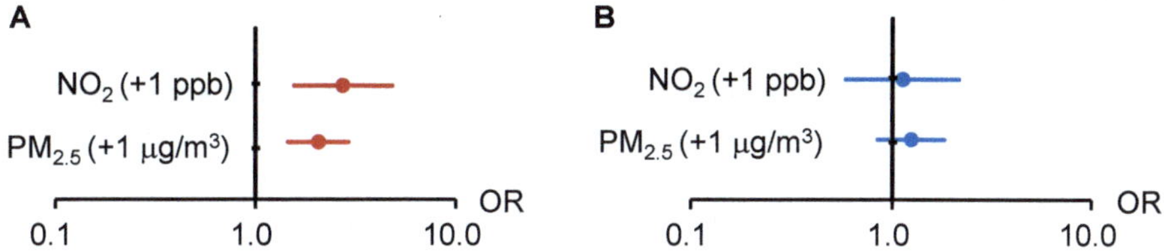

Fig. 2 Relationship between air pollutants and asthma. **a** females ($n = 847$), **b** males ($n = 650$). To investigate the relationship between the levels of each air pollutant and asthma prevalence, multivariable logistic regression analysis is employed. Results are presented as ORs and 95% confidence intervals. ORs are adjusted for age and body mass index. OR, odds ratio; $PM_{2.5}$, particulate matter with $a \leq 2.5$-mm diameter; NO_2, nitrogen dioxide

diet and antioxidant vitamin supplements supports an significant role for oxidative stress in the pathway through which outdoor air pollution affects asthma (Guarnieri and Balmes 2014). Obesity may not only increase the susceptibility to air pollution-related adverse effects but also serve as a risk factor for chronic cough (Morice et al. 2021).

Tobacco smoke is a mixture that includes gases and respirable PM and can adversely affect asthma. It would be a modifier of the effects of air pollution. PAHs are contained in tobacco smoke (WHO Regional Office for Europe 2021), and the urinary metabolites of PAHs in smokers are elevated. Clinically, a cross-sectional analysis has reported that the effects of PAHs on the risk of chronic cough are more pronounced in the US population who smoke than those who do not (Jiang and Zhao 2024).

Regarding additional susceptibility factors, genetic polymorphisms, including *GSTM1*, *GSTP1*, and *TNF*, which have been associated with the susceptibility to adverse outcomes of asthma from exposure to air pollutants, have also been linked to increased risk from secondhand smoke exposure (Guarnieri and Balmes 2014). A recent epidemiological study has reported ethnic differences in polygenic susceptibility to asthma between females from Germany and Japan; however, no significant gene–environment (ambient $PM_{2.5}$ and nitrogen dioxide levels) interactions on asthma were observed, and among the females from Germany, the polygenic risk score, which was calculated using previous genome-wide association studies, was associated with asthma (Kress et al. 2022).

Regarding biologic susceptibility, a previous study has reported that repeated high pyrene exposure is associated with an increased asthma risk, particularly in non-atopic children (Jung et al. 2012).

4 Prognosis

Air pollution from both ambient and occupational sources has been known to exacerbate and increase obstructive lung disease-related mortality (Låg et al. 2020; WHO Regional Office for Europe 2021). However, whether PAHs are associated with increases in respiratory mortality remains uncertain. A previous cohort study of male asphalt workers has suggested that PAH exposure can contribute to obstructive lung disease-related mortality (Burstyn et al. 2003).

Regarding asthma prognosis, the evidence suggests that PAH exposure can negatively impact the prognosis by increasing disease severity and exacerbations, particularly for vulnerable populations (Huang et al. 2018; Jung et al. 2012), increasing the risk of emergency department visits for asthma (Jung et al. 2012) and contributing to persistent wheezing in children (Sherris et al. 2024). The following vulnerable populations may be more susceptible to PAH-induced asthma effects: children, especially those without atopy (Jung et al. 2012; Sherris et al. 2024); females (Huang et al. 2018; Sherris et al. 2024); and individuals with occupational exposures (Huang et al. 2018).

5 Pathophysiology

PAH exposure has been shown to activate pathophysiological responses, including oxidative stress, redox imbalance, inflammation, bronchial smooth muscle contraction, epithelial and endothelial dysfunction, and dysregulated lung development (Låg et al. 2020; WHO Regional Office for Europe 2021). The risk of the development or exacerbation of asthma, COPD, and respiratory infections may be enhanced by these chemical properties of PAHs. The following are the main pathophysiological mechanisms involved in the development of PAH-induced nonmalignant respiratory diseases.

Airway inflammation and oxidative stress

PAHs can trigger oxidative stress in the respiratory system, thereby causing inflammation and tissue damage (Låg et al. 2020; WHO Regional Office for Europe 2021). This process occurs through reactive oxygen species (ROS) generation, inflammatory pathway activation, and proinflammatory cytokine release. These PAH-triggered pathways are mediated, in part, by their specific binding to cellular receptors, including the aryl hydrocarbon receptor (AhR) and the G protein-coupled receptor β_2-adrenergic receptor (β_2-AR) (Låg et al. 2020; WHO Regional Office for Europe 2021). Furthermore, along with damaging DNA, the metabolites of PAHs generated by cytochrome P450 (CYP) enzymes may form ROS and reactive electrophilic metabolites, which may inactivate essential signaling enzymes and transporters by oxidizing sulfhydryl groups and trigger inflammation (Låg et al. 2020). A recent study using a mouse model has shown that BaP exposure dose-dependently induced lung inflammatory response and that Wnt5a and downstream YAP/TAZ activation was accompanied by the enhanced release of epithelial-derived thymic stromal lymphopoietin and interleukin (IL)-33, which served as proinflammatory cytokines (Fan et al. 2022). A cross-sectional study using data from the US National Health and Nutrition Examination Survey have observed that urinary 2-hydroxyfluorene levels are associated with the systemic inflammation index, and an additional mediation analysis has shown that the association between urinary 2-hydroxyfluorene and COPD prevalence is mediated by the index (Xiao et al. 2023). Such responses may include increased neutrophil, eosinophil, and macrophage levels as well as ROS production, which are essential components of pulmonary inflammation. Sustained inflammation may exacerbate or contribute to the development of lung diseases, including asthma and COPD (WHO Regional Office for Europe 2021).

Oxidative stress plays a critical role in the pathogenesis of asthma and is regulated by protein phosphatases and intracellular antioxidant molecules, including nuclear factor erythroid 2-related factor 2, glutathione, and the peroxidase system (Cho and Moon 2010; Låg et al. 2020). Reactive PAH metabolites and ROS may interact with these systems. Moreover, PAHs can increase oxidative stress through the AhR pathway (Låg et al. 2020; Yang et al. 2017). PAHs are metabolically activated by CYP to form reactive species, including epoxides, peroxides, semiquinones, and quinones, which can generate ROS. Furthermore, redox imbalance may result in the accumulation of unfolded proteins and cause an "unfolded protein response" that interacts with different inflammatory and stress signaling pathways (Låg et al. 2020).

Furthermore, studies have suggested that PAHs can act through IgE to stimulate inflammatory responses and enhance allergic reactions (Karimi et al. 2015). In vivo PAH exposure may influence B and T helper (Th) cell differentiation by skewing immune responses toward a Th2-specific profile favoring B cell production of IgE and eosinophils, which are both hallmarks of allergic inflammation and allergic asthma (WHO Regional Office for Europe 2021). Other studies have indicated that PAHs can directly act on AhR in T cells, thereby leading to enhanced Th17 cell differentiation. As IL-17 and IL-22 sources, Th17 cells are implicated in the pathogenesis of asthma (WHO Regional Office for Europe 2021).

Airway remodeling

Chronic PAH exposure can lead to structural changes to the airways, characterized by airway wall thickening, increased smooth muscle mass, and subepithelial fibrosis. An experimental study has shown that the activation of AhR, one of the receptors for PAH, in an acute lung injury mouse model suppresses TGF-β–induced expression of genes associated with the epithelial–mesenchymal transition (Zimmerman et al. 2024). Furthermore, the knockdown of AhR in primary alveolar epithelial cells led to α-smooth muscle actin, type 1 collagen, and Fn1 upregulation as well as reduced epithelial gene expression. Additionally, the barrier function in primary murine alveolar epithelial cells following inflammatory injury was enhanced by AhR activation with an endogenous AhR ligand (Zimmerman et al. 2024). Collectively, AhR activation in alveolar epithelial cells preserves alveolar integrity by enhancing barrier function and limiting stress-induced expression of mesenchymal genes (Zimmerman et al. 2024).

Immune system modulation

PAHs can alter pulmonary immune responses, potentially leading to enhanced allergic responses, increased IgE and other asthma biomarker production, and alterations in the production of cytokines, including IL-4 and IL-10 (Al-Daghri et al. 2013). Recent studies have revealed that AhR, a receptor for PAHs, can evoke a number of other cellular functions and play a significant role in immune regulation (Låg et al. 2020). AhR is expressed in immune cells and in the parenchymal parts of the lungs, including the epithelial layer, submucosal glands, fibroblasts, and endothelial layer of the pulmonary vasculature (Låg et al. 2020). Both the innate and adaptive immune systems require AhR signaling at critical checkpoints. Ligand-induced AhR activation may suppress T cell-dependent immune responses and influence the balance between Th17 and regulatory T cells as well as dendritic cell function (Låg et al. 2020). Regarding PAHs, exposure to AhR ligands, including B[a]P, caused enhanced calcium- and ROS-dependent mast cell signaling, thereby leading to degranulation, mediator and cytokine release, and anaphylactic response in vivo (Låg et al. 2020). Conversely, some studies have suggested that ligand binding to the AhR can cause differences in the receptor conformation, resulting in different activation of downstream signaling (Låg et al. 2020). Thus, some PAHs that are not classified as AhR activators owing to limited or no effects on CYP1 expression are considered to exert other effects through interaction with AhR.

Impaired lung function

PAH exposure has been associated with reductions in lung function parameters, including forced expiratory volume in 1 s, forced vital capacity, and forced expiratory flow (Chinemerem C. Nwaozuzu et al. 2021). β_2-AR activation in the airways by the binding of catecholamines enables smooth muscle relaxation and bronchodilation, in part through intracellular calcium and cyclic adenosine monophosphate (cAMP) signaling activation; therefore, inhaled β_2-AR agonists are used in the treatment of asthma and COPD (Låg et al. 2020). Of note, a PAH mixture with > 50% pyrene has been reported to impair β_2-AR function in airway epithelial and smooth muscle cells and interfere with the treatment with β_2-AR–specific agonist procaterol (Factor et al. 2011). Other in vitro studies have suggested that B[a]P at low concentrations can directly interact with the ligand-binding pocket of the β_2-AR, thereby causing increased intracellular calcium and cAMP signaling and β_2-AR desensitization (WHO Regional Office for Europe 2021). Collectively, PAH interference with β_2-AR signaling in the airways could potentially prevent bronchodilation and may contribute to asthma and COPD development or exacerbation.

Cellular and molecular effects

At the cellular level, PAHs can cause DNA damage in lung cells, alterations in cell cycle regulation, and epigenetic changes affecting gene expression. Moreover, several studies have investigated the inflammatory responses in PAH-exposed lung cells. In primary human macrophages, B[a]P exposure induces IL-8, a well-established pulmonary inflammatory cytokine, via an AhR-dependent pathway (Låg et al. 2020). In vascular endothelial and lung epithelial cells, B[a]P and other PAHs have been reported to induce the expression of the CXC motif chemokine ligand 8 (CXCL8) and other chemokines and cytokines. For example, in addition to B[a]P, a previous in vitro study has demonstrated that various PAHs, including pyrene and benzo[e]pyrene, upregulate the intracellular calcium concentration independently of AhR in human endothelial cells (Mayati et al. 2012). Furthermore, another in vitro study has revealed that 1-nitropyrene (1-NP) induced an increase in intracellular calcium levels in human bronchial epithelial cells, at least partly through β_2-AR activation, and suggested that both β_2-AR and intracellular calcium signaling can be involved in 1-NP–induced CXCL8 responses (Låg et al. 2020; Mayati et al. 2014). Conversely, evidence has shown that the binding of ligands to AhR can induce both pro- and anti-inflammatory responses (WHO Regional Office for Europe 2021). Accordingly, a previous study has indicated that B[a]P attenuates the proinflammatory phenotype of murine macrophages by inducing anti-inflammatory IL-10 expression in an AhR-dependent manner (Fueldner et al. 2018; Låg et al. 2020). Additionally, AhR activation in alveolar epithelial cells has been reported to preserve alveolar integrity by modulating inflammatory cytokine expression while enhancing barrier function and limiting stress-induced expression of mesenchymal genes (Zimmerman et al. 2024). Collectively, differences in ligands in terms of metabolism and induced toxicity, binding to AhR, and subsequent classic or nonclassic genomic or nongenomic pathway activation may explain the multifaceted role of AhR in respiratory diseases (Låg et al. 2020).

Interaction with other environmental factors

PAHs frequently coexist with other air pollutants, and their combined effects may exacerbate asthma symptoms. For example, PAHs are frequently associated with PM, especially $PM_{2.5}$, which can independently contribute to respiratory conditions (Anyenda et al. 2016a). Furthermore, the interaction between PAHs and other pollutants, such as nitrogen dioxide (NO_2), may have synergistic effects on **asthma** development and exacerbation (Anyenda et al. 2016a). Additionally, a case–control study has demonstrated that the associations between urinary PAH metabolites and adult asthma were more pronounced in the participants who were smokers and had a history of occupational dust exposure (Huang et al. 2018). Furthermore, a recent cross-sectional analysis has reported that the effects of PAHs, measured using urinary metabolites, on the risk of chronic cough are more pronounced in the US population with higher levels of serum cotinine, a marker of environmental tobacco smoke (Jiang and Zhao 2024).

Early life exposure

Prenatal and early childhood PAH exposure have been associated with the development of asthma symptoms later in life (Jung et al. 2012). For example, a birth cohort study in Europe has observed that prenatal PAH exposure increases the risk of various respiratory symptoms, including wheezing, cough, and sore throat, in newborns (Jedrychowski et al. 2005). An in vivo study has demonstrated that hyperoxic lung injury observed in newborn mice exposed to PAH in utero is dose-dependently augmented by prenatal PAH exposure and that CYP1A1 and CYP1A2 are protective against the injury (Stading et al. 2024).

Several pieces of evidence from the epigenome-wide association studies have suggested that early life environmental events can prompt widespread and persistent variations in transcriptional profiling. Moreover, these variations arise in response to DNA damage and/or as a consequence of epigenetic modifications. Growing evidence has associated PAHs with epigenetic variations involving alterations in DNA methylation, histone modification, and micro-RNA regulation. Epigenetic alterations due to PAH exposure have been linked to pulmonary and cardiovascular diseases, nervous system disorder, and cancer. These findings offer novel perspectives into respiratory disease development and progression in response to PAH toxicity and exposure in the fetal and early postnatal periods (Das and Ravi 2022).

Respiratory neuronal reflexes and respiratory symptoms

Respiratory reflexes contribute to symptoms, including cough, and are regulated by vagal afferent nerves/C-fibers, which innervate the airways (Låg et al. 2020). Robinson et al. have recently reported that diesel exhaust particles can activate vagal C-fibers in guinea pigs (Låg et al. 2020; WHO Regional Office for Europe 2021). The organic extract, including PAHs, not the clean particles, evoked the vagal C-fiber depolarization in guinea pigs and humans. Notably, this effect was also observed following exposure to very low concentrations (1 nM) of phenanthrene, a noncarcinogenic PAH that is among the most abundant PAHs in urban air (Låg et al. 2020; WHO Regional Office for Europe 2021). The mechanisms underlying these responses appear to involve AhR activation and the subsequent calcium influx through transient receptor potential channels.

Lung epithelial barrier and respiratory infections

Experimental studies have suggested a role of PAHs in the increased susceptibility to various pathogens when exposed to air pollution. Lung epithelial barrier disruption; impaired mucociliary clearance; immune defects, including reduced pathogen recognition and killing; impaired cytokine release and surface receptor expression to pathogen products; Th2-skewed response; impaired infectious agent clearance; and delayed recovery from severe disease constitute the possible mechanistic explanations (Låg et al. 2020).

Experimental data from mouse models have revealed that AhR is involved (via nonclassical genomic and nongenomic pathways) in orchestrating appropriate immune responses in barrier organs, including the lungs (WHO Regional Office for Europe 2021). The epithelial barrier may be disrupted by PAHs by affecting tight and/or gap junctions via AhR-dependent mechanisms (Låg et al. 2020). As AhR has significant functions in immune system development and differentiation, long-term ligand binding to AhR may influence the vulnerability to respiratory infections. Along with immune cells, including macrophages, T lymphocytes, dendritic, and mast cells, both epithelial and endothelial cells are also targets of AhR-mediated changes in immune function (WHO Regional Office for Europe 2021).

Both enhancement and protection against infections have been reported following AhR ligand exposure (Låg et al. 2020; WHO Regional Office for Europe 2021). As PAHs can bind to AhR and activate several AhR pathways, they are among the environmental contaminants suggested to influence respiratory infection susceptibility. Notably, as low-molecular-weight PAHs can bind to AhR and, in part, trigger nonclassical pathways or interfere with the functions of AhR, it is speculated that they can affect host susceptibility and response to respiratory infections.

6 Public Health Implications

Considering the widespread presence of PAHs in air pollution, reducing PAH exposure is essential for preserving respiratory health (WHO Regional Office for Europe 2021). Studies have associated air pollutants, including PAHs, with cardiovascular and respiratory diseases, highlighting the significance of proactive strategies (Guarnieri and Balmes 2014; Rajagopalan and Landrigan 2021) (Table 2).

Public health service organizations

Public health organizations and specialists have an in-depth understanding of pollution in disease prevention programs, medical education, and clinical practice and acknowledge that pollution is a significant and potentially preventable risk factor for respiratory diseases. Based on this information, physicians can perform targeted interventions by obtaining pollution exposure history for each patient when relevant, assessing individual susceptibility, and providing a guide on pollution avoidance. Follow-up investigation may be recommended when the initial history suggests clinically meaningful exposure. The occupational history is key for identifying workplace exposures.

Patients exhibiting various characteristics, including asthma, who are vulnerable to the adverse effects of pollution (Table 1) should be advised to evaluate the air conditions by referring to published information, including the Environmental Protection Agency's Air Quality Index (https://www.airnow.gov/aqi-basics/) and the Atmospheric Environmental Regional Observation System (https://soramame.env.go.jp/) provided by the Ministry of the Environment, Government of Japan, when planning outdoor activities and limit such activities on days when the air quality is unsafe.

Individual-level strategies

At the individual level, lifestyle changes, including quitting smoking and avoiding secondhand smoke, reducing charred food consumption, and avoiding vehicle exhaust exposure, can significantly reduce PAH exposure (Wang et al. 2022). From an occupational perspective, workers in high-risk industries should use personal protective equipment and undergo safety training.

Table 2 Societal and governmental strategies for reducing air pollution exposure and pollution-related diseases

Public health service organizations
Tools for communicating pollution risk
Low-cost monitoring
Educational campaigns
Government
Urban infrastructure upgrades
Clean energy incentives
Emission trading
Energy
Shift to clean and renewable energy
Transportation reform
Ecosystem
Urban redesign
Green space
Walkable trails
Traffic relocation
Planned net zero communities
Lifestyle
Personal monitors
Active transport
Regular exercise
Plant-based diets

Government and societal measures

Mitigating PAH exposure necessitates systemic efforts to control pollution at its sources. A rapid government-supported shift from fossil fuels to renewable energy sources, including wind and solar power, is the most effective strategy for achieving this goal. This approach is cost-effective and benefits both health and the economy (Rajagopalan and Landrigan 2021). Tax incentives for renewable energy, elimination of subsidies for fossil fuels, and enforcement of the "polluter pays" principle to tax emissions are the suggested policy tools (Rajagopalan and Landrigan 2021). In low-income countries, household air pollution is considered to be most effectively controlled by providing low-income families with affordable access to cleaner fuels (Rajagopalan and Landrigan 2021).

At a broader societal level, PAH levels in the environment are controlled by stricter regulation and enforcement of industrial emissions, promotion of cleaner sources of energy, and regular monitoring of air quality (Li et al. 2017).

Role of healthcare professionals

Healthcare professionals are uniquely positioned to advocate for pollution control and disease prevention. By emphasizing to policymakers the well-documented associations between pollution, climate change, and health as well as by highlighting that pollution mitigation measures will also mitigate climate change and vice versa, healthcare professionals can contribute to catalyze enduring public health benefits (Rajagopalan and Landrigan 2021).

Evidence-based prevention and management strategies for asthma

A previous review has described that one strategy for reducing the exacerbation of air pollution-induced asthma is for local governments to issue smog alerts of high-pollution days (Guarnieri and Balmes 2014). Outdoor activities, especially intense exercise, during such periods should be avoided by individuals with asthma and other pre-existing cardiopulmonary disorders. Furthermore, to minimize exposure to traffic-related pollutants, individuals with asthma are recommended to reside at least 300 m away from major roadways (Guarnieri and Balmes 2014). Traffic-related air pollution can exacerbate asthma; however, concentrations of motor vehicle emissions, such as ultrafine PM and black carbon particles, substantially decrease with a 300-m distance. Moreover, in-vehicle pollutant exposure during commuting with open windows can be very high. Personalized asthma care plans can incorporate these measures to mitigate air pollution.

Advancements in pollution awareness

Technological innovations, including portable sensors and digital platforms, can enhance awareness and support personalized pollution avoidance strategies. These tools empower individuals to make informed decisions for reducing exposure and mitigating health risks (Rajagopalan and Landrigan 2021).

7 Management

Diet and lifestyle modifications

Some studies have suggested that the effects of PAH exposure can be mitigated by certain nutritional factors. A cross-sectional study of children in the US has demonstrated that urinary PAH metabolites are significantly associated with reduced lung function in children with asthma and vitamin D deficiency but not in those with asthma and vitamin D sufficiency (Han et al. 2018). These findings suggest that vitamin D sufficiency helps protect against the detrimental effects of PAH exposure on lung function in children with asthma. Another cross-sectional study of adults in the US has reported that individuals with a low antioxidant-based diet and lifestyle exhibited a marked decrease in lung function for increased urinary monohydroxy PAH levels, whereas those with a medium or high antioxidant-based diet and lifestyle showed no significant associations (Xu et al. 2024). These findings suggest the significance of maintaining an antioxidant-based diet and lifestyle in counteracting PAH exposure-induced lung dysfunction.

Medications

Although no medications are specifically formulated for counteracting PAH-induced airway diseases, several classes of medications are commonly administered for treating and managing obstructive lung disease-related symptoms, regardless of the underlying trigger. Besides avoidance of exposure on high-pollution days, these medications can help prevent exacerbations. For details on pharmacologic therapies, please refer to the national and international guidelines for each disease. The following are the main drugs used in the current treatment of asthma:

(1) *Bronchodilators*

β_2-adrenergic agonists are frequently used for treating reactive airway diseases, including asthma. However, the function of β_2-AR may be impaired by PAH exposure, potentially reducing the efficacy of these medications (Factor et al. 2011). Individuals with asthma who remain symptomatic despite a combination use of β_2-adrenergic agonists and inhaled corticosteroids are recommended adding long-acting muscarinic antagonists in their regimen (Porsbjerg et al. 2023).

(2) *Anti-inflammatory medications*

- Inhaled corticosteroids

Maintenance inhaled corticosteroid therapy can prevent asthma exacerbations. Several studies have suggested that this approach decreases adverse responses to pollutant exposures (Guarnieri and Balmes 2014). National and international guidelines have recommended inhaled corticosteroids combined with leukotriene receptor antagonists or long-acting bronchodilators as a treatment option for asthma (Porsbjerg et al. 2023). Furthermore, these guidelines have emphasized the possibility of adverse effects of inhaled corticosteroids, including sore and dry throat. Despite corticosteroid treatment, patients with asthma may be the most likely to develop respiratory symptoms with exposure to air pollutants. Owing to the potential interactive effects of air pollutants with allergens or secondhand smoke, avoidance of these exposures is essential, especially during high-pollution days (Guarnieri and Balmes 2014).

- Biologics

Several biological therapies are available for severe asthma. Examples include anti-IgE, anti-IL-5/5R, anti-IL-4/1, and antithymic stromal lymphopoietin (Porsbjerg et al. 2023). A recent study has demonstrated that benralizumab, a humanized monoclonal antibody against IL-5 receptor-alfa, improved acute eosinophilic exacerbations of asthma and COPD better than the current standard of care with prednisolone alone (Ramakrishnan et al. 2024). Another study has reported that benralizumab-treated patients achieved reductions in inhaled corticosteroid therapy while maintaining asthma control (Jackson et al. 2024a). More recently, depemokimab, an ultra-long-acting biologic therapy with enhanced binding affinity for IL-5, reduced the annualized rate of exacerbations in patients with severe asthma with an eosinophilic phenotype (Jackson et al. 2024b).

Of note, the most effective approach to PAH-induced asthma would likely involve medication management combined with efforts for reducing PAH exposure. For optimal asthma control, regular monitoring and adjustment of the treatment plan by a healthcare professional are crucial.

8 Summary and Future Perspectives

Recent human epidemiological and experimental studies have reinforced the hypothesis that exposure to PAHs, frequently associated with PM, partially explains the heightened risk of nonmalignant respiratory diseases, including asthma, observed following PM exposure (WHO Regional Office for Europe 2021). Epidemiological evidence links PAHs to adverse respiratory outcomes, including reduced lung function, asthma exacerbations, and increased morbidity and mortality from obstructive lung diseases. However, unraveling the specific contributions of PAHs from those of other combustion-derived components, including PM, remains challenging.

Experimental research has elucidated the potential mechanisms underlying these associations. PAHs appear to exert their effects through various pathways, including AhR-dependent mechanisms, PAH binding involvement, increased gene expression, and metabolic activation into reactive intermediates. These processes not only trigger proinflammatory mediators and oxidative stress but also cause subtler changes, including altered calcium signaling and membrane depolarization. Emerging evidence suggests that PAH-induced immunomodulation can also occur through nonclassical pathways, involving organic compounds attached to the carbonaceous core of combustion particles. However, the dose–response relationship, the mechanisms of action, and the relative contributions of individual PAH species have posed significant uncertainties.

Our ability to monitor PAH exposure and its health impacts has been enhanced by recent advancements in digital health technologies and portable air pollution sensors. Additionally, global efforts for reducing air pollution, through transitioning to renewable energy and implementing stricter emission regulations, provide promising pathways for mitigating PAH-associated health risks.

Future research should address key knowledge gaps, including the precise role of PAHs in respiratory diseases, their interactions with other pollutants, and the development of robust regulatory standards for specific PAH species, including B[a]P.

A comprehensive approach combining individual-level interventions with systemic pollution control strategies remains imperative for protecting public health. To advance our understanding and mitigate the health impacts of PAHs in ambient air, enhanced collaboration among researchers, policymakers, and healthcare professionals will be critical.

References

Agency for Toxic Substances and Disease Registry (ATSDR), 2009. Toxicity of Polycyclic Aromatic Hydrocarbons (PAHs). *Case Studies in Environmental Medicine*, pp. 1–68

Al-Daghri, N.M., Alokail, M.S., Abd-Alrahman, S.H., Draz, H.M., Yakout, S.M., Clerici, M., 2013. Polycyclic aromatic hydrocarbon exposure and pediatric asthma in children: A case-control study. *Environ. Heal.* 12, 2–7. https://doi.org/10.1186/1476-069X-12-1

Anyenda, E.O., Higashi, T., Kambayashi, Y., Nguyen, T.T.T., Michigami, Y., Fujimura, M., Hara, J., Tsujiguchi, H., Kitaoka, M., Asakura, H., Hori, D., Yamada, Y., Hayashi, K., Hayakawa, K., Nakamura, H., 2016a. Associations of Cough Prevalence with Ambient Polycyclic Aromatic Hydrocarbons, Nitrogen and Sulphur Dioxide: A Longitudinal Study. *Int. J. Environ. Res. Public Health* 13, 800. https://doi.org/10.3390/ijerph13080800

Anyenda, E.O., Higashi, T., Kambayashi, Y., Thao, N.T.T., Michigami, Y., Fujimura, M., Hara, J., Tsujiguchi, H., Kitaoka, M., Asakura, H., Hori, D., Yamada, Y., Hayashi, K., Hayakawa, K., Nakamura, H., 2016b. Exposure to daily ambient particulate polycyclic aromatic hydrocarbons and cough occurrence in adult chronic cough patients: A longitudinal study. *Atmos. Environ.* 140, 34–41. https://doi.org/10.1016/j.atmosenv.2016.05.042

Brauer, M., Roth, G.A., Aravkin, A.Y., Zheng, P., Abate, K.H., Abate, Y.H., Abbafati, C., Abbasgholizadeh, R., Abbasi, M.A., Abbasian, M., Abbasifard, M., Abbasi-Kangevari, M., Abd ElHafeez, S., Abd-Elsalam, S., Abdi, P., Abdollahi, M., Abdoun, M., Abdulah, D.M., Abdullahi, A., Abebe, M., Abedi, Aidin, Abedi, Armita, Abegaz, T.M., Zuñiga, R.A.A., Abiodun, O., Abiso, T.L., Aboagye, R.G., Abolhassani, H., Abouzid, M., Aboye, G.B., Abreu, L.G., Abualruz, H., Abubakar, B., Abu-Gharbieh, E., Abukhadijah, H.J.J., Aburuz, S., Abu-Zaid, A., Adane, M.M., Addo, I.Y., Addolorato, G., Adedoyin, R.A., Adekanmbi, V., Aden, B., Adetunji, J.B., Adeyeoluwa, T.E., Adha, R., Adibi, A., Adnani, Q.E.S., Adzigbli, L.A., Afolabi, A.A., Afolabi, R.F., Afshin, A., Afyouni, S., Afzal, M.S., Afzal, S., Agampodi, S.B., Agbozo, F., Aghamiri, S., Agodi, A., Agrawal, A., Agyemang-Duah, W., Ahinkorah, B.O., Ahmad, A., Ahmad, D., Ahmad, F., Ahmad, N., Ahmad, S., Ahmad, T., Ahmed, Ali, Ahmed, Anisuddin, Ahmed, Ayman, Ahmed, L.A., Ahmed, M.B., Ahmed, S., Ahmed, S.A., Ajami, M., Akalu, G.T., Akara, E.M., Akbarialiabad, H., Akhlaghi, S., Akinosoglou, K., Akinyemiju, T., Akkaif, M.A., Akkala, S., Akombi-Inyang, B., Al Awaidy, S., Al Hasan, S.M., Alahdab, F., AL-Ahdal, T.M.A., Alalalmeh, S.O., Alalwan, T.A., Al-Aly, Z., Alam, K., Alam, N., Alanezi, F.M., Alanzi, T.M., Albakri, A., AlBataineh, M.T., Aldhaleei, W.A., Aldridge, R.W., Alemayohu, M.A., Alemu, Y.M., Al-Fatly, B., Al-Gheethi, A.A.S., Al-Habbal, K., Alhabib, K.F., Alhassan, R.K., Ali, Abid, Ali, Amjad, Ali, B.A., Ali, I., Ali, L., Ali, M.U., Ali, R., Ali, S.S.S., Ali, W., Alicandro, G., Alif, S.M., Aljunid, S.M., Alla, F., Al-Marwani, S., Al-Mekhlafi, H.M., Almustanyir, S., Alomari, M.A., Alonso, J., Alqahtani, J.S., Alqutaibi, A.Y., Al-Raddadi, R.M., Alrawashdeh, A., Al-Rifai, R.H., Alrousan, S.M., Al-Sabah, S.K., Alshahrani, N.Z., Altaany, Z., Altaf, A., Al-Tawfiq, J.A., Altirkawi, K.A., Aluh, D.O., Alvis-Guzman, N., Alvis-Zakzuk, N.J., Alwafi, H., Al-Wardat, M.S., Al-Worafi, Y.M., Aly, H., Aly, S., Alzoubi, K.H., Al-Zyoud, W., Amaechi, U.A., Mohammadi, M.A., Amani, R., Amiri, S., Amirzade-Iranaq, M.H., Ammirati, E., Amu, H., Amugsi, D.A., Amusa, G.A., Ancuceanu, R., Anderlini, D., Anderson, J.A., Andrade, P.P., Andrei, C.L., Andrei, T., Anenberg, S.C., Angappan, D., Angus, C., Anil, A., Anil, S., Anjum, A., Anoushiravani, A., Antonazzo, I.C., Antony, C.M., Antriyandarti, E., Anuoluwa, B.S., Anvari, D., Anvari, S., Anwar, S., Anwar, S.L., Anwer, R., Anyabolo, E.E., Anyasodor, A.E., Apostol, G.L.C., Arabloo, J., Bahri, R.A., Arafat, M., Areda, D., Aregawi, B.B., Aremu, A., Armocida, B., Arndt, M.B., Ärnlöv, J., Arooj, M., Artamonov, A.A., Artanti, K.D., Aruleba, I.T., Arumugam, A., Asbeutah, A.M., Asgary, S., Asgedom, A.A., Ashbaugh, C., Ashemo, M.Y., Ashraf, T., Askarinejad, A., Assmus, M., Astell-Burt, T., Athar, M., Athari, S.S., Atorkey, P., Atreya, A., Aujayeb, A., Ausloos, M., Avila-Burgos, L., Awoke, A.A., Quintanilla, B.P.A., Ayatollahi, H., Portugal, C.A., Ayuso-Mateos, J.L., Azadnajafabad, S., Azevedo, R.M.S., Azhar, G.S., Azizi, H., Azzam, A.Y., Backhaus, I.L., Badar, M., Badiye, A.D., Bagga, A., Baghdadi, S., Bagheri, N., Bagherieh, S., Taghanaki, P.B., Bai, R., Baig, A.A., Baker, J.L., Bakkannavar,

S.M., Balasubramanian, M., Baltatu, O.C., Bam, K., Bandyopadhyay, S., Banik, B., Banik, P.C., Banke-Thomas, A., Bansal, H., Barchitta, M., Bardhan, M., Bardideh, E., Barker-Collo, S.L., Bärnighausen, T.W., Barone-Adesi, F., Barqawi, H.J., Barrero, L.H., Barrow, A., Barteit, S., Basharat, Z., Basiru, A., Basso, J.D., Bastan, M.M., Basu, S., Batchu, S., Batra, K., Batra, R., Baune, B.T., Bayati, M., Bayileyegn, N.S., Beaney, T., Behnoush, A.H., Beiranvand, M., Béjot, Y., Bekele, A., Belgaumi, U.I., Bell, A.W., Bell, M.L., Bello, M.B., Bello, O.O., Belo, L., Beloukas, A., Bendak, S., Bennett, D.A., Bennitt, F.B., Bensenor, I.M., Benzian, H., Beran, A., Berezvai, Z., Bernabe, E., Bernstein, R.S., Bettencourt, P.J.G., Bhagavathula, A.S., Bhala, N., Bhandari, D., Bhardwaj, N., Bhardwaj, P., Bhaskar, S., Bhat, A.N., Bhat, V., Bhatti, G.K., Bhatti, J.S., Bhatti, M.S., Bhatti, R., Bhuiyan, M.A., Bhutta, Z.A., Bikbov, B., Bishai, J.D., Bisignano, C., Biswas, A., Biswas, B., Biswas, R.K., Bjørge, T., Boachie, M.K., Boakye, H., Bockarie, M.J., Bodolica, V., Bodunrin, A.O., Bogale, E.K., Bolla, S.R., Boloor, A., Hashemi, M.B., Boppana, S.H., Basara, B.B., Borhany, H., Carvajal, A.B., Bouaoud, S., Boufous, S., Bourne, R., Boxe, C., Braithwaite, D., Brant, L.C., Brar, A., Breitborde, N.J.K., Breitner, S., Brenner, H., Briko, A.N., Britton, G., Brown, C.S., Browne, A.J., Brunoni, A.R., Bryazka, D., Bulamu, N.B., Bulto, L.N., Buonsenso, D., Burkart, K., Burns, R.A., Busse, R., Bustanji, Y., Butt, N.S., Butt, Z.A., dos Santos, F.L.C., Cagney, J., Cahuana-Hurtado, L., Calina, D., Cámera, L.A., Campos, L.A., Campos-Nonato, I.R., Cao, C., Cao, F., Cao, Y., Capodici, A., Cárdenas, R., Carr, S., Carreras, G., Carrero, J.J., Carugno, A., Carvalho, F., Carvalho, M., Castaldelli-Maia, J.M., Castañeda-Orjuela, C.A., Castelpietra, G., Catalá-López, F., Catapano, A.L., Cattaruzza, M.S., Caye, A., Cederroth, C.R., Cegolon, L., Cenderadewi, M., Cercy, K.M., Cerin, E., Chadwick, J., Chakraborty, C., Chakraborty, P.A., Chakraborty, S., Chan, J.S.K., Chan, R.N.C., Chandan, J.S., Chandika, R.M., Chaturvedi, P., Chen, A.T., Chen, C.S., Chen, H., Chen, M.X., Chen, M., Chen, S., Cheng, C.Y., Cheng, E.T.W., Cherbuin, N., Chi, G., Chichagi, F., Chimed-Ochir, O., Chimoriya, R., Ching, P.R., Chirinos-Caceres, J.L., Chitheer, A., Cho, W.C.S., Chong, B., Chopra, H., Chowdhury, R., Christopher, D.J., Chu, D.T., Chukwu, I.S., Chung, E., Chung, S.C., Chutiyami, M., Cioffi, I., Cogen, R.M., Cohen, A.J., Columbus, A., Conde, J., Corlateanu, A., Cortese, S., Cortesi, P.A., Costa, V.M., Costanzo, S., Criqui, M.H., Cruz, J.A., Cruz-Martins, N., Culbreth, G.T., da Silva, A.G., Dadras, O., Dai, X., Dai, Z., Daikwo, P.U., Dalli, L.L., Damiani, G., D'Amico, E., D'Anna, L., Darwesh, A.M., Das, J.K., Das, S., Dash, N.R., Dashti, M., Dávila-Cervantes, C.A., Weaver, N.D., Davitoiu, D.V., De la Hoz, F.P., de la Torre-Luque, A., De Leo, D., Debopadhaya, S., Degenhardt, L., Bo, C. Del, Delgado-Enciso, I., Delgado-Saborit, J.M., Demoze, C.K., Denova-Gutiérrez, E., Dervenis, N., Dervišević, E., Desai, H.D., Desai, R., Devanbu, V.G.C., Dewan, S.M.R., Dhali, A., Dhama, K., Dhane, A.S., Dhimal, M.L., Dhimal, M., Dhingra, S., Dhulipala, V.R., Dhungana, R.R., da Silva, D.D., Diaz, D., Diaz, L.A., Diaz, M.J., Dima, A., Ding, D.D., Dinu, M., Djalalinia, S., Do, T.C., Do, T.H.P., do Prado, C.B., Dodangeh, M., Dohare, S., Dokova, K.G., Dong, W., Dongarwar, D., D'Oria, M., Dorostkar, F., Dorsey, E.R., Doshi, R., Doshmangir, L., Dowou, R.K., Driscoll, T.R., Dsouza, A.C., Dsouza, H.L., Dumith, S.C., Duncan, B.B., Duraes, A.R., Duraisamy, S., Dushpanova, A., Dzianach, P.A., Dziedzic, A.M., Ebrahimi, A., Echieh, C.P., Ed-Dra, A., Edinur, H.A., Edvardsson, D., Edvardsson, K., Efendi, F., Eftekharimehrabad, A., Eini, E., Ekholuenetale, M., Ekundayo, T.C., El Arab, R.A., El Sayed Zaki, M., El-Dahiyat, F., Elemam, N.M., Elgar, F.J., ElGohary, G.M.T., Elhabashy, H.R., Elhadi, M., Elmehrath, A.O., Elmeligy, O.A.A., Elshaer, M., Elsohaby, I., Emeto, T.I., Esfandiari, N., Eshrati, B., Eslami, M., Esmaeili, S.V., Estep, K., Etaee, F., Fabin, N., Fagbamigbe, A.F., Fagbule, O.F., Fahimi, S., Falzone, L., Fareed, M., e Sá Farinha, C.S., Faris, M.A.I.E.M., Faris, P.S., Faro, A., Fasina, F.O., Fatehizadeh, A., Fauk, N.K., Fazylov, T., Feigin, V.L., Feng, X., Fereshtehnejad, S.M., Feroze, A.H., Ferrara, P., Ferrari, A.J., Ferreira, N., Fetensa, G., Feyisa, B.R., Filip, I., Fischer, F., Fitriana, I., Flavel, J., Flohr, C., Flood, D., Flor, L.S., Foigt, N.A., Folayan, M.O., Force, L.M., Fortuna, D., Foschi, M., Franklin, R.C., Freitas, A., Friedman, S.D., Fux, B., Sridevi, G., Gaal, P.A., Gaihre, S., Gajdács, M., Galali, Y., Gallus, S., Gandhi, A.P., Ganesan, B., Ganiyani, M.A., Garcia, V., Gardner, W.M., Garg, R.K., Gautam, R.K., Gebi, T.G., Gebregergis, M.W., Gebrehiwot, M., Gebremariam, T.B.B., Gebremeskel, T.G., Gerema, U., Getacher, L., a Getahun, G.K., Getie, M., Ghadirian, F., Ghafarian, S., Jolfayi,

A.G., Ghailan, K.Y., Ghajar, A., Ghasemi, M.R., Dabaghi, G.G., Ghasemzadeh, A., Ghassemi, F., Ghazy, R.M., Gholami, A., Gholamrezanezhad, A., Gholizadeh, N., Ghorbani, M., Gil, A.U., Gil, G.F., Gilbertson, N.M., Gill, P.S., Gill, T.K., Gindaba, E.Z., Girmay, A., Glasbey, J.C., Gnedovskaya, E. V., Göbölös, L., Godinho, M.A., Goel, A., Golechha, M., Goleij, P., Golinelli, D., Gomes, N.G.M., Gopalani, S.V., Gorini, G., Goudarzi, H., Goulart, A.C., Gouravani, M., Goyal, A., Graham, S.M., Grivna, M., Grosso, G., Guan, S.Y., Guarducci, G., Gubari, M.I.M., Guha, A., Guicciardi, S., Gulati, S., Gulisashvili, D., Gunawardane, D.A., Guo, C., Gupta, A.K., Gupta, B., Gupta, M., Gupta, Rahul, Gupta, Rajat Das, Gupta, Rajeev, Gupta, S., Gupta, V.B., Gupta, Vijai Kumar, Gupta, Vivek Kumar, Habibzadeh, F., Habibzadeh, P., Hadaro, T.S., Hadian, Z., Haep, N., Haghi-Aminjan, H., Haghmorad, D., Hagins, H., Haile, D., Hailu, A., Ali, A.H., Halboub, E.S., Halimi, A., Hall, B.J., Haller, S., Halwani, R., Hamadeh, R.R., Hamdy, N.M., Hameed, S., Hamidi, S., Hammoud, A., Hanif, A., Hanifi, N., Haq, Z.A., Haque, M.R., Harapan, H., Hargono, A., Haro, J.M., Hasaballah, A.I., Hasan, I., Hasan, M.J., Hasan, S.M.M., Hasani, H., Hasanian, M., Hashmeh, N., Hasnain, M.S., Hassan, A., Hassan, I., Hassan Zadeh Tabatabaei, M.S., Hassani, S., Hassanipour, S., Hassankhani, H., Haubold, J., Havmoeller, R.J., Hay, S.I., Hebert, J.J., Hegazi, O.E., Hegena, T.Y., Heidari, G., Heidari, M., Helfer, B., Hemmati, M., Henson, C.A., Herbert, M.E., Herteliu, C., Heuer, A., Hezam, K., Hinneh, T.K., Hiraike, Y., Hoan, N.Q., Holla, R., Hon, J., Hoque, M.E., Horita, N., Hossain, S., Hosseini, S.E., Hosseinzadeh, H., Hosseinzadeh, M., Hostiuc, M., Hostiuc, S., Hoven, H., Hsairi, M., Hsu, J.M., Hu, C., Huang, J., Huda, M.N., Hulland, E.N., Hultström, M., Hushmandi, K., Hussain, J., Hussein, N.R., Huynh, C.K., Huynh, H.H., Ibitoye, S.E., Idowu, O.O., Ihler, A.L., Ikeda, N., Ikuta, K.S., Ilesanmi, O.S., Ilic, I.M., Ilic, M.D., Imam, M.T., Immurana, M., Inbaraj, L.R., Irham, L.M., Isa, M.A., Islam, M.R., Ismail, F., Ismail, N.E., Iso, H., Isola, G., Iwagami, M., Iwu, C.C.D., Iwu-Jaja, C.J., Vinothini, J., Jaafari, J., Jacob, L., Jacobsen, K.H., Jadidi-Niaragh, F., Jahankhani, K., Jahanmehr, N., Jahrami, H., Jain, A., Jain, N., Jairoun, A.A., Jaiswal, A., Jakovljevic, M., Yengejeh, R.J., Jamora, R.D.G., Jatau, A.I., Javadov, S., Javaheri, T., Jayaram, S., Jeganathan, J., Jeswani, B.M., Jiang, H., Johnson, C.O., Jokar, M., Jomehzadeh, N., Jonas, J.B., Joo, T., Joseph, A., Joseph, N., Joshi, V., Joshua, C.E., Jozwiak, J.J., Jürisson, M., Kaambwa, B., Kabir, A., Kabir, Z., Kadashetti, V., Kahn, E.M., Kalani, R., Kaliyadan, F., Kalra, S., Kamath, R., Kanagasabai, T., Kanchan, T., Kandel, H., Kanmiki, E.W., Kanmodi, K.K., Kansal, S.K., Kapner, D.J., Kapoor, N., Karagiannidis, E., Karajizadeh, M., Karakasis, P., Karanth, S.D., Karaye, I.M., Karch, A., Karim, A., Karimi, H., Karmakar, S., Kashoo, F.Z., Kasraci, H., Kassahun, W.D., Kassebaum, N.J., Kassel, M.B., Katikireddi, S.V., Kauppila, J.H., Kawakami, N., Kaydi, N., Kayode, G.A., Kazemi, F., Keiyoro, P.N., Kemmer, L., Kempen, J.H., Kerr, J.A., Kesse-Guyot, E., Khader, Y.S., Khafaie, M.A., Khajuria, H., Khalaji, A., Khalil, M., Khalilian, A., Khamesipour, F., Khan, A., Khan, M.N., Khan, M., Khan, M.J., Khan, M.A.B., Khanmohammadi, S., Khatab, K., Khatatbeh, H., Khatatbeh, M.M., Khatib, M.N., Khavandegar, A., Kashani, H.R.K., Khidri, F.F., Khodadoust, E., Khormali, M., Khorrami, Z., Khosla, A.A., Khosrowjerdi, M., Khreis, H., Khusun, H., Kifle, Z.D., Kim, K., Kim, M.S., Kim, Y.J., Kimokoti, R.W., Kisa, A., Kisa, S., Knibbs, L.D., Knudsen, A.K.S., Koh, D.S.Q., Kolahi, A.A., Kompani, F., Kong, J., Koren, G., Korja, M., Korshunov, V.A., Korzh, O., Kosen, S., Kothari, N., Koul, P.A., Laxminarayana, S.L.K., Krishan, K., Krishnamoorthy, V., Krishnamoorthy, Y., Krishnan, B., Krohn, K.J., Defo, B.K., Bicer, B.K., Kuddus, M.A., Kuddus, M., Kugbey, N., Kuitunen, I., Kulimbet, M., Kulkarni, V., Kumar, A., Kumar, N., Kumar, V., Kundu, S., Kurmi, O.P., Kusnali, A., Kusuma, D., Kutluk, T., Vecchia, C. La, Ladan, M.A., Laflamme, L., Lahariya, C., Lai, D.T.C., Lal, D.K., Lallukka, T., Lám, J., Lan, Q., Lan, T., Landires, I., Lanfranchi, F., Langguth, B., Lansingh, V.C., Laplante-Lévesque, A., Larijani, B., Larsson, A.O., Lasrado, S., Lauriola, P., Le, H.H., Le, L.K.D., Le, N.H.H., Le, T.T.T., Leasher, J.L., Ledda, C., Lee, M., Lee, P.H., Lee, S.W., Lee, S.W.H., Lee, Y.H., LeGrand, K.E., Leigh, J., Leong, E., Lerango, T.L., Lescinsky, H., Leung, J., Li, M.C., Li, W.Z., Li, W., Li, Y., Li, Z., Ligade, V.S., Lim, L.L., Lim, S.S., Lin, R.T., Lin, S., Liu, C., Liu, G., Liu, Jinli, Liu, Jue, Liu, R.T., Liu, S., Liu, W., Liu, Xiaofeng, Liu, Xuefeng, Livingstone, K.M., Llanaj, E., Lohiya, A., López-Bueno, R., Lopukhov, P.D., Lorkowski, S., Lotufo, P.A., Lozano, R., Lubinda, J., Lucchetti, G., Luo, L., Lv,

H., Amin, H.I.M., Ma, Z.F., Maass, K.L., Mabrok, M., Machairas, N., Machoy, M., Mafhoumi, A., El Razek, M.M.A., Maghazachi, A.A., Prasad, D.R.M., Maharaj, S.B., Mahmoud, M.A., Mahmoudi, E., Majeed, A., Makram, O.M., Makris, K.C., Malasala, S., Maled, V., Malhotra, K., Malik, A.A., Malik, I., Malinga, L.A., Malta, D.C., Mamun, A.A., Manda, A.L., Manla, Y., Mansour, A., Mansouri, B., Mansouri, P., Mansourian, M., Mansournia, M.A., Mantovani, L.G., Manu, E., Marateb, H.R., Maravilla, J.C., Marsh, E., Martinez, G., Martinez-Piedra, R., Martini, S., Martins-Melo, F.R., Martorell, M., Marx, W., Maryam, S., Mathangasinghe, Y., Mathioudakis, A.G., Matozinhos, F.P., Mattumpuram, J., Maugeri, A., Maulik, P.K., Mayeli, M., Mazidi, M., Mazzotti, A., McGrath, J.J., McKee, M., McKowen, A.L.W., McLaughlin, S.A., McPhail, M.A., McPhail, S.M., Mechili, E.A., Mehmood, A., Mehmood, K., Mehrabani-Zeinabad, K., Nasab, E.M., Meier, T., Mejia-Rodriguez, F., Meto, T.M., Mekonnen, B.D., Menezes, R.G., Mengist, B., Mensah, G.A., Mensah, L.G., Mentis, A.F.A., Meo, S.A., Meretoja, A., Meretoja, T.J., Mersha, A.M., Mesfin, B.A., Mestrovic, T., Mettananda, K.C.D., Mettananda, S., Miazgowski, T., Micha, G., Michalek, I.M., de Sá, A.C.M.G.N., Miller, T.R., Mirarefin, M., Mirghafourvand, M., Mirica, A., Mirijello, A., Mirrakhimov, E.M., Mirshahi, A., Mirzaei, M., Mishra, A.K., Mishra, V., Mitchell, P.B., Mithra, P., Mittal, C., Moazen, B., Moberg, M.E., Mocciaro, G., Mohamadkhani, A., Mohamed, A.Z., Mohamed, A.I., Mohamed, J., Mohamed, M.F.H., Mohamed, N.S., Mohammadi, E., Mohammadi, S., Mohammadian-Hafshejani, A., Mohammadifard, N., Mohammed, H., Mohammed, M., Mohammed, Salahuddin, Mohammed, Shafiu, Mokdad, A.H., Monasta, L., Mondello, S., Moni, M.A., Ghalibaf, A.A.M., Moore, C.E., Moradi, M., Moradi, Y., Moraga, P., Morawska, L., Moreira, R.S., Morovatdar, N., Morrison, S.D., Morze, J., Heris, R.M., Mossialos, E., Motappa, R., Mougin, V., Mousavi, P., Msherghi, A., Mubarik, S., Muccioli, L., Mueller, U.O., Mulita, F., Mullany, E.C., Munjal, K., Murillo-Zamora, E., Murlimanju, B. V., Musina, A.M., Mustafa, G., Muthu, S., Muthupandian, S., Muthusamy, R., Muzaffar, M., Myung, W., Nafei, A., Nagarajan, A.J., Nagaraju, S.P., Nagel, G., Naghavi, M., Naghavi, P., Naik, G.R., Naik, G., Nainu, F., Nair, T.S., Najdaghi, S., Ansari, N.N., Nanavaty, D.P., Nangia, V., Swamy, S.N., Davani, D.N., Nascimento, B.R., Nascimento, G.G., Nashwan, A.J., Natto, Z.S., Nauman, J., Navaratna, S.N.K., Naveed, M., Nayak, B.P., Nayak, V.C., Ndejjo, R., Nduaguba, S.O., Negash, H., Negoi, I., Negoi, R.I., Nejadghaderi, S.A., Nejjari, C., Nematollahi, M.H., Nepal, S., Neupane, S., Ng, M., Nguefack-Tsague, G., Ngunjiri, J.W., Nguyen, D.H., Nguyen, N.N.Y., Nguyen, Phat Tuan, Nguyen, Phuong The, Nguyen, V.T., Minh, D.N.T., Niazi, R.K., Nicholson, S.I., Nie, J., Nikoobar, A., Nikpoor, A.R., Ningrum, D.N.A., Nnaji, C.A., Noman, E.A., Nomura, S., Noroozi, N., Norrving, B., Noubiap, J.J., Nri-Ezedi, C.A., Ntaios, G., Ntsekhe, M., Nunemo, M.H., Nurrika, D., Nutor, J.J., Oancea, B., O'Connell, E.M., Odetokun, I.A., O'Donnell, M.J., Oduro, M.S., Ogunfowokan, A.A., Ogunkoya, A., Oh, I.H., Okati-Aliabad, H., Okeke, S.R., Okekunle, A.P., Okonji, O.C., Olagunju, A.T., Olasupo, O.O., Olatubi, M.I., Oliveira, A.B., Oliveira, G.M.M., Olorukooba, A.A., Olufadewa, I.I., Olusanya, B.O., Olusanya, J.O., Oluwafemi, Y.D., Omar, H.A., Bali, A.O., Omer, G.L., Ong, K.L., Ong, S., Onwujekwe, O.E., Onyedibe, K.I., Oppong, A.F., Ordak, M., Orish, V.N., Ornello, R., Orpana, H.M., Ortiz, A., Ortiz-Prado, E., Osman, W.M.S., Ostroff, S.M., Osuagwu, U.L., Otoiu, A., Otstavnov, N., Otstavnov, S.S., Ouyahia, A., Owolabi, M.O., Oyeyemi, I.T., Oyeyemi, O.T., Padukudru, M.P.A., Pacheco-Barrios, K., Padron-Monedero, A., Padubidri, J.R., Pal, P.K., Palicz, T., Pan, F., Pan, H.F., Pana, A., Panda, S.K., Panda-Jonas, S., Pandey, A., Pandi-Perumal, S.R., Pangaribuan, H.U., Pantazopoulos, I., Stoian, A.M.P., Papadopoulou, P., Parent, M.C., Parija, P.P., Parikh, R.R., Park, Seoyeon, Park, Sungchul, Parsons, N., Pashaei, A., Pasovic, M., Passera, R., Patil, S., Patoulias, D., Patthipati, V.S., Paudel, U., Pawar, S., Toroudi, H.P., Peden, A.E., Pedersini, P., Peng, M., Pensato, U., Pepito, V.C.F., Peprah, E.K., Peprah, P., Peres, M.F.P., Perianayagam, A., Perico, N., Perna, S., Pesudovs, K., Petcu, I.R., Petermann-Rocha, F.E., Pham, H.T., Philip, A.K., Phillips, M.R., Pickering, B. V., Pierannunzio, D., Pigeolet, M., Pigott, D.M., Piracha, Z.Z., Piradov, M.A., Pisoni, E., Piyasena, M.P., Plass, D., Plotnikov, E., Poddighe, D., Polkinghorne, K.R., Poluru, R., Pond, C.D., Popovic, D.S., Porru, F., Postma, M.J., Poudel, G.R., Pour-Rashidi, A., Pourshams, A., Pourtaheri, N., Prabhu, D., Prada, S.I., Pradhan, J., Pradhan, P.M.S., Prasad, M., Prates, E.J.S., Purnobasuki, H., Purohit, B.M., Puvvula,

J., Qasim, N.H., Qattea, I., Qazi, A.S., Qian, G., Qiu, S., Rad, M.R., Radfar, A., Radhakrishnan, R.A., Radhakrishnan, V., Shahraki, H.R., Rafferty, Q., Rafiei, A., Raggi, A., Raghav, P.R., Raheem, N., Rahim, F., Rahim, M.J., Rahimifard, M., Rahimi-Movaghar, V., Rahman, M.O., Rahman, M.A., Rahmani, A.M., Rahmani, B., Rahmanian, M., Rahmanian, N., Rahmanian, V., Rahmati, M., Rahmawaty, S., Raimondo, D., Rajaa, S., Rajendran, V., Rajput, P., Ramadan, M.M., Ramasamy, S.K., Ramasubramani, P., Ramazanu, S., Ramteke, P.W., Rana, J., Rana, K., Ranabhat, C.L., Rane, A., Rani, U., Ranta, A., Rao, C.R., Rao, M., Rao, P.C., Rao, S.J., Rasella, D., Rashedi, S., Rashedi, V., Rashidi, M., Rashidi, M.M., Rasouli-Saravani, A., Ratan, Z.A., Babu, G.R., Rauniyar, S.K., Rautalin, I., Rawaf, D.L., Rawaf, S., Rawassizadeh, R., Razo, C., Reda, Z.F.F., Reddy, M.M.R.K., Redwan, E.M.M., Reifels, L., Reitsma, M.B., Remuzzi, G., Reshmi, B., Resnikoff, S., Restaino, S., Reyes, L.F., Rezaei, M., Rezaei, Nazila, Rezaei, Negar, Rezaeian, M., Rhee, T.G., Riaz, M.A., Ribeiro, A.L.P., Rickard, J., Robinson-Oden, H.E., Rodrigues, C.F., Rodrigues, M., Rodriguez, J.A.B., Roever, L., Romadlon, D.S., Ronfani, L., Rosauer, J.J., Roshandel, G., Rostamian, M., Rotimi, K., Rout, H.S., Roy, B., Roy, N., Rubagotti, E., de Andrade Ruela, G., Rumisha, S.F., Runghien, T., Russo, M., Ruzzante, S.W., Chandan, S.N., Saad, A.M.A., Saber, K., Saber-Ayad, M.M., Sabour, S., Sacco, S., Sachdev, P.S., Sachdeva, R., Saddik, B., Saddler, A., Sadee, B.A., Sadeghi, E., Sadeghi, M., Majd, E.S., Saeb, M.R., Saeed, U., Safari, M., Safi, S., Safi, S.Z., Sagar, R., Sagoe, D., Sharif-Askari, F.S., Sharif-Askari, N.S., Sahebkar, A., Sahoo, S.S., Sahu, M., Saif, Z., Sajid, M.R., Sakshaug, J.W., Salam, N., Salamati, P., Salami, A.A., Salaroli, L.B., Salehi, L., Salehi, S., Salem, M.R., Salem, M.Z.Y., Salihu, D., Salimi, S., Salum, G.A., Kafil, H.S., Samadzadeh, S., Samodra, Y.L., Samuel, V.P., Samy, A.M., Sanabria, J., Sanjeev, R.K., Sanna, F., Santomauro, D.F., Santric-Milicevic, M.M., Sarasmita, M.A., Saraswathy, S.Y.I., Saravanan, A., Saravi, B., Sarikhani, Y., Sarmiento-Suárez, R., Sarode, G.S., Sarode, S.C., Sartorius, B., Sarveazad, A., Sathian, B., Sattin, D., Sawhney, M., Saya, G.K., Sayeed, A., Sayeed, M.A., Sayyah, M., Schinckus, C., Schmidt, M.I., Schuermans, A., Schumacher, A.E., Schutte, A.E., Schwarzinger, M., Schwebel, D.C., Schwendicke, F., Selvaraj, S., Semreen, M.H., Senthilkumaran, S., Serban, D., Serre, M.L., Sethi, Y., Shafie, M., Shah, H., Shah, N.S., Shah, P.A., Shah, S.M., Shahbandi, A., Shaheen, A.A., Shahid, S., Shahid, W., Shahsavari, H.R., Shahwan, M.J., Shaikh, M.A., Shaikh, S.Z., Shalash, A.S., Sham, S., Shamim, M.A., Shams-Beyranvand, M., Shamshirgaran, M.A., Shamsi, M.A., Shanawaz, M., Shankar, A., Sharfaei, S., Sharifan, A., Sharifi-Rad, J., Sharma, M., Sharma, U., Sharma, V., Shastry, R.P., Shavandi, A., Shehabeldine, A.M.E., Shehzadi, S., Sheikh, A., Shen, J., Shetty, A., Shetty, B.S.K., Shetty, P.H., Shiani, A., Shiferaw, D., Shigematsu, M., Shin, M.J., Shiri, R., Shittu, A., Shiue, I., Shivakumar, K.M., Shivarov, V., Shool, S., Shorofi, S.A., Shrestha, R., Shrestha, S., Shuja, K.H., Shuval, K., Si, Y., Siddig, E.E., Silva, D.A.S., Silva, L.M.L.R., Silva, S., Silva, T.P.R., Simpson, C.R., Singh, A., Singh, B.B., Singh, B., Singh, G., Singh, H., Singh, J.A., Singh, M., Singh, N.P., Singh, P., Singh, S., Sinto, R., Sivakumar, S., Siwal, S.S., Skhvitaridze, N., Skou, S.T., Sleet, D.A., Sobia, F., Soboka, M., Socea, B., Solaimanian, S., Solanki, R., Solanki, S., Soliman, S.S.M., Somayaji, R., Song, Y., Sorensen, R.J.D., Soriano, J.B., Soyiri, I.N., Spartalis, M., Spearman, S., Spencer, C.N., Sreeramareddy, C.T., Stachteas, P., Stafford, L.K., Stanaway, J.D., Stanikzai, M.H., Stein, C., Stein, D.J., Steinbeis, F., Steiner, C., Steinke, S., Steiropoulos, P., Stockfelt, L., Stokes, M.A., Straif, K., Stranges, S., Subedi, N., Subramaniyan, V., Suleman, M., Abdulkader, R.S., Sundström, J., Sunkersing, D., Sunnerhagen, K.S., Suresh, V., Swain, C.K., Szarpak, L., Szeto, M.D., Damavandi, P.T., Tabarés-Seisdedos, R., Tabatabaei, S.M., Malazy, O.T., Tabatabaeizadeh, S.A., Tabatabai, S., Tabche, C., Tabish, M., Tadakamadla, S.K., Abkenar, Y.T., Soodejani, M.T., Taherkhani, A., Taiba, J., Takahashi, K., Talaat, I.M., Tamuzi, J.L., Tan, K.K., Tang, H., Tat, N.Y., Taveira, N., Tefera, Y.M., Tehrani-Banihashemi, A., Temesgen, W.A., Temsah, M.H., Teramoto, M., Terefa, D.R., Teye-Kwadjo, E., Thakur, R., Thangaraju, P., Thankappan, K.R., Thapar, R., Thayakaran, R., Thirunavukkarasu, S., Thomas, N., Thomas, N.K., Tian, J., Tichopad, A., Ticoalu, J.H.V., Tiruye, T.Y., Tobe-Gai, R., Tolani, M.A., Tolossa, T., Tonelli, M., Topor-Madry, R., Topouzis, F., Touvier, M., Tovani-Palone, M.R., Trabelsi, K., Tran, J.T., Tran, M.T.N., Tran, N.M., Trico, D., Trihandini, I., Troeger, C.E., Tromans, S.J., Truyen, T.T.T.T., Tsatsakis, A., Tsermpini, E.E.,

Tumurkhuu, M., Udoakang, A.J., Udoh, A., Ullah, A., Ullah, Saeed, Ullah, Sana, Umair, M., Umakanthan, S., Unim, B., Unnikrishnan, B., Upadhyay, E., Urso, D., Usman, J.S., Vaithinathan, A.G., Vakili, O., Valenti, M., Valizadeh, R., Van den Eynde, J., van Donkelaar, A., Varga, O., Vart, P., Varthya, S.B., Vasankari, T.J., Vasic, M., Vaziri, S., Venketasubramanian, N., Verghese, N.A., Verma, M., Veroux, M., Verras, G.I., Vervoort, D., Villafañe, J.H., Villalobos-Daniel, V.E., Villani, L., Villanueva, G.I., Vinayak, M., Violante, F.S., Vlassov, V., Vo, B., Vollset, S.E., Volovat, S.R., Vos, T., Vujcic, I.S., Waheed, Y., Wang, C., Wang, F., Wang, S., Wang, Y., Wang, Y.P., Wanjau, M.N., Waqas, M., Ward, P., Waris, A., Wassie, E.G., Weerakoon, K.G., Weintraub, R.G., Weiss, D.J., Weiss, E.J., Weldetinsaa, H.L.L., Wells, K.M., Wen, Y.F., Wiangkham, T., Wickramasinghe, N.D., Wilkerson, C., Willeit, P., Wilson, S., Wong, Y.J., Wongsin, U., Wozniak, S., Wu, C., Wu, D., Wu, F., Wu, Z., Xia, J., Xiao, H., Xu, S., Xu, X., Xu, Y.Y., Yadav, M.K., Yaghoubi, S., Yamagishi, K., Yang, L., Yano, Y., Yaribeygi, H., Yasufuku, Y., Ye, P., Yesodharan, R., Yesuf, S.A., Yezli, S., Yi, S., Yiğit, A., Yigzaw, Z.A., Yin, D., Yip, P., Yismaw, M.B., Yon, D.K., Yonemoto, N., You, Y., Younis, M.Z., Yousefi, Z., Yu, C., Yu, Y., Zadey, S., Zadnik, V., Zakham, F., Zaki, N., Zakzuk, J., Zamagni, G., Zaman, S. Bin, Zandieh, G.G.Z., Zanghì, A., Zar, H.J., Zare, I., Zarimeidani, F., Zastrozhin, M.S., Zeng, Y., Zhai, C., Zhang, A.L., Zhang, H., Zhang, L., Zhang, M., Zhang, Y., Zhang, Z., Zhang, Z.J., Zhao, H., Zhao, J.T., Zhao, X.J.G., Zhao, Yang, Zhao, Yong, Zhong, C., Zhou, Jingjing, Zhou, Juexiao, Zhou, S., Zhu, B., Zhu, L., Zhu, Z., Ziaeian, B., Ziafati, M., Zielińska, M., Zimsen, S.R.M., Zoghi, G., Zoller, T., Zumla, A., Zyoud, Saed H., Zyoud, Samer H., Murray, C.J.L., Gakidou, E., 2024. Global burden and strength of evidence for 88 risk factors in 204 countries and 811 subnational locations, 1990–2021: a systematic analysis for the Global Burden of Disease Study 2021. *Lancet* 403, 2162–2203. https://doi.org/10.1016/S0140-6736(24)00933-4

Burstyn, I., Boffetta, P., Heederik, D., Partanen, T., Kromhout, H., Svane, O., Langård, S., Frentzel-Beyme, R., Kauppinen, T., Stücker, I., Shaham, J., Ahrens, W., Cenée, S., Ferro, G., Heikkilä, P., Hooiveld, M., Johansen, C., Randem, B.G., Schill, W., 2003. Mortality from obstructive lung diseases and exposure to polycyclic aromatic hydrocarbons among asphalt workers. *Am. J. Epidemiol.* 158, 468–78. https://doi.org/10.1093/aje/kwg180

Cho, Y.S., Moon, H.-B., 2010. The role of oxidative stress in the pathogenesis of asthma. *Allergy. Asthma Immunol. Res.* 2, 183–7. https://doi.org/10.4168/aair.2010.2.3.183

Das, D.N., Ravi, N., 2022. Influences of polycyclic aromatic hydrocarbon on the epigenome toxicity and its applicability in human health risk assessment. *Environ. Res.* 213, 113677. https://doi.org/10.1016/j.envres.2022.113677

Factor, P., Akhmedov, A.T., McDonald, J.D., Qu, A., Wu, J., Jiang, H., Dasgupta, T., Panettieri, R.A., Perera, F., Miller, R.L., 2011. Polycyclic aromatic hydrocarbons impair function of β2-adrenergic receptors in airway epithelial and smooth muscle cells. *Am. J. Respir. Cell Mol. Biol.* 45, 1045–9. https://doi.org/10.1165/rcmb.2010-0499OC

Fan, L., Li, W., Ma, J., Cheng, M., Xie, L., Ye, Z., Xie, Y., Wang, B., Yu, L., Zhou, Y., Chen, W., 2022. Benzo(a)pyrene induces airway epithelial injury through Wnt5a-mediated non-canonical Wnt-YAP/TAZ signaling. *Sci. Total Environ.* 815, 151965. https://doi.org/10.1016/j.scitotenv.2021.151965

Fueldner, C., Kohlschmidt, J., Riemschneider, S., Schulze, F., Zoldan, K., Esser, C., Hauschildt, S., Lehmann, J., 2018. Benzo(a)pyrene attenuates the pattern-recognition-receptor induced proinflammatory phenotype of murine macrophages by inducing IL-10 expression in an aryl hydrocarbon receptor-dependent manner. *Toxicology* 409, 80–90. https://doi.org/10.1016/j.tox.2018.07.011

Guarnieri, M., Balmes, J.R., 2014. Outdoor air pollution and asthma. *Lancet* 383, 1581–1592. https://doi.org/10.1016/S0140-6736(14)60617-6

Han, Y.-Y., Rosser, F., Forno, E., Celedón, J.C., 2018. Exposure to polycyclic aromatic hydrocarbons, vitamin D, and lung function in children with asthma. *Pediatr. Pulmonol.* 53, 1362–1368. https://doi.org/10.1002/ppul.24084

Hara, A., Odajima, H., Matsuzaki, H., Fujimura, M., Toma, T., Wada, T., Ohkura, N., Zhao, J., Pham, K.O., Suzuki, K., Tsujiguchi, H., Takami, A., Hayakawa, K., Nakamura, H., 2022. Association between Cough and Ambient Polycyclic Aromatic Hydrocarbons in Patients with Chronic Cough: An Observational Study in Two Regions of Japan. *Appl. Sci.* 12, 12505. https://doi.org/10.3390/app122412505

Hara, A., Sato, T., Kress, S., Suzuki, K., Pham, K.-O., Tajima, A., Schikowski, T., Nakamura, H., 2024. Sex-specific associations between air pollutants and asthma prevalence in Japanese adults: a population-based study. *Int. J. Environ. Health Res.* 1–9. https://doi.org/10.1080/09603123.2024.2352597

Huang, X., Zhou, Y., Cui, X., Wu, X., Yuan, J., Xie, J., Chen, W., 2018. Urinary polycyclic aromatic hydrocarbon metabolites and adult asthma: A case-control study. *Sci. Rep.* 8, 7658. https://doi.org/10.1038/s41598-018-26021-3

Jackson, D.J., Heaney, L.G., Humbert, M., Kent, B.D., Shavit, A., Hiljemark, L., Olinger, L., Cohen, D., Menzies-Gow, A., Korn, S., SHAMAL Investigators, 2024a. Reduction of daily maintenance inhaled corticosteroids in patients with severe eosinophilic asthma treated with benralizumab (SHAMAL): a randomised, multicentre, open-label, phase 4 study. *Lancet* 403, 271–281. https://doi.org/10.1016/S0140-6736(23)02284-5

Jackson, David J, Wechsler, M.E., Jackson, Daniel J, Bernstein, D., Korn, S., Pfeffer, P.E., Chen, R., Saito, J., de Luíz Martinez, G., Dymek, L., Jacques, L., Bird, N., Schalkwijk, S., Smith, D., Howarth, P., Pavord, I.D., SWIFT-1 and SWIFT-2 Investigators, SWIFT-1 Investigators, SWIFT-2 Investigators, 2024b. Twice-Yearly Depemokimab in Severe Asthma with an Eosinophilic Phenotype. *N. Engl. J. Med.* 391, 2337–2349. https://doi.org/10.1056/NEJMoa2406673

Jedrychowski, W., Galas, A., Pac, A., Flak, E., Camman, D., Rauh, V., Perera, F., 2005. Prenatal ambient air exposure to polycyclic aromatic hydrocarbons and the occurrence of respiratory symptoms over the first year of life. *Eur. J. Epidemiol.* 20, 775–82. https://doi.org/10.1007/s10654-005-1048-1

Jiang, M., Zhao, H., 2024. Association of chronic cough with exposure to polycyclic aromatic hydrocarbons in the US population. *Heliyon* 10, e23413. https://doi.org/10.1016/j.heliyon.2023.e23413

Jung, K.H., Yan, B., Moors, K., Chillrud, S.N., Perzanowski, M.S., Whyatt, R.M., Hoepner, L., Goldstein, I., Zhang, B., Camann, D., Kinney, P.L., Perera, F.P., Miller, R.L., 2012. Repeated exposure to polycyclic aromatic hydrocarbons and asthma: effect of seroatopy. *Ann. allergy, asthma Immunol.* 109, 249–54. https://doi.org/10.1016/j.anai.2012.07.019

Karimi, P., Peters, K.O., Bidad, K., Strickland, P.T., 2015. Polycyclic aromatic hydrocarbons and childhood asthma. *Eur. J. Epidemiol.* 30, 91–101. https://doi.org/10.1007/s10654-015-9988-6

Kress, S., Hara, A., Wigmann, C., Sato, T., Suzuki, K., Pham, K.-O., Zhao, Q., Areal, A., Tajima, A., Schwender, H., Nakamura, H., Schikowski, T., 2022. The Role of Polygenic Susceptibility on Air Pollution-Associated Asthma between German and Japanese Elderly Women. *Int. J. Environ. Res. Public Health* 19, 9869. https://doi.org/10.3390/ijerph19169869

Låg, M., Øvrevik, J., Refsnes, M., Holme, J.A., 2020. Potential role of polycyclic aromatic hydrocarbons in air pollution-induced non-malignant respiratory diseases. *Respir. Res.* 21, 299. https://doi.org/10.1186/s12931-020-01563-1

Li, D., Wang, B., Feng, G., Xie, M., Wang, L., Gao, R., 2017. Effect of the GSTM1 genotype on the biomarkers of exposure to polycyclic aromatic hydrocarbons: Meta-analysis. *Int. J. Occup. Med. Environ. Health* 30, 177–201. https://doi.org/10.13075/ijomeh.1896.01054

Mayati, A., Le Ferrec, E., Holme, J.A., Fardel, O., Lagadic-Gossmann, D., Ovrevik, J., 2014. Calcium signaling and β2-adrenergic receptors regulate 1-nitropyrene induced CXCL8 responses in BEAS-2B cells. *Toxicol. Vitr.* 28, 1153–7. https://doi.org/10.1016/j.tiv.2014.05.012

Mayati, A., Le Ferrec, E., Lagadic-Gossmann, D., Fardel, O., 2012. Aryl hydrocarbon receptor-independent up-regulation of intracellular calcium concentration by environmental polycyclic aromatic hydrocarbons in human endothelial HMEC-1 cells. *Environ. Toxicol.* 27, 556–62. https://doi.org/10.1002/tox.20675

Miller, R.L., Garfinkel, R., Lendor, C., Hoepner, L., Li, Z., Romanoff, L., Sjodin, A., Needham, L., Perera, F.P., Whyatt, R.M., 2010. Polycyclic aromatic hydrocarbon metabolite levels and pediatric allergy and asthma in an inner-city cohort. *Pediatr. allergy Immunol.* 21, 260–7. https://doi.org/10.1111/j.1399-3038.2009.00980.x

Morice, A., Dicpinigaitis, P., McGarvey, L., Birring, S.S., 2021. Chronic cough: new insights and future prospects. *Eur. Respir. Rev.* 30, 210127. https://doi.org/10.1183/16000617.0127-2021

Nwaozuzu, Chinemerem C, Partick-Iwuanyanwu, K.C., Abah, S.O., 2021. Systematic Review of Exposure to Polycyclic Aromatic Hydrocarbons and Obstructive Lung Disease. *J. Heal. Pollut.* 11, 210903. https://doi.org/10.5696/2156-9614-11.31.210903

Padula, A.M., Balmes, J.R., Eisen, E.A., Mann, J., Noth, E.M., Lurmann, F.W., Pratt, B., Tager, I.B., Nadeau, K., Hammond, S.K., 2015. Ambient polycyclic aromatic hydrocarbons and pulmonary function in children. *J. Expo. Sci. Environ. Epidemiol.* 25, 295–302. https://doi.org/10.1038/jes.2014.42

Perera, F., Tang, W., Herbstman, J., Tang, D., Levin, L., Miller, R., Ho, S., 2009. Relation of DNA methylation of 5'-CpG island of ACSL3 to transplacental exposure to airborne polycyclic aromatic hydrocarbons and childhood asthma. *PLoS One* 4, e4488. https://doi.org/10.1371/journal.pone.0004488

Porsbjerg, C., Melén, E., Lehtimäki, L., Shaw, D., 2023. Asthma. *Lancet* 401, 858–873. https://doi.org/10.1016/S0140-6736(22)02125-0

Rajagopalan, S., Landrigan, P.J., 2021. Pollution and the Heart. *N. Engl. J. Med.* 385, 1881–1892. https://doi.org/10.1056/NEJMra2030281

Ramakrishnan, S., Russell, R.E.K., Mahmood, H.R., Krassowska, K., Melhorn, J., Mwasuku, C., Pavord, I.D., Bermejo-Sanchez, L., Howell, I., Mahdi, M., Peterson, S., Bengtsson, T., Bafadhel, M., 2024. Treating eosinophilic exacerbations of asthma and COPD with benralizumab (ABRA): a double-blind, double-dummy, active placebo-controlled randomised trial. *Lancet Respir. Med.* https://doi.org/10.1016/S2213-2600(24)00299-6

Sherris, A.R., Loftus, C.T., Szpiro, A.A., Dearborn, L.C., Hazlehurst, M.F., Carroll, K.N., Moore, P.E., Adgent, M.A., Barrett, E.S., Bush, N.R., Day, D.B., Kannan, K., LeWinn, K.Z., Nguyen, R.H.N., Ni, Y., Riederer, A.M., Robinson, M., Sathyanarayana, S., Zhao, Q., Karr, C.J., 2024. Prenatal polycyclic aromatic hydrocarbon exposure and asthma at age 8-9 years in a multi-site longitudinal study. *Environ. Heal.* 23, 26. https://doi.org/10.1186/s12940-024-01066-2

Smargiassi, A., Goldberg, M.S., Wheeler, A.J., Plante, C., Valois, M.-F., Mallach, G., Kauri, L.M., Shutt, R., Bartlett, S., Raphoz, M., Liu, L., 2014. Associations between personal exposure to air pollutants and lung function tests and cardiovascular indices among children with asthma living near an industrial complex and petroleum refineries. *Environ. Res.* 132, 38–45. https://doi.org/10.1016/j.envres.2014.03.030

Stading, R., Swanson, L., Xia, G., Jiang, W., Wang, L., Couroucli, X., Lingappan, K., Moorthy, B., 2024. Prenatal exposure to polycyclic aromatic hydrocarbons (PAHs) augments neonatal hyperoxic lung injury: Role of cytochrome P450 (CYP)1A1, 1A2, and 1B1. *Free Radic. Biol. Med.* 211, 35–46. https://doi.org/10.1016/j.freeradbiomed.2023.12.001

Vohra, K., Vodonos, A., Schwartz, J., Marais, E.A., Sulprizio, M.P., Mickley, L.J., 2021. Global mortality from outdoor fine particle pollution generated by fossil fuel combustion: Results from GEOS-Chem. *Environ. Res.* 195, 110754. https://doi.org/10.1016/j.envres.2021.110754

Wang, I.-J., Karmaus, W.J.J., Yang, C.-C., 2017. Polycyclic aromatic hydrocarbons exposure, oxidative stress, and asthma in children. *Int. Arch. Occup. Environ. Health* 90, 297–303. https://doi.org/10.1007/s00420-017-1198-y

Wang, Z., Ng, K., Warner, R.D., Stockmann, R., Fang, Z., 2022. Reduction strategies for polycyclic aromatic hydrocarbons in processed foods. *Compr. Rev. food Sci. food Saf.* 21, 1598–1626. https://doi.org/10.1111/1541-4337.12905

WHO Regional Office for Europe 2021. Human health effects of polycyclic aromatic hydrocarbons as ambient air pollutants: report of the Working Group on Polycyclic Aromatic Hydrocarbons of the Joint Task Force on the Health Aspects of Air Pollution. *World Health Organization*

Woodruff, T.J., 2024. Health Effects of Fossil Fuel-Derived Endocrine Disruptors. *N. Engl. J. Med.* 390, 922–933. https://doi.org/10.1056/NEJMra2300476

Xiao, Y., Zhang, L., Liu, H., Huang, W., 2023. Systemic inflammation mediates environmental polycyclic aromatic hydrocarbons to increase chronic obstructive pulmonary disease risk in United States adults: a cross-sectional NHANES study. *Front. public Heal.* 11, 1248812. https://doi.org/10.3389/fpubh.2023.1248812

Xu, Z., Lei, X., Chen, C., 2024. Antioxidant diet/lifestyle could mitigate the adverse impacts of urinary polycyclic aromatic hydrocarbons on lung function. *Environ. Res.* 246, 118099. https://doi.org/10.1016/j.envres.2024.118099

Yang, L., Wang, W.-C., Lung, S.-C.C., Sun, Z., Chen, C., Chen, J.-K., Zou, Q., Lin, Y.-H., Lin, C.-H., 2017. Polycyclic aromatic hydrocarbons are associated with increased risk of chronic obstructive pulmonary disease during haze events in China. *Sci. Total Environ.* 574, 1649–1658. https://doi.org/10.1016/j.scitotenv.2016.08.211

Zimmerman, E., Sturrock, A., Reilly, C.A., Burrell-Gerbers, K.L., Warren, K., Mir-Kasimov, M., Zhang, M.A., Pierce, M.S., Helms, M.N., Paine, R., 2024. Aryl Hydrocarbon Receptor Activation in Pulmonary Alveolar Epithelial Cells Limits Inflammation and Preserves Lung Epithelial Cell Integrity. *J. Immunol.* 213, 600–611. https://doi.org/10.4049/jimmunol.2300325

Analytical Methods

Oxidative Derivatives of Polycyclic Aromatic Hydrocarbons and Their Metabolites

Akira Toriba

Abstract Oxygenated PAHs (oxy-PAHs), such as hydroxylated PAHs (OHPAHs), PAH quinones (PAHQs) and carbonyl PAHs (oxo-PAHs), have been found in the atmosphere as well as in soot from automobile, soil, sediment, seawater and waste water. Gas chromatography with mass spectrometry (GC–MS) was preferred in the analysis of oxy-PAHs in the environmental samples and almost methods need acetyl derivatization to improve sensitivity. On the other hand, an advantage of LC-bases techniques is their good sensitivity without derivatization. The metabolites of PAHs and their derivatives in biological fluids have been investigated as potential biomarkers for assessing human exposure to them, and, particularly, urinary 1-hydroxypyrene and other OHPAHs have been used in various studies. Urinary OHPAHs have been frequently analyzed by HPLC with fluorescence detection and LC–MS/MS. The urinary metabolites of PAHQs (9,10-Phenanthrenequinone) and nitro-PAHs (1-nitropyrene) were identified by LC–MS/MS and suggested as a new type of biomarker.

Keywords Polycyclic aromatic hydrocarbons (PAHs) · Oxygenated PAH derivatives · Reactive oxygen species · Metabolite · Urinary biomarker

1 Introduction

Polycyclic aromatic hydrocarbons (PAHs), such as benzo[*a*]pyrene (BaP), are molecules that contain over two aromatic rings are an important group of carcinogenic/mutagenic pollutants in the environment (Lewtas 2007). Recently, oxygenated PAHs (oxy-PAHs), such as hydroxylated PAHs (OHPAHs), PAH quinones (PAHQs) and carbonyl PAHs (oxo-PAHs), and hydroxylated PAHQs and oxo-PAHs, have been found in the atmosphere as well as in soot from automobile, soil, sediment, seawater and waste water. Both OHPAHs and PAHQs showed endocrine disruption (Hayakawa et al. 2007, 2011; Motoyama et al. 2009) and overproduction of reactive

A. Toriba (✉)
Graduate School of Biomedical Sciences, Nagasaki University, Nagasaki, Japan
e-mail: toriba@nagasaki-u.ac.jp

K. Hayakawa and Y. Aoki (eds.), *A New Era of Polycyclic Aromatic Hydrocarbons*,
https://doi.org/10.1007/978-981-96-7695-8_10

oxygen species (ROS) (Bolton et al. 2000; Toriba et al. 2016), and these derivatives may induce similar or strong adverse health effects compared to parent PAHs (Lewtas 2007), indicating the necessity of environmental monitoring and to determine these exposure amounts to human. The structures of representative oxy-PAHs are shown in Fig. 1. On the other hand, oxy-PAHs such as OHPAHs and PAHQs are produced by the metabolism of PAHs in human and animal body and existed in biological fluids such as urine (Santos et al. 2019; Toriba and Hayakawa 2007). In particular, urinary 1-hydroxypyrene has been widely used as a biomarker to evaluate exposure to PAHs in epidemiological studies (Hansen et al. 2008).

According to different measures of carcinogenicity (genotoxicity, mutagenicity, tumor promotion or enzyme induction), the most hazardous PAHQs and oxo-PAHs

Fig. 1 Structures of representative oxygenated PAHs (Oxy-PAHs)

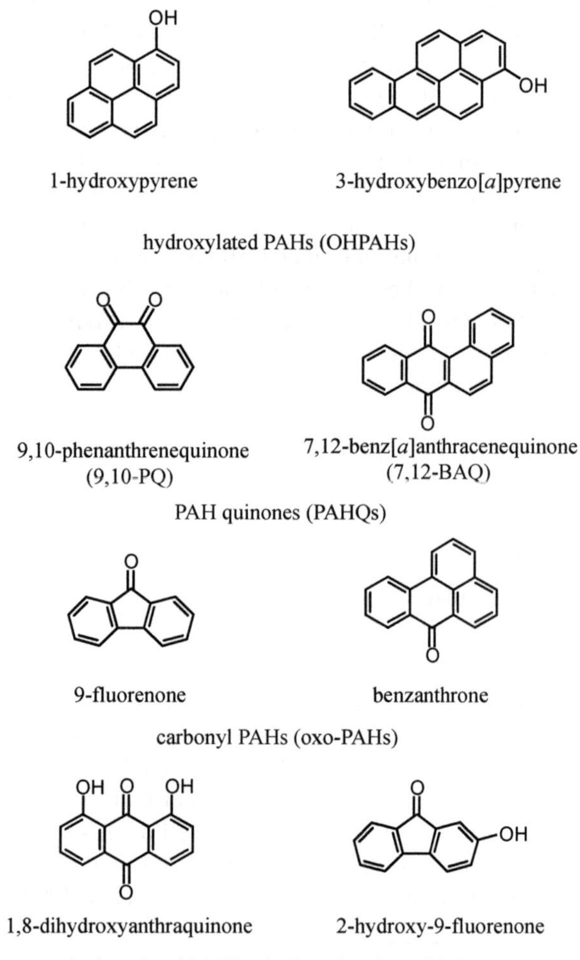

are 7,12-benz[*a*]anthracenequinone (7,12-BAQ), 5,12-naphthacenequinone (5,12-NapQ), 6 *H*-benzo[*cd*]pyren-6-one (BcdP-6-one) and 11 *H*-benzo[*b*]fluoren-11-one (BbF-11-one) (Clergé et al. 2019). PAHQs such as 9,10-phenanthrenequinone (9,10-PQ) can overproduce ROS through a redox cycle (Bolton et al. 2000). Enzymatic and non-enzymatic redox cycling produces their corresponding semiquinone radicals and subsequently generates ROS including superoxide, hydrogen peroxide and hydroxyl radical. The ROS production can cause oxidative stresses. *o*-PAHQs overproduced ROS and significantly decreased the cell viability, while *p*-PAHQs did not produce ROS and had little cytotoxicity (Motoyama et al. 2009). OHPAHs are known to get estrogenic and antiestrogenic activities especially for 4 rings-OHPAHs such as 3-, 4- and 10-hydroxybenz[*a*]anthracenes (3-, 4-, and 10-OHBaAs) and 2-hydroxychrysene (2-OHChr) (Hayakawa et al. 2007). Environmental monitoring data for oxy-PAHs are needed to evaluate their potential human health risk.

2 Analytical Methods for Oxy-PAHs in Environmental Samples

OHPAHs have been analyzed mainly by gas chromatography with mass spectrometry (GC–MS) with electron impact (EI) ionization after trimethylsilyl (TMS) derivatization using *N,O*-bis(trimethylsilyl)trifluoroacetamide (BSTFA) with 1% trimethylsilyl chloride (TMCS) (Wang et al. 2007; Cochran et al. 2012; Bandowe et al. 2014b). GC-EI-MS methods with TMS derivatization was applied to analyze OHPAHs in ambient PM samples (Wang et al. 2007; Bandowe et al. 2014b), PM samples emitted from combustion sources (Cochran et al. 2012), sea water (Wang et al. 2017), waste water (Pojana and Marcomini 2007), soil (Bandowe et al. 2011) and sediment (Wang et al. 2012, 2015). GC–MS with negative chemical ionization (NCI) mode has been also used for analyzing OHPAHs after pentafluorobenzoyl derivatization (Ma et al. 2016; Lin et al. 2015).

Liquid chromatography (LC) based methods have been widely developed for OHPAH analysis. An advantage of LC-bases techniques is their good sensitivity without derivatization. Through method development using liquid chromatography with mass spectrometry (LC–MS) or LC (UPLC) with tandem mass spectrometry (LC–MS/MS), appropriate ionization methods have been discussed. The atmospheric pressure chemical ionization (APCI) and atmospheric pressure photo ionization (APPI) modes gave high ionization efficiency for OHPAHs, whereas the electrospray ionization (ESI) mode was applicable to their detection (Grosse and Letzel, 2007). Several OHPAHs in wood smoke PM (Avagyan et al. 2015, 2016) and ambient PM samples (Honda et al. 2022) were quantified by LC or UPLC-MS/MS methods. In addition to mass spectrometric detection, HPLC-based techniques with electrochemical detection (ECD) (Galceran and Moyano 1995), fluorescence detection (FL) (Barrado et al. 2012; Kamiya et al. 2005) and chemiluminescence detection (Li et al.

2011) have also been applied to a few compounds in atmospheric particulates and cigarette smoke condensate.

To monitor atmospheric concentrations of PAHQs, GC methods were preferred in the analysis of environmental samples. PAHQs can be directly analyzed by GC-EI-MS, even though most GC-EI-MS methods need acetyl derivatization to improve sensitivity (Cho et al. 2004; Delgado-Saborit et al. 2013). The GC-EI-MS methods were applied to analyze PAHQs in ambient PM (Cho et al. 2004; Andreou and Rapsomanikis 2009; Delgado-Saborit et al. 2013; Layshock et al. 2010), soil (Bandowe et al. 2010, 2011, 2014a, b; Obrist et al. 2015) and sediment (Han et al. 2015; Wang et al. 2015). GC-NCI-MS and GC-APCI-MS provided a practical direct detection of PAHQs without derivatization (Albinet et al. 2006, 2014; Keyte et al. 2016; Lammel et al. 2020). A sensitive and selective method for simultaneously determining 37 PAHQs with GC–MS/MS has been developed for the analysis of PM samples (Toriba et al. 2016). Analyses of standard atmospheric samples with this method detected 33 PAHQs, 14 of which have not previously been reported in environmental samples (Fig. 2). This strongly suggests that little-known derivatives may be important contaminants in PM toxicity.

Analytical methods based on LC–MS or LC–MS/MS have typically been accomplished through PAHQ ionization in the APCI mode (Jakober et al. 2006; Mirivel et al. 2010; O'Connell et al. 2013; Lintelmann et al. 2006; Delhomme et al. 2008; Walgraeve et al. 2012). In HPLC methods using optical detections, PAHQs were converted to fluorescence derivatives (Kishikawa et al. 2004), and the ROS production ability of PAHQs was applied to their chemiluminescence detection (Ahmed et al. 2009). However, HPLC methods have only been applied to limited kinds of PAHQs.

Oxo-PAHs such as 9-fluorenone (9-Flo), benzanthrone and 6H-benzo[*c,d*]pyrene-6-one (6H-BcdPO) are other major monitored components and have been determined by GC-NCI-MS (Albinet et al. 2006, 2014; Wang et al. 2016; Liang et al. 2023) and GC-APCI-MS (Lammel et al. 2020). The most abundant oxidized derivatives in emissions of oxo-PAHs were 9-Flo and 6H-BcdPO from coal combustion (Wang et al. 2016), 9-Flo emitted from ships (Liang et al. 2023) and 9-Flo and benzanthrone in the aerosol in central Europe (Lammel et al. 2020). Oxo-PAHs in soil (Bandowe et al. 2010, 2011, 2014a, b; Wei et al. 2024) and sediment samples (Wang et al. 2015; Li et al. 2025) were analyzed by GC-EI-MS and the composition of oxo-PAHs was dominated by 9-Flo and 1-indanone.

Hydroxylated PAHQs (2-hydroxy-1,4-naphthoquinone, 1,8-dihydroxyanthraquinone) and hydroxylated oxo-PAHs (2-hydroxy-9-fluorenone) were also detected in wood smoke PM (Cochran et al. 2012) and ambient PM samples (Wang et al. 2007). Their hydroxy group(s) were converted into their TMS ethers and then analyzed by GC–MS.

In addition to analytical methods dedicated to each PAH derivative, the comprehensive (non-targeted) measurement of PAHs and their derivatives should be helpful for characterizing air pollution and better understanding the potential health risks. Sixteen PAHs and 10 oxy-PAHs in ambient PM in central Beijing were quantified using GC with quadrupole time-of-flight MS (GC-qTOFMS), which achieves

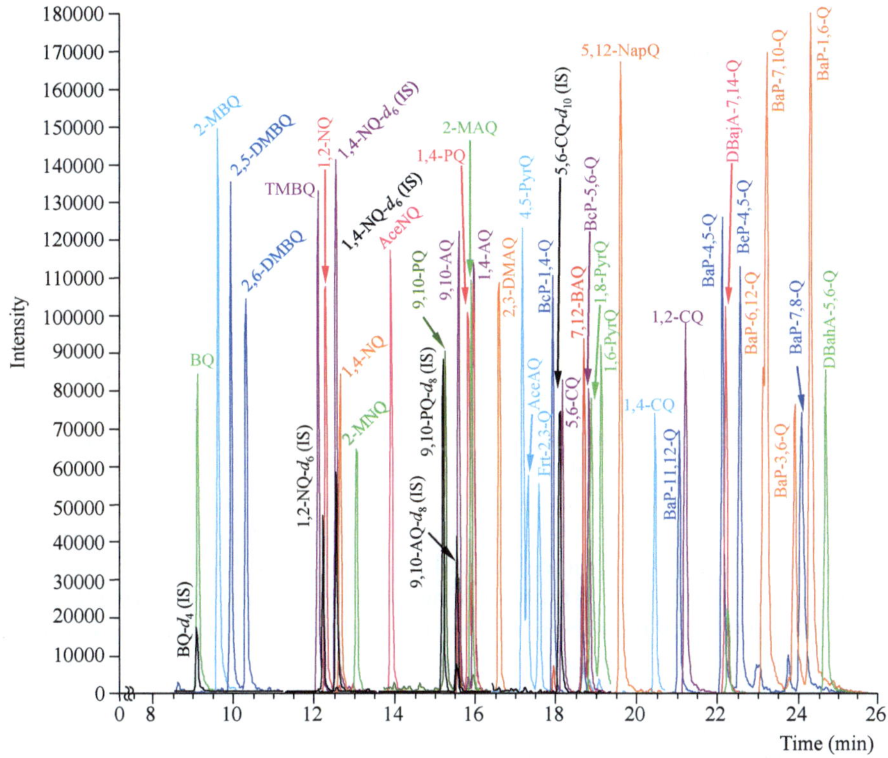

Fig. 2 Representative SRM chromatogram of TMS-derivatized PAHQ standards by the GC–MS/MS method (Toriba et al. 2016). The abbreviations of each PAHQ are as follows; 1,4-benzoquinone (BQ), 2-methyl-1,4-benzoquinone (2-MBQ), 2,5-dimethylbenzoquinone (2,5-DMBQ), 2,6-dimethylbenzoquinone (2,6-DMBQ), tetramethylbenzoquinone (TMBQ), 1,2-naphthoquinone (1,2-NQ), 1,4-naphtoquinone (1,4-NQ), 2-methyl-1,4-naphthoquinone (2-MNQ), acenaphthoquinone (AceNQ), 9,10-phenanthrenequinone (9,10-PQ), 9,10-anthraquinone (9,10-AQ), 1,4-anthraquinone (1,4-AQ), 1,4-phenanthrenequinone (1,4-PQ), aceanthraquinone (AceAQ), 5,12-naphthacenequinone (5,12-NapQ), fluoranthene-2,3-quinone (Frt-2,3-Q), 1,6-pyrenequinone (1,6-PyrQ), 1,8-pyrenequinone (1,8-PyrQ), 4,5-pyrenequinone (4,5-PyrQ), 7,12-benzanthraquinone (7,12-BAQ), benzo[*c*]phenanthrene-1,4-quinone (BcP-1,4-Q), benzo[*c*]phenanthrene-5,6-quinone (BcP-5,6-Q), 1,2-chrysenequinone (1,2-CQ), 1,4-chrysenequinone (1,4-CQ), 5,6-chrysenequinone (5,6-CQ), benzo[*e*]pyrene-4,5-quinone (BeP-4,5-Q), benzo[*a*]pyrene-1,6-quinone (BaP-1,6-Q), benzo[*a*]pyrene-3,6-quinone (BaP-3,6-Q), benzo[*a*]pyrene-4,5-quinone (BaP-4,5-Q), benzo[*a*]pyrene-6,12-quinone (BaP-6,12-Q), benzo[*a*]pyrene-7,8-quinone (BaP-7,8-Q), benzo[*a*]pyrene-7,10-quinone (BaP-7,10-Q) and benzo[*a*]pyrene-11,12-quinone (BaP-11,12-Q), dibenzo[*a,h*]anthracene-5,6-quinone (DBahA-5,6-Q) and dibenzo[*a,j*]anthracene-7,14-quinone (DBajA-7,14-Q)

high resolving power and mass accuracy (Elzein et al. 2019). Two-dimensional gas chromatography (GC × GC) is available to separate and identify a variety of compounds compared to normal GC. Comprehensive GC × GC with TOFMS (GC × GC-TOFMS) was recently employed to determine PAHs and oxy-PAHs in air samples (Geldenhuys et al. 2015).

3 Metabolites of PAHs and Their Derivatives

In order to evaluate the exposure of humans to PAHs and nitro-derivatives of PAHs (NPAHs) via multiple routes, PAH or NPAH concentrations in the atmosphere and in inhaled particulate matter (PM) of individuals collected using a personal air sampler have been measured in many areas. However, there are a wide variety of PAH and NPAH sources and individual differences exist in lifestyle factors such as smoking and dietary habits and in the functions of the human body, such as absorption, distribution, and excretion. Owing to this, a more precise way of evaluating human exposure was needed and, to that end, biological markers or biomarkers have been developed to evaluate the exposure level on the basis of measurements of metabolites in the tissues and body fluids. Such biomarkers can be broadly defined as measurable changes in a biological system that are caused by exposure to an exogenous chemical (Metcalf and Orloff 2004).

Recently, a variety of PAH and NPAH metabolites in biological fluids have been investigated as potential biomarkers of exposure to these compounds. Since humans are usually exposed to quite complex mixtures of PAHs and NPAHs, it is necessary to choose suitable metabolites, besides a number of relevant parent compounds. In biological fluids, urine is an excellent candidate sample for determining the metabolites owing to the non-invasiveness and convenience of collecting samples.

Incorporated PAHs into the human body are converted to phenols, dihydrodiols, and tetrols by cytochrome P450-dependent monooxygenases. In addition, one part of the dihydrodiols is converted into dihydrodiol epoxides by additional oxidation. As an example, BaP forms BaP-7,8-diol-9,10-epoxide, which can covalently bind to nucleophilic sites (guanine and adenine bases) in DNA to form BaP-DNA adducts and induces a mutation in the DNA. These types of metabolites are considered as carcinogenic intermediates of BaP and general PAHs. On the other hand, the largest fraction of the metabolites, the phenols and dihydrodiols, are excreted in urine or bile in the form of glucuronide and sulfate conjugates (Jongeneelen 1997).

To date, measuring PAH-DNA adducts in white blood cells, albumin adduct in plasma and measuring PAH metabolites (monohydroxylated PAHs) in urine have been presented as methods that could provide biomarkers for evaluating PAH exposure (Jongeneelen 1997; Jacob and Seidel 2002). Owing to the simplicity and convenience of collecting samples and of the experimental procedures, urinary metabolites have been widely used as biomarkers and 1-hydroxypyrene (1-OHP) is one of the most common markers.

An HPLC-FL method for determining 1-OHP in human urine has been established by Jongeneelen et al. (1987) and the method became the most commonly used method. This requires an enzymatic hydrolysis step to deconjugate the glucuronide and sulfate of 1-OHP. For the quantitative determination, a standard addition method has generally been used. However, more than two measurements for each sample were needed and it took time to conduct the analysis. To avoid the inconvenience, an HPLC method that uses deuterated 1-OHP as an internal standard has been developed (Chetiyanukornkul et al. 2002). To measure multiple PAH metabolites in human urine simultaneously, many analytical procedures using HPLC with fluorescence detection (Toriba et al. 2003; Chetiyanukornkul et al. 2006; Barbeau et al. 2011), GC-(HR)MS(/MS) with silylation (Woudneh et al. 2016; Li et al. 2014; Bao et al. 2025) or LC(UPLC)-MS(/MS) with (Jacob et al. 2007; Luo et al. 2015) or without derivatization (Onyemauwa et al. 2009; Lankova et al. 2016; Huang et al. 2022) have been developed to determine urinary PAH metabolites having 4–5 rings including 1-OHP which are important for PAH exposure assessment.

Urinary metabolites of BaP have been of particular interest as a biomarker for PAH exposure because a carcinogenic PAH is preferable to non-carcinogenic pyrene. However, environmental exposure to BaP is lower than pyrene and there are various metabolic and excretion pathways. Therefore, the levels of urinary 3-OHBaP were expected to be of the order of ng/L and there is greater difficulty in analyzing its metabolites. 3-hydroxybenzo[*a*]pyrene (3-OHBaP), a major metabolite of BaP, was identified and quantified at low ng/g creatinine concentration in human urine (Förster et al. 2008). 3-OHBaP in human urine samples has been successfully determined by HPLC with fluorescence detection (Barbeau et al. 2011) and LC–MS/MS with derivatization (Luo et al. 2015; Sarkar et al. 2009; Yao et al. 2014).

The above mentioned GC–MS and some LC–MS/MS methods need enzymatic hydrolysis of its conjugates to 1-OHP for generally 2–16 h (overnight) (Chetiyanukornkul et al. 2006; Jongeneelen et al. 1987). The hydrolysis step is time-consuming, and its possible incomplete hydrolysis should be taken into consideration. Furthermore, GC–MS methods and some LC–MS/MS methods require a derivatization step in addition to the hydrolysis. Thus, direct measurement of 1-hydroxypyrene-glucuronide (1-OHP-G) would be ideal for the analysis of urine samples. 1-OHP-G levels account for more than 80% of total pyrene metabolites in human urine (Singh et al. 1995). Since 1-OHP-G yields threefold to fivefold higher fluorescence than unconjugated 1-OHP, an HPLC-FL method to detect 1-OHP-G has been developed by using the fluorescence advantage (Singh et al. 1995). An LC–MS/MS using a specific fragmentation of the glucuronide sugar moiety on ESI mode has been developed for determining 1-OHP-G. (Kakimoto et al. 2008). The proposed LC–MS/MS method avoids the problems associated with interfering substances, hydrolysis and derivatization (Fig. 3).

There are a few studies on the measurement of PAHQ metabolite(s) in biological samples. One PAHQ, 9,10-Phenanthrenequinone (9,10-PQ), has been identified in diesel exhaust particulates and PM (Cho et al. 2004; Jakober et al. 2007; Eiguren-Fernandez et al. 2008; Andreou and Rapsomanikis 2009). The metabolic pathway of 9,10-PQ plays an important role in the production of ROS and 9,10-PQ is metabolized

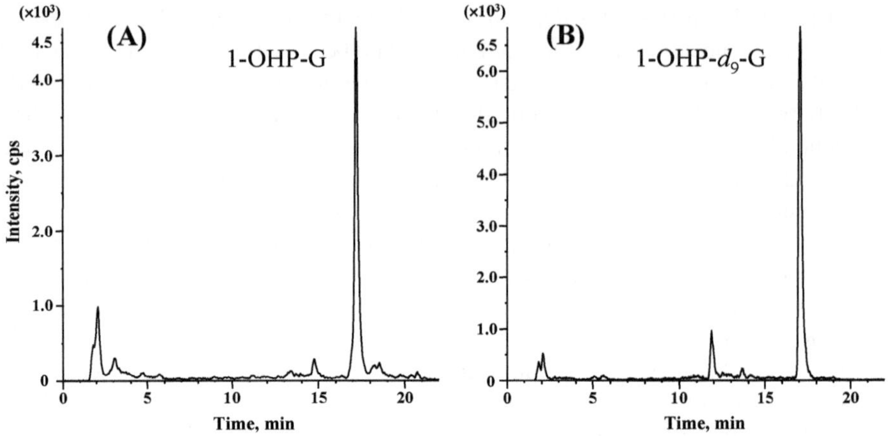

Fig. 3 Representative SRM chromatograms (transition *m/z* 393.1 → 217.1 for 1-OHP-G and *m/z* 402.1 → 226.1 for the internal standard (1-OHP-d_9-G)) of a non-smoker urine sample (A: 1-OHP-G and B: internal standard) (Kakimoto et al. 2008)

to 9, 10-dihydroxyphenanthrene (9, 10-PQH_2) by NAD(P)H quinone oxidoreductase (NQO1) and aldo–keto reductase (AKR) through two-electron reduction pathways. In vitro, once 9, 10-PQH_2 is formed by the reduction, its conjugation by UGTs could terminate redox cycling of the quinone and the monoglucuronide of 9, 10-PQH_2 (9, 10-PQHG) is transported into extracellular space (Taguchi et al. 2008). Recently, an LC–MS/MS method for determining urinary 9,10-PQHG has been reported and have identified the metabolite as a metabolite of 9,10-PQ in human urine, directly confirming PAHQ exposure through respiration (Asahi et al. 2014). Urinary 9,10-PQHG should be useful to evaluate 9,10-PQ (PAHQs) exposure and the health effects of ROS upon exposure.

Among NPAHs, 1-nitropyrene (1-NP) is one of the most abundant NPAHs in diesel exhaust particulates (DEP) and has been proposed as a chemical marker for diesel exhaust (Scheepers et al. 1995; Bamford et al. 2003). Therefore, the metabolites of 1-NP are expected to be a specific biomarker of exposure to DEP because of 1-NP's strong association with diesel exhaust (Zielinska et al. 2004). The metabolism of 1-NP has been studied using various tissues and species (Ball et al. 1984; El-Bayoumy and Hecht 1984; Howard et al. 1995). 1-NP is metabolized essentially through two routes: cytochrome P450-mediated C oxidation and the reduction of nitro group. Some intermediates have been shown to bind to DNA or proteins. Urinary or fecal metabolites that have typically been observed in in vivo studies are hydroxy-1-nitropyrenes (3-, 6-, and 8-OHNP), hydroxy-*N*-acetyl-1-aminopyrenes (3-, 6-, and 8-OHNAAP), *trans*-4,5-dihydro-4,5-dihydroxy-1-nitropyrene, *N*-acetyl-1-aminopyrene (NAAP), and 1-aminopyrene (1-AP).

In humans, only a few studies have reported methods for determining urinary metabolites of 1-NP or other NPAHs. Although 1-AP in human urine has been measured (Grimmer et al. 2000; Seidel et al. 2002; Laumbach et al. 2009), 1-NP metabolites such as OHNPs and OHNAAPs, which are expected to be the major metabolites from in vivo and in vitro studies, have never been determined in human urine. A sensitive and specific LC–MS/MS method and an effective pretreatment method for the simultaneous measurement of urinary 1-NP metabolites has been reported (Toriba et al. 2007). This is intended to provide biomarkers of low-level 1-NP (DEP) exposure in humans. Since Almost all the 1-NP metabolites in human urine are present as glucuronide and/or sulfate conjugates, urine samples were hydrolyzed enzymatically. The use of blue rayon, which has a high specificity for polycyclic aromatic compounds, has made it possible to enrich the analytes selectively from a large volume of urine (100 mL). The 1-NP metabolite concentrations were determined in urine samples of healthy subjects who were non-occupationally exposed to 1-NP (DEP). Among the 3 structural isomers, 6- and 8-OHNAAPs and 6- and 8-OHNPs were abundant in human urine, while 3-OHNAAP and 3-OHNP were hardly detected in the urine samples (Fig. 4). This report is the first to demonstrate the presence of OHNAAPs and OHNPs in human urine, in agreement with previous studies that predicted that these metabolites should be excreted into human urine (Howard et al. 1990; Silvers et al. 1992; Chae et al. 1999). The measurement of urinary 1-NP metabolites was applied to taxi drivers' urine samples collected in China, and their concentrations were greater than metabolite levels previously reported in non-occupationally exposed subjects (Miller-Schulze et al. 2013). The sum of urinary 1-NP metabolites was associated with personal exposure to 1-NP and the results suggested that the metabolites reflect recent exposure (Galaviz et al. 2017).

4 Conclusion

The determinations of oxy-PAHs in environment are not easy and this is a reason why the development of research on the effects of PAHOHs and PAHQs to human has been slow. From the point of view of the health risk, several methods recently developed will be useful for future monitoring studies. On the other hand, the evaluation of exposure to PAHs and their derivatives using urinary metabolites as biomarkers is effective when considering the impact of the chemicals on health. Much attention has focused on the development of biomarkers for pyrene, BaP, 1-NP and 9,10-PQ. Specific and sensitive analytical methods for their metabolites into human urine should accelerate studies of the assessment of human exposure to them.

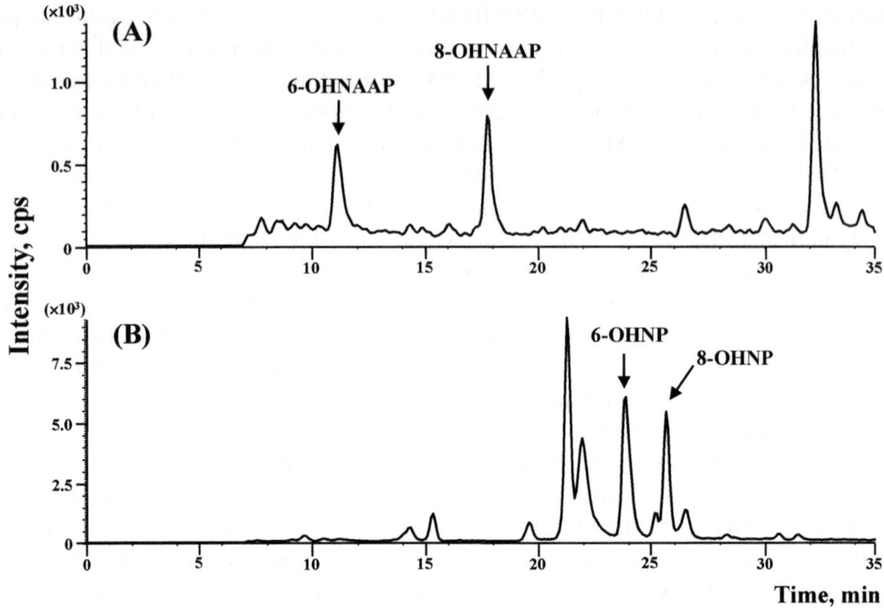

Fig. 4 Representative SRM chromatograms of 1-NP metabolites in a human urine sample. **a** OHNAAPs (*m/z* 274 → 231), **b** OHNPs (*m/z* 262 → 232). Each peak of the metabolites corresponds to 3.76 pmol/L of urine (6-OHNAAP), 1.84 pmol/L of urine (8-OHNAAP), 5.78 pmol/L of urine (6-OHNP) and 2.86 pmol/L of urine (8-OHNP) (Toriba et al. 2007)

References

Adeline, Clergé Jérémie, Le Goff Claire, Lopez Jérôme, Ledauphin Raphaël, Delépée (2019) Oxy-PAHs: occurrence in the environment and potential genotoxic/mutagenic risk assessment for human health. Critical Reviews in Toxicology 49(4), 302–328. https://doi.org/10.1080/10408444.2019.1605333

Ahmed S, Kishikawa N, Ohyama K, Maki T, Kurosaki H, Nakashima K, Kuroda N (2009) An ultrasensitive and highly selective determination method for quinones by high-performance liquid chromatography with photochemically initiated luminol chemiluminescence. J. Chromatogr. A 1216: 3977–3984.

Albinet A, Leoz-Garziandia E, Budzinski H, Villenave E (2006) Simultaneous analysis of oxygenated and nitrated polycyclic aromatic hydrocarbons on standard reference material 1649a (urban dust) and on natural ambient air samples by gas chromatography-mass spectrometry with negative ion chemical ionisation. J. Chromatogr. A 1121: 106–113.

Albinet A, Nalin F, Tomaz S, Beaumont J, Lestremau F (2014) A simple QuEChERS-like extraction approach for molecular chemical characterization of organic aerosols: Application to nitrated and oxygenated PAH derivatives (NPAH and OPAH) quantified by GC-NICIMS. Anal. Bioanal. Chem. 406: 3131–3148.

Andreou G, Rapsomanikis S (2009) Polycyclic aromatic hydrocarbons and their oxygenated derivatives in the urban atmosphere of Athens. J. Hazard. Mater. 172: 363–373.

Asahi M, Kawai M, Toyama T, Kumagai Y, Chuesaard T, Tang N, Kameda T, Hayakawa K, Toriba A (2014) Identification and quantification of in vivo metabolites of 9,10-phenanthrenequinone in human urine associated with producing reactive oxygen species. Chem. Res. Toxicol. 27: 76–85.

Avagyan R, Nyström R, Boman C, Westerholm R (2015) Determination of hydroxylated polycyclic aromatic hydrocarbons by HPLC-photoionization tandem mass spectrometry in wood smoke particles and soil samples. Anal. Bioanal. Chem. 407: 4523–4534.

Avagyan R, Nyström R, Lindgren R, Boman C, Westerholm R (2016) Particulate hydroxy-PAH emissions from a residential wood log stove using different fuels and burning conditions. Atmos. Environ. 140: 1–9.

Ball LM, Kohan MJ, Inmon JP, Claxton LD, Lewtas J (1984) Metabolism of 1-nitro[^{14}C]pyrene in vitro in the rat and mutagenicity of urinary metabolites. Carcinogenesis, 5: 1557–1564.

Bamford HA, Bezabeh DZ, Schantz S, Wise SA, Baker JE (2003) Determination and comparison of nitrated-polycyclic aromatic hydrocarbons measured in air and diesel particulate reference materials. Chemosphere, 50, 575–587.

Bandowe BAM, Shukurov N, Kersten M, Wilcke W (2010) Polycyclic aromatic hydrocarbons (PAHs) and their oxygen-containing derivatives (OPAHs) in soils from the Angren industrial area, Uzbekistan. Environ. Pollut. 158: 2888–2899.

Bandowe BAM, Sobocka J, Wilcke W (2011) Oxygen-containing polycyclic aromatic hydrocarbons (OPAHs) in urban soils of Bratislava, Slovakia: Patterns, relation to PAHs and vertical distribution. Environ. Pollut. 159: 539–549.

Bandowe BAM, Lueso MG, Wilcke W (2014) Oxygenated polycyclic aromatic hydrocarbons and azaarenes in urban soils: A comparison of a tropical city (Bangkok) with two temperate cities (Bratislava and Gothenburg). Chemosphere 107: 407–414.

Bandowe BAM, Meusel H, Huang R, Ho K, Cao J, Hoffmann T, Wilcke W (2014b) $PM_{2.5}$-bound oxygenated PAHs, nitro-PAHs and parent-PAHs from the atmosphere of a Chinese megacity: Seasonal variation, sources and cancer risk assessment. Sci. Total Environ. 473–474: 77–87.

Bao S, Qian J, Qiu T, Jiang W, Gu W, Qu Y, Bai X, Yu X, Jiang Y, Tang S, Lv Y, Shi X (2025) Evaluation of various mass spectrometry mode based on gas chromatography for quantifying the polycyclic aromatic hydrocarbon metabolites in human urine. J. Chromatogr. A 1739: 465521.

Barbeau D, Maître A, Marques M (2011) Highly sensitive routine method for urinary 3-hydroxybenzo[*a*]pyrene quantitation using liquid chromatography-fluorescence detection and automated off-line solid phase extraction. Analyst 136: 1183–1191.

Barrado AI, Garcia S, Castrillejo Y, Perez RM (2012) Hydroxy-PAH levels in atmospheric PM10 aerosol samples correlated with season, physical factors and chemical indicators of pollution. Atmos. Pollut. Res. 3: 81–87.

Bolton JL, Trush MA, Penning TM, Dryhurst G, Monks TJ (2000) Role of Quinones in Toxicology. Role of quinones in toxicology. Chem. Res. Toxicol. 13: 35–160.

Chae YH, Thomas T, Guengerich FP, Fu PP, El-Bayoumy K. (1999) Comparative metabolism of 1-, 2-, and 4-nitropyrene by human hepatic and pulmonary microsomes. Cancer Res. 59: 1473–1480.

Chetiyanukornkul T, Toriba A, Kizu R, Makino T, Nakazawa H, Hayakawa K (2002) Determination of 1-hydroxypyrene in human urine by high-performance liquid chromatography with fluorescence detection using a deuterated internal standard. J. Chromatogr. A 961: 107–112.

Chetiyanukornkul T, Toriba A, Kameda T, Tang N, Hayakawa K (2006) Simultaneous determination of urinary hydroxylated metabolites of naphthalene, fluorene, phenanthrene, fluoranthene and pyrene as multiple biomarkers of exposure to polycyclic aromatic hydrocarbons. Anal. Bioanal. Chem. 386: 712–718.

Cho AK, Stefano DE, You Y, Rodriguez CE, Schmitz DA, Kumagai, Y, Miguel AH, Eiguren-Fernandez A, Kobayashi T, Avol E, Froines JR (2004) Determination of four quinones in diesel exhaust particles, SRM 1649a, and atmospheric $PM_{2.5}$. Aerosol Sci. Techonol. 38: 68–81.

Cochran RE, Dongari N, Jeong H, Beránek J, Haddadi S, Shipp J, Kubátová, A (2012) Determination of polycyclic aromatic hydrocarbons and their oxy-, nitro-, and hydroxy-oxidation products. Anal. Chim. Acta 740: 93–103.

Delgado-Saborit JM, Alam MS, Godri Pollitt KJ, Stark C, Harrison RM (2013) Analysis of atmospheric concentrations of quinones and polycyclic aromatic hydrocarbons in vapour and particulate phases. Atomos. Environ. 77: 974–982.

Delhomme O, Millet M, Herckes P (2008) Determination of oxygenated polycyclic aromatic hydrocarbons in atmospheric aerosol samples by liquid chromatography-tandem mass spectrometry. Talanta 74: 703–710.

Eiguren-Fernandez A, Miguel AH, Lu R, Purvis K, Grant B, Mayo P, Stefanoa ED, Cho AK, Froines J (2008) Atmospheric formation of 9,10-phenanthraquinone in the Los Angeles air basin. Atmos. Environ. 42: 2312–2319.

El-Bayoumy K, Hecht SS (1984) Metabolism of 1-nitro[U-4,5,9,10-^{14}C]pyrene in the F344 Rat. Cancer Res. 44: 4317–4322.

Elzein A, Dunmore, RE, Ward MW, Hamilton JF, Lewis AC (2019). Variability of polycyclic aromatic hydrocarbons and their oxidative derivatives in wintertime Beijing, China. Atmos. Chem. Phys., 19: 8741–8758.

Förster K, Preuss R, Roßbach B, Brüning T, Angerer J, Simon P (2008) 3-Hydroxybenzo[*a*]pyrene in the urine of workers with occupational exposure to polycyclic aromatic hydrocarbons in different industries. Occup. Environ. Med., 65: 224–229.

Galaviz VE, Quintana PJE, Yost MG, Sheppard L, Paulsen MH, Camp JE, Simpson CD (2017) Urinary metabolites of 1-nitropyrene in US–Mexico border residents who frequently cross the San Ysidro Port of Entry. J. Expo. Sci. Environ. Epidemiol. 27: 84–89.

Galceran MT, Moyano E (1995) Determination of hydroxyl-substituted polycyclic aromatic-hydrocarbons by high-performance liquid chromatography with electrochemical detection. J. Chromatogr. A 715: 41–48.

Geldenhuys, G., Rohwer, E. R., Naudé, Y., Forbes, P. B. C. (2015). Monitoring of atmospheric gaseous and particulate polycyclicaromatic hydrocarbons in South African platinum mines utilisingportable denuder sampling with analysis by thermaldesorption–comprehensive gas chromatography–massspectrometry. J. Chromatogr. A, 1380, 17–28.

Grimmer G, Dettbarn G, Seidel A, Jacob J (2000) Detection of carcinogenic aromatic amines in the urine of non-smokers. J. Sci. Total. Environ. 247, 81–90.

Grosse S, Letzel T (2007) Liquid chromatography/atmospheric pressure ionization mass spectrometry with post-column liquid mixing for the efficient determination of partially oxidized polycyclic aromatic hydrocarbons. J. Chromatogr. A 1139: 75–83.

Han YM, Wei C, Bandowe BAM, Wilcke W, Cao JJ, Xu BQ, Gao SP, Tie XX, Li GH, Jin ZD, An ZS (2015) Elemental carbon and polycyclic aromatic compounds in a 150-year sediment core from Lake Qinghai, Tibetan Plateau, China: Influence of regional and local sources and transport pathways. Environ. Sci. Technol. 49: 4176–4183.

Hansen ÅM, Mathiesen L, Pedersen M, Knudsen LE (2008) Urinary 1-hydroxypyrene (1-HP) in environmental and occupational studies-A review. Int. J. Hyg. Health 211: 471–503.

Hayakawa K, Onoda Y, Tachikawa C, Hosoi S, Yoshida M, Chung SW, Kizu R, Toriba A, Kameda T, Tang N. (2007) Estrogenic/antiestrogenic activities of polycyclic aromatic hydrocarbons and their monohydroxylated derivatives by yeast two-hybrid assay. J. Health Sci., 53: 562–570.

Hayakawa K, Bekki K, Yoshita M, Tachikawa C, Kameda T, Tang N, Toriba A Hosoi S (2011) Estrogenic/antiestrogenic activities of quinoid polycyclic aromatic hydrocarbons. J. Health Sci., 57: 274–280.

Honda M, Hayakawa K, Zhang L, Tang N, Nakamura, H (2022) Seasonal Variability and Risk Assessment of Atmospheric Polycyclic Aromatic Hydrocarbons and Hydroxylated Polycyclic Aromatic Hydrocarbons in Kanazawa, Japan. Applied Sci., 12: 9469.

Howard PC, Aoyama T, Bauer SL, Gelboin HV, Gonzalez FJ (1990) The metabolism of 1-nitropyrene by human cytochromes P450. Carcinogenesis, 11: 1539–1542.

Howard PC, Consolo MC, Dooley KL, Beland FA (1995) Metabolism of 1-nitropyrne in mice: transport across the placenta and mammary tissues. Chem.-Biol. Interact. 95: 309–325.

Huang S, Lic Q, Liua H, Ma S, Long C, Li G, Yu Y (2022) Urinary monohydroxylated polycyclic aromatic hydrocarbons in the general population from 26 provincial capital cities in China: Levels, influencing factors, and health risks. Environ Intern. 160: 107074.

Jacob J, Seidel A (2002) Biomonitoring of polycyclic aromatic hydrocarbons in human urine. J. Chromatogr., 778: 31–47.

Jacob P, Wilson M, Benowitz NL (2007) Determination of phenolic metabolites of polycyclic aromatic hydrocarbons in human urine as their pentafluorobenzyl ether derivatives using liquid chromatography–tandem mass spectrometry. Anal. Chem. 79: 587–98.

Jakober CA, Charles MJ, Kleeman MJ, Green PG (2006) LC-MS Analysis of Carbonyl Compounds and Their Occurrence in Diesel Emissions. Anal. Chem. 78: 5086–5093

Jakober CA, Riddle SG, Robert MA, Destaillats H, Charles MJ, Green PG, Kleeman MJ (2007) Quinone emissions from gasoline and diesel motor vehicles. Environ. Sci. Technol. 41: 4548–4554.

Jongeneelen FJ, Anzion RBM, Henderson PT (1987) Determination of hydroxylated metabolites of polycyclic aromatic hydrocarbons in urine. J. Chromatogr. 413: 227–232.

Jongeneelen FJ (1997) Methods for routine biological monitoring of carcinogenic PAH-mixtures. Sci. Total Environ. 199: 141–149.

Kakimoto K, Toriba A, Ohno T, Ueno M, Kameda T, Tang N, Hayakawa K (2008) Direct measurement of the glucuronide conjugate of 1-hydroxypyrene in human urine by using liquid chromatography with tandem mass spectrometry. J. Chromatogr. B 867: 259–263.

Kamiya M, Toriba A., Onoda Y, Kizu, R, Hayakawa K (2005) Evaluation of estrogenic activities of hydroxylated polycyclic aromatic hydrocarbons in cigarette smoke condensate. Food Chem. Toxicol. 43: 1017–1027.

Keyte IJ, Albinet A, Harrison RM (2016) On-road traffic emissions of polycyclic aromatic hydrocarbons and their oxy- and nitro- derivative compounds measured in road tunnel environments. Sci. Total Environ. 566–567: 1131–1142.

Kishikawa N, Wada M, Ohba Y, Nakashima K, Kuroda N (2004) Highly sensitive and selective determination of 9,10-phenanthrenequinone in airborne particulates using high-performance liquid chromatography with pre-column derivatization and fluorescence detection. J. Chromatogr. A 1057: 83–88.

Lammel G, Kitanovski Z, Kukučka P, Novák J, Arangio AM, Codling GP, Filippi A, Hovorka J, Kuta J, Leoni C, Přibylová P, Prokeš R, Sáňka O, Shahpoury P, Tong H, Wietzoreck M (2020) Oxygenated and nitrated polycyclic aromatic hydrocarbons inambient air-levels, phase partitioning, mass size distributions, andinhalation bioaccessibility. Environ. Sci. Technol. 54 (5), 2615–2625.

Laumbach R, Tong J, Zhang L, Ohman-Strickland P, Stern A, Fiedler N, Kipen H, Kelly-McNeil K, Lioy P, Zhang J (2009) Quantification of 1-aminopyrene in human urine after a controlled exposure to diesel exhaust. J. Environ. Monitor. 11: 153–159.

Layshock JA, Wilson G, Anderson KA (2010) Ketone and quinone-substituted polycyclic aromatic hydrocarbons in Mussel Tissue, sediment, urban dust, and diesel particulate matrices. Environ. Toxicol. Chem. 29: 2450–2460.

Lankova D, Urbancova K, Sram RJ, Hajslova J, Pulkrabova J (2016) A novel strategy for the determination of polycyclic aromatic hydrocarbon monohydroxylated metabolites in urine using ultra-high-performance liquid chromatography with tandem mass spectrometry. Anal. Bioanal. Chem. 408: 2515–2525.

Lewtas J (2007) Air pollution combustion emissions: Characterization of causative agents and mechanisms associated with cancer, reproductive, and cardiovascular effects. Mutat. Res. 636: 95–133.

Li R, Kameda T, Li Y, Toriba A, Tang N, Hayakawa K, Lin JM (2011) Hydrogen peroxide-sodium hydrosulfite chemiluminescence system combined with high-performance liquid chromatography for determination of 1-hydroxypyrene in airborne particulates. Talanta 85 2711–2714.

Li S, Chena W, Liu Y, Wang H, Li Y, Zhang Z, Xingjian Yang X (2025) Unraveling the spatiotemporal trends and source attribution of polycyclic aromatic hydrocarbons and oxygenated derivatives in Guangzhou agricultural ditch sediment. Ecotoxicol. Environ. Saf. 289, 117425.

Li Z, Romanoff LC, Trinidad DA, Pittman EN, Hilton D, Hubbard K, Carmichael H, Parker J, Calafat AM, Sjödin A (2014) Quantification of 21 metabolites of methylnaphthalenes and polycyclic aromatic hydrocarbons in human urine. Anal. Bioanal. Chem. 406: 3119–3129.

Liang X, Wang L, Du W, Chen Y, Yun X, Chen Y, Shen G, Shen H, Yang X, Tao S (2023) Emission factors of oxygenated polycyclic aromatic hydrocarbons from ships in China. Environ. Pollut. 337: 122483.

Lin Y, Ma Y, Qiu X, Li R, Fang Y, Wang J, Zhu Y, Hu D (2015) Sources, transformation, and health implications of PAHs and their nitrated, hydroxylated, and oxygenated derivatives in $PM_{2.5}$ in Beijing. J. Geophys. Res. Atmos 14: 7219–7228.

Lintelmann J, Fischer K, Matuschek G (2006) Determination of oxygenated polycyclic aromatic hydrocarbons in particulate matter using high-performance liquid chromatography-tandem mass spectrometry. J. Chromatogr. A 1133: 241–247.

Luo K, Gao Q, Hu J (2015) Derivatization method for sensitive determination of 3-hydroxybenzo[*a*]pyrene in human urine by liquid chromatography-electrospray tandem mass spectrometry. J. Chromatogr. A 1379: 51–55.

Ma Y, Cheng Y, Qiu X, Lin Y, Cao J and Hu D (2016) A quantitative assessment of source contributions to fine particulate matter ($PM_{2.5}$)-bound polycyclic aromatic hydrocarbons (PAHs) and their nitrated and hydroxylated derivatives in Hong Kong. Environ. Pollut. 219: 742–749.

Metcalf SW and Orloff KG (2004) Biomarkers of exposure in community setting. J. Toxicol. Environ. Health, Part A, 67: 715–726.

Miller-Schulze JP, Paulsen M, Kameda T, Toriba A, Tang N, Tamura K, Dong L, Zhang X, Hayakawa K, Yost MG, Simpson CD (2013) Evaluation of urinary metabolites of 1-nitropyrene as biomarkers for exposure to diesel exhaust in taxi drivers of Shenyang, China. J. Expo. Sci. Environ. Epidemiol. 23: 170–175.

Mirivel G, Riffault V, Galloo JC (2010) Simultaneous determination by ultra-performance liquid chromatography-atmospheric pressure chemical ionization time-of-flight mass spectrometry of nitrated and oxygenated PAHs found in air and soot particles. Anal. Bioanal. Chem. 397: 243–256.

Motoyama, Y., Bekki, K., Chung, S.-W., Tang, N., Kameda, T., Toriba, A., Taguchi, K., Hayakawa, K. (2009). Oxidative stress more strongly induced by ortho- than para-quinoid polycyclic aromatic hydrocarbons in A549 cells. J. Health Sci., 55, 845–850.

Obrist D, Zielinska B, Perlinger JA (2015) Accumulation of polycyclic aromatic hydrocarbons (PAHs) and oxygenated PAHs (OPAHs) in organic and mineral soil horizons from four U.S. remote forests. Chemosphere 134: 98–105.

O'Connell SG, Haigh T, Wilson G, Anderson KA (2013) An analytical investigation of 24 oxygenated-PAHs (OPAHs) using liquid and gas chromatography-mass spectrometry. Anal. Bioanal. Chem. 405: 8885–8896.

Onyemauwa F, Rappaport SM, Sobus JR, Gajdošová D, Wu R, Waidyanatha S (2009) Using liquid chromatography-tandem mass spectrometry to quantify monohydroxylated metabolites of polycyclic aromatic hydrocarbons in urine. J. Chromatogr. B 877: 1117–1125.

Pojana G, Marcomini A (2007) Determination of monohydroxylated metabolites of polycyclic aromatic hydrocarbons (OH-PAHs) from wastewater-treatment plants. Intern J. Environ. Anal. Chem. 87: 627–636.

Santos PM, Sánchez MdN., Pavón JLP, B.M. Cordero, BM (2019) Determination of polycyclic aromatic hydrocarbons in human biological samples: A critical review. Trends Anal. Chem. 113: 194–209.

Sarkar M, Liu J, Koval T, Wang J, Feng S, Serafin R, Jin Y, Xie Y, Newland K, Roethig HJ (2009) Evaluation of biomarkers of exposure in adult cigarette smokers using Marlboro Snus. Nocotine Tob. Res. 12: 105–116.

Seidel A, Dahman D, Krekeler H, Jacob J (2002) Biomonitoring of polycyclic aromatic compounds in the urine of mining workers occupationally exposed to diesel exhaust. Int. J. Hyg. Environ. Health. 204, 333–338.

Scheepers PTJ, Martens MH, Velders DD, Fijneman P, van Kerkhoven M, Noordhoek J, Bos RP (1995) 1-Nitropyrene as a marker for the mutagenicity of diesel exhaust-derived particulate matter in workplace atmospheres. Environ. Mol. Mutagen. 25: 134–147.

Silvers KJ, Chazinski T, McManus ME, Bauer SL, Gonzalez FJ, Gelboin HV, Maurel P, Howard PC (1992) Cytochrome P-450 3A4 (nifedipine oxidase) is responsible for the C-oxidative metabolism of 1-nitropyrene in human liver microsomal samples. Cancer Res., 52: 6237–6243.

Singh R, Tucek M, Maxa K, Tenglerova J, Weyand EH (1995) A rapid and simple method for the analysis of 1-hydroxypyrene glucuronide: a potential biomarker for poly cyclic aromatic hydrocarbon exposure. Carcinogenesis 16: 2909–2915.

Taguchi K, Shimada M, Fujii S, Sumi D, Pan X, Yamano S, Nishiyama T, Hirastuka A, Yamamoto M, Cho AK, Froines JR, Kumagai Y (2008) Redox cycling of 9,10-phenanthraquinone to cause oxidative stress is terminated through its monoglucuronide conjugation in human pulmonary epithelial A549 cells. Free Radic. Biol. Med. 44: 1645–1655.

Toriba A, Chetiyanukornkul T, Kizu R, Hayakawa, K (2003) Quantification of 2-hydroxyfluorene in human urine by column-switching high performance liquid chromatography with fluorescence detection. Analyst 128: 605–610.

Toriba A, Kitaoka H, Dills R, Mizukami S, Tanabe K, Takeuchi N, Ueno M, Kameda T, Tang N, Hayakawa K, Simpson CD (2007) Identification and quantification of 1-nitropyrene metabolites in human urine as a proposed biomarker for exposure to diesel exhaust. Chem. Res. Toxicol. 20: 999–1007.

Toriba A, Hayakawa K (2007) Biomarkers of exposure to polycyclic aromatic hydrocarbons and related compounds. J Health Sci. 53:631–638.

Toriba A, Homma C, Kita M, Uozaki W, Boongla Y, Orakij W, Tang N, Kameda T, Hayakawa K (2016) Simultaneous determination of polycyclic aromatic hydrocarbon quinones by gas chromatography-tandem mass spectrometry, following a one-pot reductive trimethylsilyl derivatization. J. Chromatogr. A 1459: 89–100.

Walgraeve C, Demeestere K, De Wispelaere P, Dewulf J, Lintelmann J, Fischer K, Van Langenhove H (2012) Selective accurate-mass-based analysis of 11 oxy-PAHs on atmospheric particulate matter by pressurized liquid extraction followed by high-performance liquid chromatography and magnetic sector mass spectrometry Anal. Bioanal. Chem. 402: 1697–1711.

Wang G, Kawamur, K, Zhao X, Li Q, Dai Z, Niu H (2007) Identification, abundance and seasonal variation of anthropogenic organic aerosols from a mega-city in China. Atmos. Environ. 41: 407–416.

Wang X, Lin L, Luan T, Yang L, Tam NFY (2012) Determination of hydroxylated metabolites of polycyclic aromatic hydrocarbons in sediment samples by combining subcritical water extraction and dispersive liquid-liquid microextraction with derivatization. Anal. Chim. Acta 753: 57–63.

Wang X, Yuan K, Yang L, Lin L, Nora Tam NFY, Chen B, Luan T (2015) Characterizing the parent and oxygenated polycyclic aromatic hydrocarbons in mangrove sediments of Hong Kong. Mar. Pollut. Bull. 98: 335–340.

Wang X, Li Y, Cai F, Qing Q, Yuan K, Chen B (2017) Fully automatic single-drop microextraction with one-setp extraction and derivatization and its application for rapid analysis of hydroxylated polycyclic aromatic hydrocarbons in seawaters. Talanta 164: 727–734.

Wei L, Lv J, Zuo P, Li Y, Yang R, Zhang Q, Jiang, G (2024) The occurrence and sources of PAHs, oxygenated PAHs (OPAHs), and nitrated PAHs (NPAHs) in soil and vegetation from the Antarctic, Arctic, and Tibetan Plateau. Sci. Total Environ. 912, 169394.

Woudneh MB, Benskin JP, Grace R, Hamilton MC, Magee BH, Hoeger GC, Forsberg ND, Cosgrove JR (2016) Quantitative determination of hydroxy polycylic aromatic hydrocarbons as a biomarker of exposure to carcinogenic polycyclic aromatic hydrocarbons. J. Chromatogr. A 1454: 93–100.

Yan, Wang Yue, Xu Yingjun, Chen Chongguo, Tian Yanli, Feng Tian, Chen Jun, Li Gan, Zhang (2016) Influence of different types of coals and stoves on the emissions of parent and oxygenated PAHs from residential coal combustion in China. Environmental Pollution 212, 1–8. https://doi.org/10.1016/j.envpol.2016.01.041

Yao L, Yang J, Liu B, Zheng S, Wang W, Zhu X, Qian X (2014) Development of a sensitive method for the quantification of urinary 3-hydroxybenzo[*a*]pyrene by solid phase extraction, dansyl chloride derivatization and liquid chromatography-tandem mass spectrometry detection. Anal. Methods 6: 6488–6493.

Zielinska B, McDonald JD, Whitney K, Lawson DR (2004) Emission rates and comparative chemical composition from selected in-use diesel and gasoline-fueled vehicles. J. Air Waste Manag. Assoc. 54, 1138–1150.

Instrumental Analysis of Polycyclic Aromatic Hydrocarbons and Nitropolycyclic Aromatic Hydrocarbons

Kazuichi Hayakawa

Abstract Polycyclic aromatic hydrocarbons (PAHs) are oxidized to hydroxy-PAHs (OHPAHs), quinoid-PAHs [PAH quinones (PAHQs)], and nitro-PAHs (NPAHs). Quantitative analysis is essential for understanding their emissions, environmental behaviors, exposures, and health risks. Although NPAHs exhibit far stronger mutagenicity compared with PAHs, fewer studies on NPAHs exist owing to the previous lack of highly sensitive analytical methods. Atmospheric concentrations follow the order PAHs > PAHQs > OHPAHs > NPAHs. However, highly sensitive methods for analyzing NPAHs have now been developed, increasing the number of studies focusing on atmospheric NPAHs. Additionally, source analysis methods have emerged, with NPAHs and PAHs as critical monitoring targets. This chapter covers sampling and cleanup procedures for PAHs and NPAHs in atmospheric samples, as well as instrumental analysis techniques, including high-performance liquid chromatography–fluorescence detection/chemiluminescence detection and gas chromatography–mass spectrometry.

Keywords Polycyclic aromatic hydrocarbon · Nitropolycyclic aromatic hydrocarbon · High-performance liquid chromatography–fluorescence detection · High-performance liquid chromatography–chemiluminescence detection · Gas chromatography–mass spectrometry · Sample treatment · Sampling

1 Introduction

Polycyclic aromatic hydrocarbons (PAHs) and their oxidized derivatives, including nitro-PAHs (NPAHs), hydroxy-PAHs (OHPAHs), and quinoid-PAHs [PAH quinones (PAHQs)], are present in the atmosphere. Many of these compounds (all abbreviations are listed in Table 1) exhibit harmful biological activities, including

K. Hayakawa (✉)
Institute of Nature and Environmental Technology, Kanazawa University, Kakuma-Machi, Kanazawa, Japan
e-mail: hayakawa@p.kanazawa-u.ac.jp

K. Hayakawa and Y. Aoki (eds.), *A New Era of Polycyclic Aromatic Hydrocarbons*,
https://doi.org/10.1007/978-981-96-7695-8_11

carcinogenicity, mutagenicity, endocrine-disrupting effects, and reactive oxygen species production (Hayakawa 2022a, b). In this chapter, PAHQs and OHPAHs are collectively referred to as OPAHs.

Among these oxidized PAHs, NPAHs show stronger mutagenicity compared with PAHs; however, fewer studies address NPAHs compared to PAHs. Additionally, many NPAH sources and environmental behaviors remain unknown, primarily owing to the lack of sensitive, reliable methods available for NPAH detection. Figure 1 shows the atmospheric concentrations of selected PAHs, NPAHs, OHPAHs, and PAHQs in Kanazawa during winter 2018, with particulate-bound concentrations following the order PAHs > PAHQs > OHPAHs > NPAHs. Analytical methods must measure

Table 1 Abbreviations of PAHs and their oxidized derivatives

Polycyclic aromatic hydrocarbons (PAHs)	Naphthalene (Nap), acenaphthene (Ace), acenaphthylene (Acy), fluorene (Fle), phenanthrene (Phe), anthracene (Ant), Fluoranthene (Flu), pyrene (Pyr), benz[*a*]anthracene (BaA), chrysene (Chr), benzo[*b*]fluoranthene (BbF), benzo[*k*]fluoranthene (BkF), benzo[*a*]pyrene (BaP), dibenz[*a,h*]anthracene (DBA), benzo[*ghi*]perylene (BgPe), indeno[1,2,3-*cd*]pyrene (IDP)
Nitro-PAHs (NPAHs)	1-, 2-Nitronaphthalene (1-, 2-NNap), 5-niroacenaphthene (5-NAce), 1-, 2-, 4-nitropyrene (1-, 2-, 4-NP), 1,3-, 1,6-, 1,8-dinitropyrene (1,3-, 1,6-, 1,8-DNP), 1-, 10-nitrobenzanthrone (1-, 10-NBA), 2-nitrofluorene (2-NF), 2-, 9-nitroanthracene (2-, 9-NA), 5-nitroacenaphthene (5-NAc), 4-, 9-nitrophenanthrene (2-, 9-NPh), 3-nitrofluoranthene (3-NFL), 7-nitrobenz[*a*]anthracene (7-NBaA), 2-nitrotriphenylene (2-NTP), 6-nitrochrysene (6-NC), 6-nitrobenzo[*a*]pyrene (6-NBaP), 1-, 3-nitroperylene (1-, 3-NPer)
Hydroxylated PAHs (OHPAHs)	1-, 2-Hydroxynaphthalene (1-, 2-OHNap), 1-, 2-, 3-hydroxyfluoranthene (1-, 2-, 3-OHFlu), 1-, 2-, 3-, 9-hydroxyphenanthrene (1-, 2-, 3-, 9-OHPhe), 1-, 2-, 3-hydroxypyrene (1-, 2-, 3-OHPyr), 3-hydroxybenzo[*a*]pyrene (3-OHBaP), 6-hydroxychrysene (6-OHChr), 3-hydroxybenzo[*a*]pyrene (3-OHBaP)
PAH quinones (PAHQs)	1,4-Benzoquinone (BQ), 2-methyl-1,4-benzoquinone (2-MBQ), 2,5- and 2,6-dimethylbenzoquinones (2,5- and 2,6-DMBQ), tetramethylbenzoquinone (TMBQ), 1,2- and 1,4-naphtoquinones (1,2- and 1,4-NQ), 2-methyl-1,4-naphthoquinone (2-MNQ), acenaphthoquinone (AceNQ), 1,4- and 9,10-phenanthrenequinones (1,4- and 9,10-PQ), 1,4- and 9,10-anthraquinones (1,4- and 9,10-AQ), 1,6-, 1,8- and 4,5-pyrenequinones (1,6-,1,8- and 4,5-PyrQ), fluoranthene-2,3-quinone (2,3- FluQ), aceanthraquinone (AceAQ), 1,4- and 5,6-chrysenequinones (1,4- and 5,6-CQ), 5,12-naphthacenequinone (5,12-NapQ), benzo[*c*]phenanthrene-5,6-quinone (5,6-BcPQ), 7,12-benzanthraquinone (7,12-BAQ), benzo[*e*]pyrene-4,5-quinone (4,5- BePQ), benzo[*a*]pyrene-1,6-, -3,6-, -4,5-, -6,12- and -11,12-quinones (1,6-, -3,6-, -4,5-, -6,12- and -11,12- BaPQ), dibenzo[*a,j*]anthracene-7,14-quinone (7,14-DBajAQ) and dibenzo[*a,h*]anthracene-5,6-quinone (5,6-DBahAQ)

NPAHs at concentrations approximately two orders of magnitude lower than their parent PAHs (Hayakawa 2023).

NPAHs mainly form through two processes: (1) electrophilic nucleation of PAHs with NO_2^+ occurs directly during combustion, with the rate of formation varying by combustion temperature (Westbald et al. 2002; Zajemska et al. 2015), where the concentration ratio of NPAHs to parent PAHs ([NPAH]/[PAH]) rises with increasing combustion temperature. This results in the larger [NPAH]/[PAH] ratios in vehicle engine emissions than in coal or biomass combustion (Hayakawa et al. 2020); (2) gas-phase reactions of parent PAHs with radicals, such as OH^- and $\cdot NO_3^-$, and NO_2, in the atmosphere (Atkinson and Arey 1994). Additionally, metal ions also influence secondary NPAH formation (Daellenbach et al. 2020; Kameda et al. 2016).

PAHs and NPAHs have many combustion sources, including transportation, industry, heating, cooking, forest fire, and slash-and-burn agriculture. Varying combustion conditions, particularly temperature, influence the amount of particulate combustion-derived material (soot, P_c) and the concentrations of PAHs and NPAHs in P_c. Accordingly, a new source analysis method (the NP method) has been developed to quantify the contribution of specific sources not only to PAHs and NPAHs but also to P_c (Hayakawa et al. 2020, 2021), demonstrating the value of tracking PAHs and NPAHs simultaneously to understand their occurrence, dynamics, and exposure in the environment. The NP method is described in Part IV, Chap. 13.

Although studies on NPAHs are less common than those on PAHs, several reviews have addressed NPAHs in airborne and emission particulates as well as NPAH

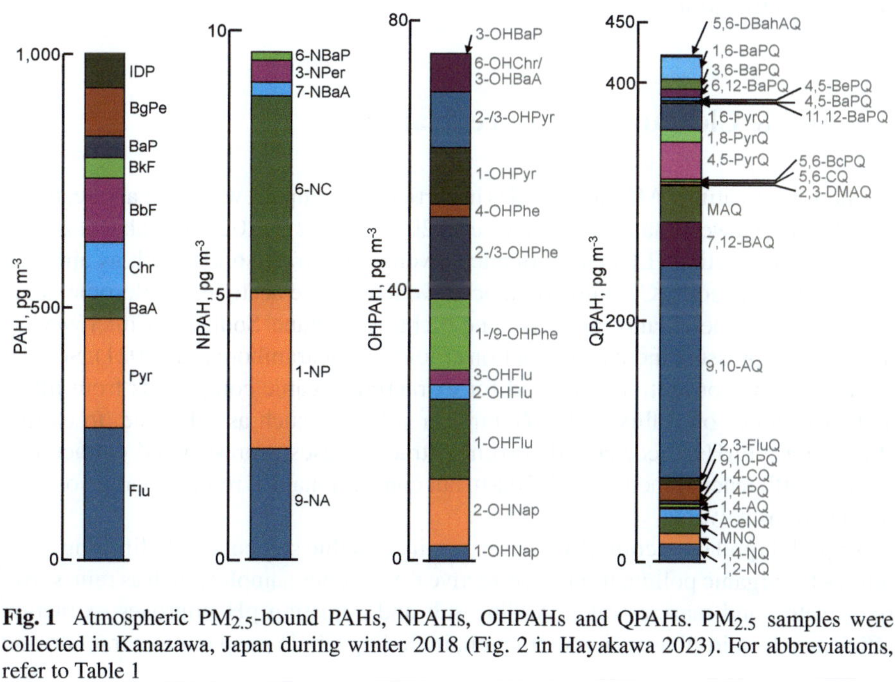

Fig. 1 Atmospheric $PM_{2.5}$-bound PAHs, NPAHs, OHPAHs and QPAHs. $PM_{2.5}$ samples were collected in Kanazawa, Japan during winter 2018 (Fig. 2 in Hayakawa 2023). For abbreviations, refer to Table 1

sampling, cleanup/concentration, and analytical methods (Bandowe and Meusel 2017; Abbas et al. 2018; Yan et al. 2021; Krzyszczak and Czech 2021; Nowakowski et al. 2022; Hayakawa 2000; Hayakawa 2016; Hayakawa 2022a, b). This chapter examines instrumental analysis methods applied to both PAHs and NPAHs.

2 Sampling

Table 2 lists the number of rings, molecular weights, vapor pressures, and water solubilities of selected PAHs and NPAHs. Adding a nitro-group decreases both the vapor pressure and water solubility of PAHs. In the atmosphere, PAHs and NPAHs exist in gas and particulate phases, depending on their vapor pressures. Under typical conditions, two- and three-ring PAHs with high vapor pressure exist in the gas phase, five- and six-ring PAHs with low vapor pressure are bound to particulates, and four-ring PAHs appear in both phases. High- and low-volume air samplers equipped with quartz or glass fiber filters are commonly used to collect particulate-bound chemicals. To capture gas-phase chemicals, a polyurethane-foam (PUF) layer is added to the air sampler after the filter, although a single PUF layer may not capture all two-ring PAHs, such as Nap. Additionally, the atmospheric concentrations of major NPAHs, including 1-NP, range from sub- to several pg m^{-3}, which are much lower than that of Pyr, as shown in Fig. 1. Using a high-volume air sampler (with a sampling speed around 1 m^3 min^{-1}), several hours are typically required to collect sufficient P_c for NPAH quantification.

3 Sample Cleanup and Concentration

Methods for treating PAHs and NPAHs in particulate samples, with their applications, have been reviewed (Bandowe and Meusel 2017; Yan et al. 2021; Hayakawa 2022a, b). Liquid extraction (LE) uses organic solvents to extract solids, such as airborne particulates and soils. Common organic solvents include acetone, dichloromethane, *n*-hexane, benzene-ethanol, and *n*-hexane-dichloromethane. Sonication, microwaves, heating, and pressure increase extraction efficiency (Scaramboni et al. 2021). Soxhlet extraction, the conventional method for extracting organic compounds from filters and PUF, relies on reflux and siphoning in solvents, such as *n*-hexane, to extract PAHs and NPAHs. Accelerated solvent extraction uses a pressurized extractor to increase efficiency (Gbeddy et al. 2020), but conventional LE requires large volumes of hazardous solvents.

Liquid–liquid extraction (LLE) uses two immiscible solvents with differing solubilities for organic pollutants and is effective for aqueous samples, such as rain, snow, river water, and marine water. Adding salt and adjusting pH improves extraction efficiency. Liquid-phase microextraction, including single-drop microextraction and

Table 2 Vapor pressures and solubilities of PAHs and NPAHs in the water

PAH abbreviation	Number of rings	Molecular weight	Vapor pressure (hPa, 25 °C)	Solubility in water (μg/L)
Nap	2	128.2	11.4	31
Ace	3	154.21	3.07	38
Acy	3	152.20	3.80	16
Fle	3	166.22	1.66	1.9
Phe	3	178.23	1.6×10^{-1}	1.1
Ant	3	178.23	8.6×10^{-4}	4.5×10^{-2}
FR	4	202.26	8.61×10^{-4}	2.6×10^{-1}
Pyr	4	202.25	5×10^{-5}	1.3×10^{-1}
BaA	4	252.31	5.43×10^{-5}	1.1×10^{-2}
Chr	4	228.29	4×10^{-6}	2×10^{-4}
BbF	5	252.31	5×10^{-7}	1.5×10^{-3}
BkF	5	252.32	5.2×10^{-8}	8×10^{-3}
BaP	5	252.31	6×10^{-8}	3.8×10^{-3}
DBA	5	302.38	1.33×10^{-8}	6×10^{-4}
BgPe	6	276.34	1.27×10^{-7}	2.6×10^{-4}
IDP	6	276.34	1.38×10^{-8}	1.9×10^{-3}
NPAH abbreviation	**Number of rings**	**Molecular weight**	**Vapor pressure (hPa, 20 °C)**	**Solubility in water (μg/L)**
1-NNap	2	173.17	6.3×10^{-4}	9.18×10^{-3}
2-NNap	2	173.17	3.7×10^{-4}	9.18×10^{-3}
5-NAce	3	199.21	5.46×10^{-5}	9.1×10^{-4}
2-NF	3	211.22	5.9×10^{-6}	2.16×10^{-4}
4-NPh	3	223.23	1.79×10^{-6}	2.915×10^{-4}
9-NA	3	223.23	1.6×10^{-6}	1.14×10^{-4}
1-NP	4	247.25	1.11×10^{-7}	1.18×10^{-5}
2-NP	4	247.25	7.36×10^{-8}	6.79×10^{-5}
4-NP	4	247.25	5.52×10^{-10}	6.79×10^{-5}
1,3-DNP	4	292.25	1.2×10^{-9}	5.4×10^{-5}
1,6-DNP	4	292.25	1.2×10^{-9}	5.4×10^{-5}
1,8-DNP	4	292.25	1.2×10^{-9}	5.4×10^{-5}
2-NFR	4	247.25	3.93×10^{-6}	2.1×10^{-3}
3-NFR	4	247.25	7.36×10^{-8}	1.95×10^{-5}
6-NC	4	273.29	1.01×10^{-8}	1.53×10^{-5}
6-NBaP	5	297.31	4.14×10^{-10}	3.49×10^{-6}

Vapor pressure and water solubility data were obtained from Abbas et al. (2018). For abbreviations, refer to Table 1

dispersive liquid–liquid microextraction (DLLME), is a miniaturized LLE. This technique uses a small volume (sub-mL) of hydrophobic organic solvent as an extractant, which is immiscible with water (Santos et al. 2018). The enrichment factor increases as the extractant volume decreases. An advanced form of DLLME, solidification of floating organic drop–DLLME, uses an organic solvent with a lower density relative to water and a melting point near room temperature. After mixing with the sample solution, organic pollutants are extracted into the organic solvent, and the organic phase becomes solid and floats on the water phase's surface due to cooling. These methods have been effectively used to extract trace levels of PAHs, NPAHs, and OPAHs (Guiñez et al. 2017; Santos et al. 2017; Borges et al. 2018).

Solid-phase extraction (SPE) employs a stationary phase to purify and concentrate organic chemicals in environmental and biological samples. Many solid phases are available; small columns with reversed-phase resins, such as C_{18}, are popular (Sun et al. 2020). Solid-phase microextraction is garnering attention because it requires no sample preparation before gas chromatography–mass spectrometry (GC–MS) analysis. Polydimethylsiloxane/divinylbenzene, cold fiber—a stainless-steel wire coated with zeolitic imidazolate frameworks—and metal–organic nanotubes have been developed for the solid phase (Kong et al. 2018, 2019; dos Santos et al. 2018; Li et al. 2017; Wang et al. 2018). These techniques have been applied to high-performance liquid chromatography (HPLC)–tandem mass spectrometry (MS/MS) analysis of NPAHs. The quick, easy, cheap, effective, rugged, and safe (i.e., QuEChERS) method, a combination of LLE and SPE, includes sample crushing, acetonitrile extraction, drying with salts (e.g., $MgSO_4$), and purification with ethylenediamine-N-propylsilane and other adsorbents. This method is used for simultaneous determination of NPAHs and OPAHs in environmental, biological, and food samples prior to GC–MS analysis (Albinet et al. 2014; Deng and Chan 2017).

Supercritical fluid extraction uses the state shift of a supercritical fluid between gas and liquid, depending on critical temperature and pressure. Common supercritical fluid extractants include carbon dioxide, nitrous oxide, ammonia, sulfur hexafluoride, ethane, and heptane, with carbon dioxide being the most widely used owing to its near-room-temperature critical point. This method simplifies organic pollutant concentration through solvent volatilization.

4 Instrumental Analysis

4.1 PAHs

The United States Environmental Protection Agency (USEPA) recommends prioritizing measurement of 16 two- to six-ring PAHs in environment assessments. The chemical structures and abbreviations of these PAHs are shown in Fig. 2. For simultaneous determination, GC–MS and HPLC–fluorescence detection (FLD) or –MS are recommended methods (USEPA, 1984). These techniques are also recommended by

Japan's Ministry of Environment (Ministry of Environment, Government of Japan 2014) and widely used globally.

Using GC–MS and HPLC-FLD in previous studies, several PAH isomer pairs, including BaA and Chr (4-ring), BbF and BkF (5-ring), and BgPe and IDP (6-ring), showed insufficient resolution. To improve separation, various stationary phases have been developed. For capillary GC columns, stationary phases (liquid phases) with varying degrees of polarity enhance separation through π–π interactions. The durability of PAHs has also improved with stationary phases chemically bonded to the column's inner surface (Hayakawa 2022a, b). Figure 3 shows a typical GC–MS chromatogram of standard PAHs (Ministry of Environment, Government of Japan 2014).

In PAH mass spectrometry, the molecular ion is the base peak; therefore, *m*/*z* values (m represents the ion mass divided by the unified electron mass unit; *z* is the ion charge) are used in quantitative GC–MS analysis. Additionally, several fragment ions are used to confirm identification. Deuterated PAHs serve as internal standards to estimate recovery during sample treatment and GC injection. Table 3 shows the *m*/*z* values used for PAH measurement. Many environmental PAH studies employ single MS with selective ion monitoring (SIM).

In the HPLC analysis of PAHs and their oxidized derivatives, reversed-phase columns, such as the C_{18} stationary phase, with gradient elution in a mixture of organic solvent and aqueous solution is effective. However, PAHs and their oxidized derivatives can interact with residual silanol on silica-based C_{18} stationary phases, causing peak broadening. This issue has been addressed using polymeric rather than

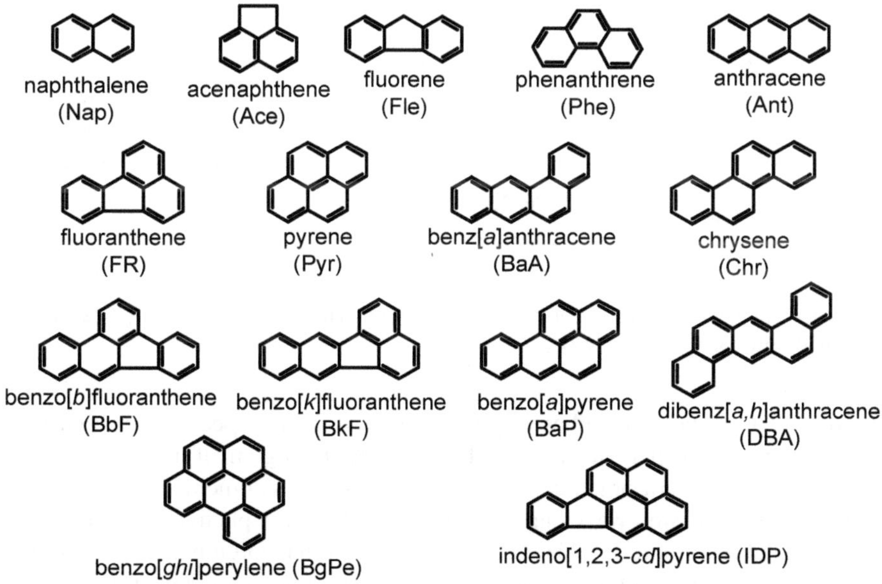

Fig. 2 USEPA's 16 PAHs recommended for measurement

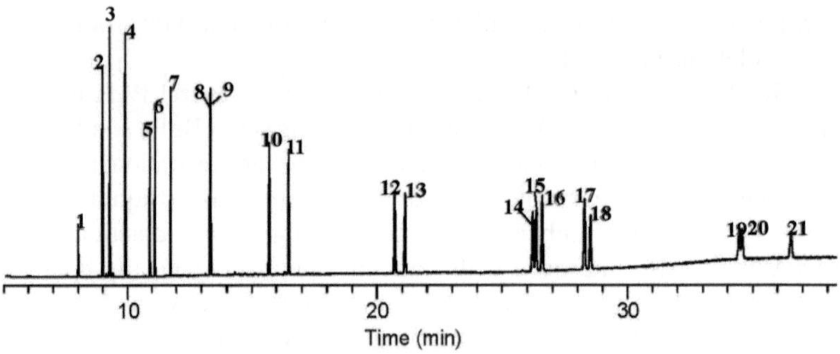

Fig. 3 Typical GC–MS chromatogram of standard PAHs. Column: SPB-50 (30 m × 0.25 mm I.D., 0.25 μm). Carrier gas: helium. Scan range: *m*/*z* 40–450. Peaks (each 10 ppm): (1) naphthalene; (2) 2-methyl naphthalene; (3) 1-methyl naphthalene; (4) biphenyl; (5) acenaphthylene; (6) acenaphthene; (7) fluorene; (8) phenanthrene; (9) anthracene; (10) fluoranthene; (11) pyrene; (12) benz[*a*]anthracene; (13) chrysene; (14) benzo[*b*]fluoranthene; (15) benzo[*k*]fluoranthene; (16) benzo[*j*]fluoranthene; (17) benzo[*e*]pyrene; (18) benzo[*a*]pyrene; (19) indeno[1,2,3-*cd*]pyrene; (20) dibenz[*a,h*]anthracene; (21) benzo[*ghi*]perylene (Figure in page 21 in Ministry of Environment, Government of Japan 2014)

Table 3 PAH *m*/*z* values for GC–MS analysis

PAH abbreviation	*m*/*z* for Q	*m*/*z* for C	PAH abbreviation	*m*/*z* for Q	*m*/*z* for C
Nap	128	102	BaP	252	253, 126
Ace	152	150	IDP	276	138, 227
Fle	166	165	BgPe	276	139, 279
Phe	178	179, 176	DBA	278	139, 279
Ant	178	179, 176	Nap-d_8	136	
FR	202	101, 203	Phe-d_{10}	188	94, 189
Pyr	202	101, 203	Pyr-d_{10}	212	106, 213
BaA	228	229, 226	FR-d_{10}	212	106, 213
Chr	228	226, 229	Chr-d_{12}	240	120, 241
BbF	252	253, 126	BaP-d_{12}	264	132, 265
BkF	252	253, 126	perylene-d_{12}	264	260

Q, quantification; C, confirmation. *m*/*z* value data were obtained from Hayakawa (2022a, b). For PAH abbreviations, refer to Table 1.

monomeric C_{18} stationary phases, producing well-defined peaks with high plate numbers. Figure 4 displays a typical HPLC-FLD chromatogram of standard reference material (SRM) 1941a organics extracted from marine sediment, with changes in excitation and emission wavelengths (Wise et al. 2012). A phenyl stationary phase is also available for unique PAH separations. For optimal column and mobile-phase combinations, users should consult manufacturers' guidelines. Given that the 16 USEPA PAHs, except acenaphthylene, are fluorescent, FLD is widely used in HPLC

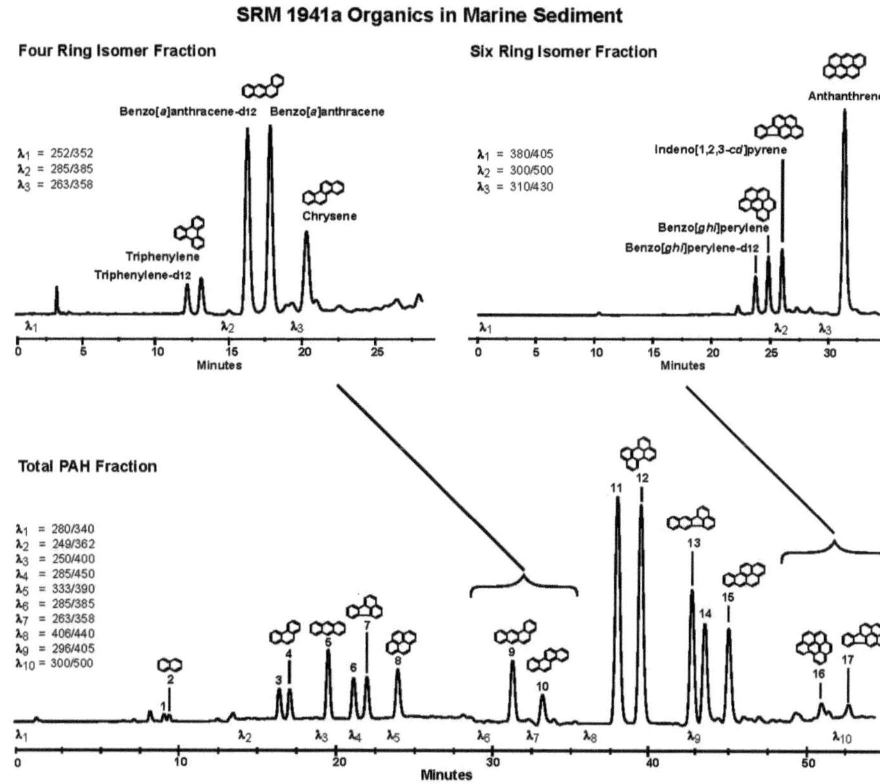

Fig. 4 Chromatograms of PAHs in Soxhlet extracts of SRM 1941a from marine sediments, obtained via HPLC-FLD. Column: polymeric C_{18} (25 cm × 0.46 cm I.D., 5 μm particle size). Excitation and fluorescence wavelengths, see in the chromatograms. Peak numbers 1, 3, 6, 11 and 14: deuterated PAHs used as internal standards for quantification (Fig. 2 in Wise et al. 2012)

for PAHs. During chromatographic analysis, excitation and emission wavelengths are adjusted to maximize sensitivity and selectivity for each PAH.

4.2 NPAHs

Unlike the 16 USEPA-recommended PAHs, priority NPAHs have not been established for environmental studies. Most studies have focused on mono- and di-NPAHs with two to five rings. Figure 5 shows the chemical structures and abbreviations of NPAHs targeted for environmental analysis in the author's laboratory. Although PAH standards and solutions, including deuterated compounds, are commercially available, crystals of several NPAHs and their solutions are less accessible owing to the limited number of manufacturers.

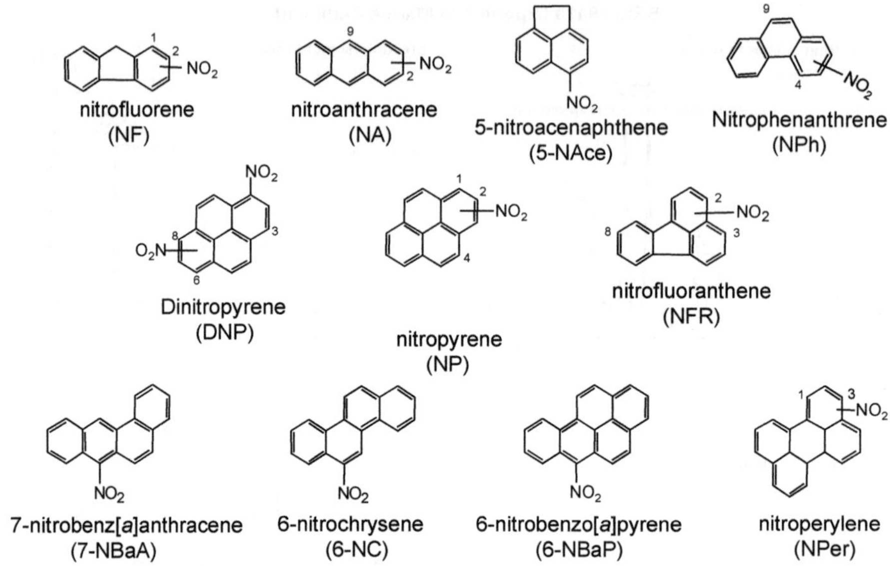

Fig. 5 NPAHs targeted in the author's laboratory

The nitro group of NPAHs can be reduced electrochemically or chemically to nitroso-PAHs, hydroxyamino-PAHs (OHPAHs), and amino-PAHs (APAHs), with both OHPAHs and APAHs being fluorescent and detectable via HPLC-FLD. Table 4 shows the excitation and emission wavelengths for APAHs. Common reduction reagents or catalysts for NPAHs before HPLC analysis include sodium hydroxide, zinc–copper, and cadmium–copper. An on-column reduction of NPAHs has also been developed (Tejada et al. 1986), and the FLD time is adjusted to detect each compound at optimal wavelengths.

Off-line chemical reduction of NPAHs requires large sample volumes (mL) and prolonged processing, such as refluxing a mixture of NPAHs and sodium hydrosulfide in methanol for 1.5 h. To address this, mini-columns packed with metal catalysts, such as zinc, have been integrated with HPLC-FLD as on-line reducers (Hayakawa et al. 1993), although these columns have limited lifespans. To improve durability, short columns packed with platinum (Pt)- and Pt/rhodium (Rh)-coated alumina, originally designed as three-way catalysts for vehicle exhaust gas (Matthey 2024), were developed, enabling rapid reduction of NPAHs to APAHs in a hot methanol eluent. This system was successfully combined with FLD to give sensitive HPLC determination of NPAHs (Delhomme et al. 2007; Ohno et al. 2009).

Chemiluminescence detection (CLD), sensitive to specific fluorophores, employs energy from chemical reactions to excite fluorophores from the ground to excited stages, with the released light having the same fluorescence spectrum. Notably, CLD does not require a light source, unlike FLD (Fig. 6).

CLD using peroxyoxalate (PO) esters, such as bis(2,4,6-trichlorophenyl)oxalate, was found to be more sensitive to APAHs compared with FLD (Sigvardson et al.

Table 4 Excitation and emission wavelengths used for HPLC-FLD analysis of NPAHs[a]

NPAH abbreviation[b]	Excitation wavelength[c] (nm)	Emission wavelength[c] (nm)
1-NF	285	370
2-NF	285	370
2-NA	260	495
9-NA	263	505
9-NPh	247	430
1-NP	360	430
1,3-DNP	395	445
1,6-DNP	369	442
1,8-DNP	395	454
3-NFR	300	530
8-NFR	300	550
6-NC	273	437
6-NBaP	420	475
3-NPer	227	540

[a]NPAHs were chemically reduced before their injection into HPLC-FLD. [b]Abbreviations and chemical compositions of NPAHs, refer to Fig. 3. [c]Wavelength data were obtained from Tejada et al. (1986) and Liu and Robbat (1991)

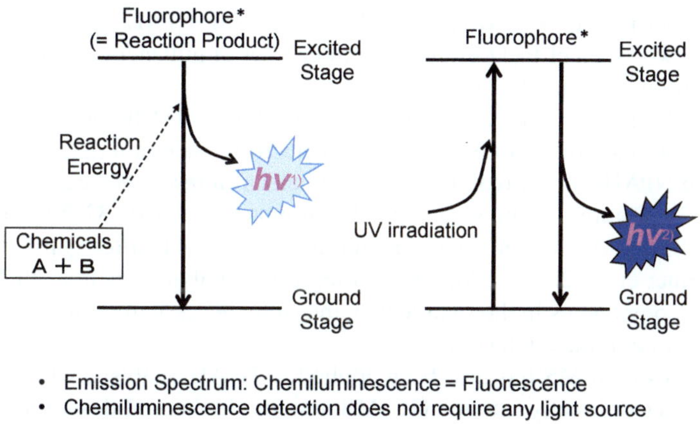

Fig. 6 Chemiluminescence and fluorescence detection

1984). Figure 7 shows the PO-CLD detection mechanism for NPAHs. HPLC-CLD detection limits ($S/N = 3$) for 1,3-, 1,6-, and 1,8-DNPs as well as 1-NP were in the sub-fmol range, approximately two orders of magnitude lower than those of FLD (Hayakawa et al. 1995, 2001). Introducing reduction and concentration columns packed with Pt/Rh-coated alumina and octadecylsilane, respectively, has led to an

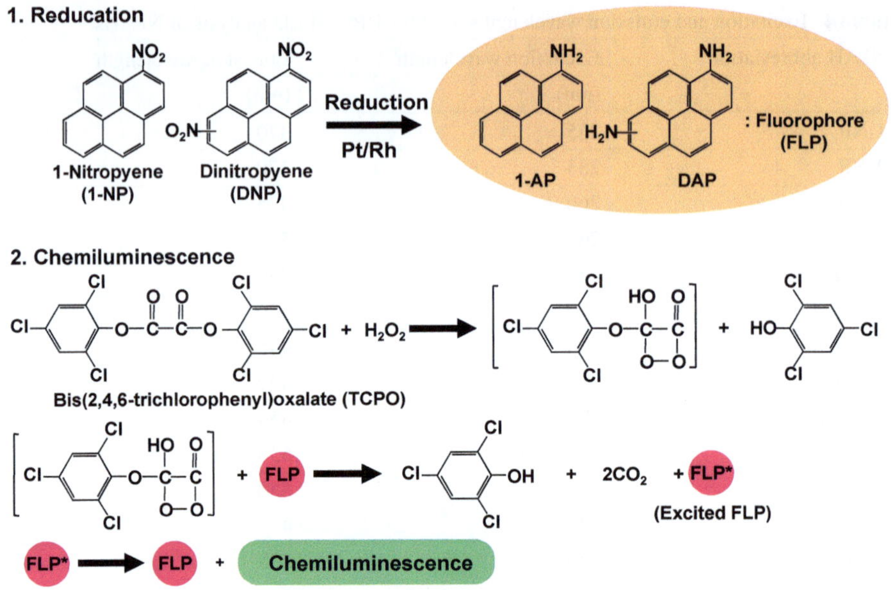

Fig. 7 Chemiluminescence detection mechanism for NPAHs

HPLC-CLD system for on-line reduction/concentration (Fig. 8), with typical chromatograms of airborne particulate extracts and NPAH standards detected via the system shown in Fig. 9 (Hayakawa 2016), enhancing NPAH analysis. Replacing CLD with two-channel FLD in this system enables simultaneous PAH and NPAH analysis (Boongla et al. 2017).

For NPAH analysis via GC–MS, negative ion chemical (NIC) ionization mode is two orders of magnitude more sensitive than electron ionization (EI) mode, as electronegative NPAHs ionize readily via thermal electron resonance capture (Bezabeh et al. 2003). Figure 10 shows a typical chromatogram with 32 NPAHs and 32 other oxidized PAH derivatives in organic aerosol extracts after QuEChER treatment (Albinet et al. 2014). When measuring trace analytes, such as NPAHs, ions from other components in the sample may interfere, necessitating the use of an MS/MS system to minimize interference.

Laser ionization MS has also been applied to NPAH analysis, with small solid samples injected directly into the GC–MS system via thermal desorption (Li et al. 2018). Sensitivity improvements of two- to eightfold were achieved by introducing an on-line chemical reducer that converts NPAHs to their aminated analogs (APAHs) using hydrazine monohydrate in GC coupled with multiphoton ionization time-of-flight–MS (Tang et al. 2019). In single MS with selective ion monitoring mode, ions from sample co-components can interfere with measurement. MS/MS systems provide an effective means to selectively monitor analyte ions, reducing this interference.

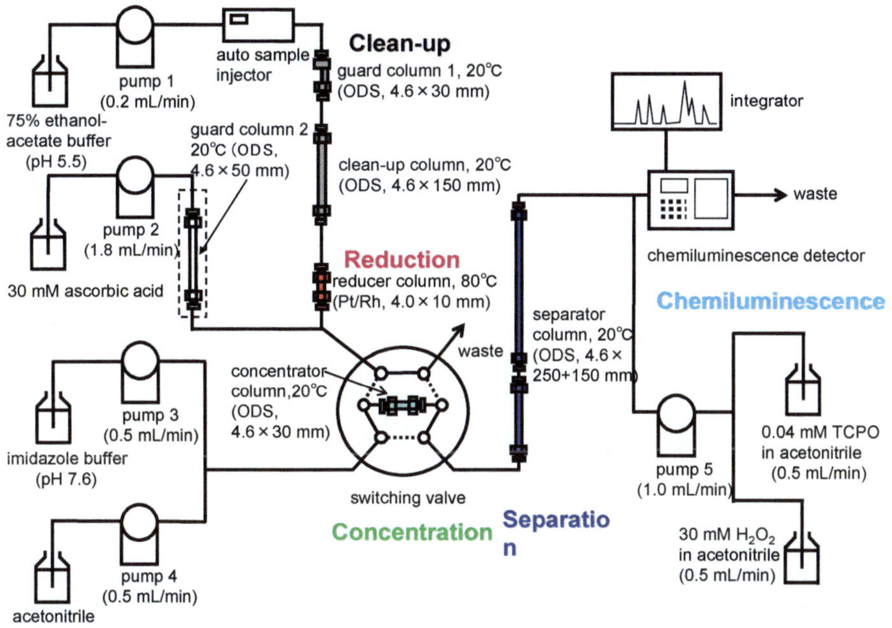

Fig. 8 Schematic diagram representing the on-line reduction/concentration HPLC- CLD system for NPAHs (Fig. 2 in Hayakawa 2016, with small modification)

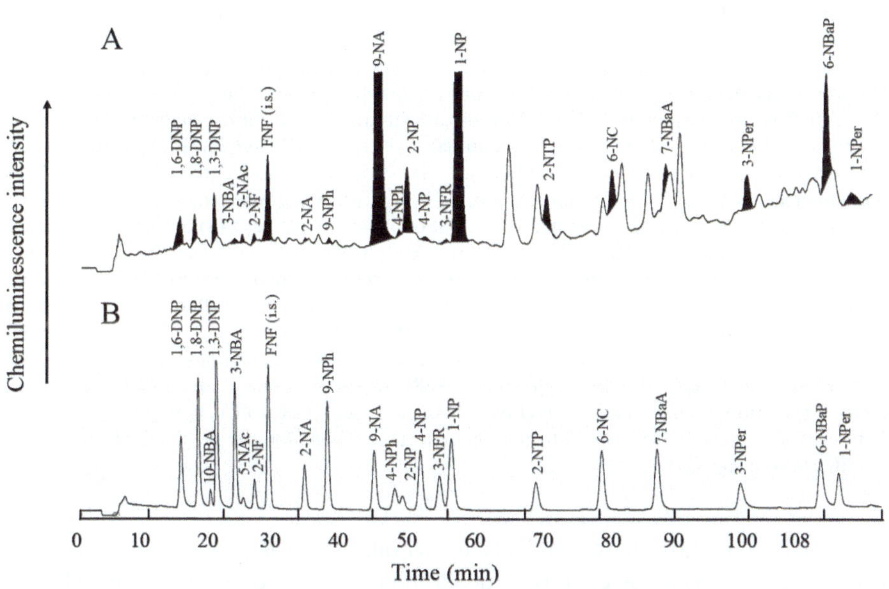

Fig. 9 HPLC-CLD chromatograms for **a** airborne particulates in Kanazawa and **b** the NPAH standard. HPLC-CLD system is illustrated in Fig. 8 (Fig. 3 in Hayakawa 2016)

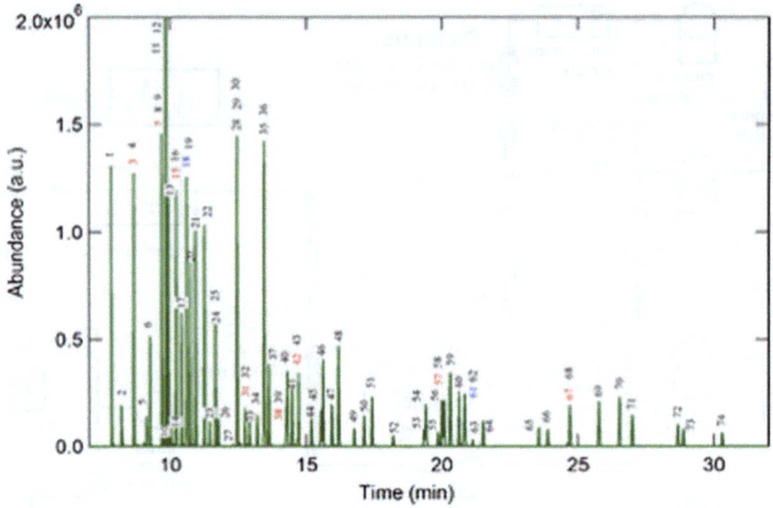

Fig. 10 Typical reconstituted GC–MS total ion chromatogram obtained from a standard solution of NPAH and OPAH (500 pg μL^{-1}, with 1 μL injected, in SIM acquisition mode). Peaks: (1) Phthaldialdehyde; (2) Phthalic anhydride; (3) 1,4-Naphthoquinone-d_6; (4) 1,4-Naphthoquinone; (5) 1-Naphthaldehyde; (6) 2-Formyl-*trans*-cinnamaldehyde; (7) 1-Nitronaphthalene-d_7; (8) 1-Nitronaphthalene; (9) 1,2-Naphthoquinone; (10) Benzophenone; (11 and 12) 2-Methyl-1-NN + 1-Methyl-5-NN[a], (13) 2-Nitronaphthalene; (14) 1-Acenaphthenone; (15) 2-Nitrobiphenyl-d_9; (16) 2-Nitrobiphenyl; (17) 2-Methyl-4-nitronaphthalene; (18) 9-Fluorenone-d_8; (19) 9-Fluorenone; (20) 1-Methyl-4-nitronaphthalene; (21) 1-Methyl-6-nitronaphthalene, (22) 3-Nitrobiphenyl; (23) 1,2-Naphthalic anhydride; (24) Biphenyl-2–2′-dicarboxaldehyde; (25) Xantone; (26) Acenaphthenequinone; (27) 2,3-Naphthalenedicarboxylic anhydride; (28) 1,5-Dinitronaphthalene; (29) Anthrone; (30) 6*H*-Dibenzo[*b*,*d*]pyran-6-one; (31) Anthraquinone-d_8; (32) 9,10-Anthraquinone; (33) 3-Nitrodibenzofuran; (34) 5-Nitroacenaphthene; (35) 1,8-Naphthalic anhydride; (36) 1,4-Anthraquinone; (37) 4,4'-Biphenyldicarboxaldehyde; (39) 2-Methylanthraquinone; (40) 2-Nitrofluorene; (41) 9-Phenanthrene carboxaldehyde; (42) 9-Nitroanthracene-d_9; (43) 9-Nitroanthracene; (44) 9,10-Phenanthrenequinone; (45) 2-Nitro-9-fluorenone; (46) 9-Nitrophenanthrene; (47) 2-Nitrodibenzothiophene; (48) 3-Nitrophenanthrene; (49) 2-Nitroanthracene; (50) 9-Methyl-10-nitroanthracene; (51) Benzo[*a*]fluorenone; (52) Benzo[*b*]fluorenone; (53) Benzanthrone; (54) 1-Pyrene carboxaldehyde; (55) Aceanthrenequinone; (56) 2-Nitrofluoranthene; (57) 3-Nitrofluoranthene-d_9; (58) 3-Nitrofluoranthene; (59) 4-Nitropyrene; (60) Benz[*a*]anthracene-7,12-dione; (61) 1-Nitropyrene-d_9; (62) 1-Nitropyrene; (63) 2-Nitropyrene; (64) 1,4-Chrysenequinone; (65) 7-Nitrobenz[*a*]anthracene; (66) 5,6-Chrysenequinone; (67) 6-Nitrochrysene-d_{11}; (68) 6-Nitrochrysene; (69) 1,3-Dinitropyrene; (70) 1,6-Dinitropyrene; (71) 1,8-Dinitropyrene; (72) 1-Nitrobenzo[*e*]pyrene; (73) 6-Nitrobenzo[*a*]pyrene; (74), 3-Nitrobenzo[*e*]pyrene (Fig. 2 in Albinet et al. 2014, with small modification in caption)

In analyses of hazardous chemicals in environmental samples, tandem mass spectrometry (MS/MS) by using a triple quadrupole mass spectrometer (QqQ-MS) has become attractive since its high sensitivity and specificity (Sun et al. 2020). As described above, two common ionization methods, electron impact (EI) and chemical ionization (CI), are commonly used for GC/MS. The EI method has advantages

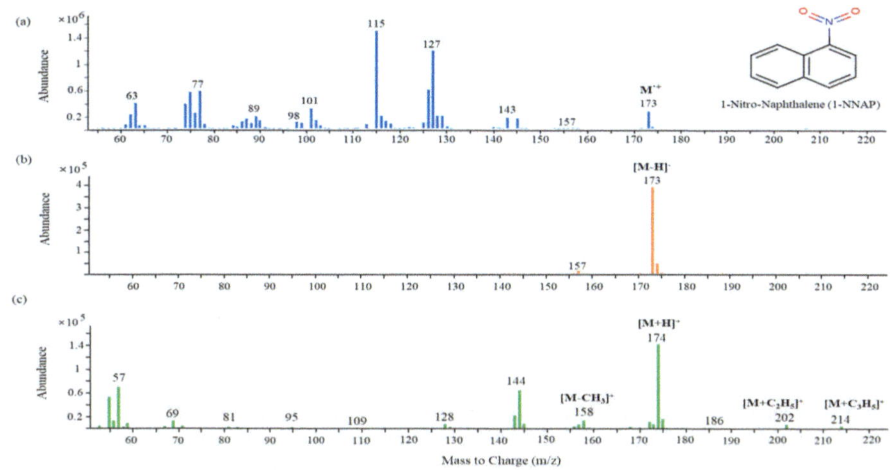

Fig. 11 Mass spectra of 1-nitrophenanthrene obtained from each ionization mode. **a** EI; **b** NCI; **c** PCI (Fig. 1 in Jo et al. 2023)

such as a mass spectral library search to assist compound identification (Vu-Duc et al. 2021). Conversely, the CI method is selective and sensitive to several compounds, such as NPAHs and OPAHs. However, it depends on the selection of negative chemical ionization (NCI) and positive chemical ionization (PCI) (Bezabeh et al. 2003). In recent years, detailed comparisons of these detection modes for the analysis of NPAHs and OPAHs in environmental samples using GC–MS and GC–MS/MS has been studied (Gramblicka et al 2023; Jo et al. 2023) .

Figure 11 compares the mass spectra of 1-NNap obtained by three ionization methods: EI, NCI, and PCI. EI has the lowest signal ion abundance. In addition, in NCI, only a molecular ion ($[M - H]^-$) is observed, and the mass spectrum is simple and stable. In PCI, in addition to a molecular ion ($[M + H]^+$), $[M + CH_3]^+$ and $[M + C_2H_5]^+$ are also observed. Therefore, for high-sensitivity analysis of NPAH and OPAH, it is more appropriate to use SIM mode under the NCI condition or selected reaction monitoring (SRC) mode under the PCI condition. Figure 12 shows chromatograms of NPAHs and OPAHs under NCI and PCI conditions. The difference in the abundance of NPAH is small between NCI and PCI conditions, but the abundance of OHPAH is larger for OPAH by the PCI, suggesting it more suitable.

4.3 *Metabolites of PAHs and NPAHs*

In animal body, PAHs oxidatively metabolized to hydroxyl derivatives in the presence of P450 in animals, and then conjugated. Therefore, after hydrolyzing these conjugates, the resulting hydroxylated PAHs can be analyzed by HPLC/FLD or CLD system. The urinary excretion of PAHs, such as fluorene, phenanthrene, fluoranthene

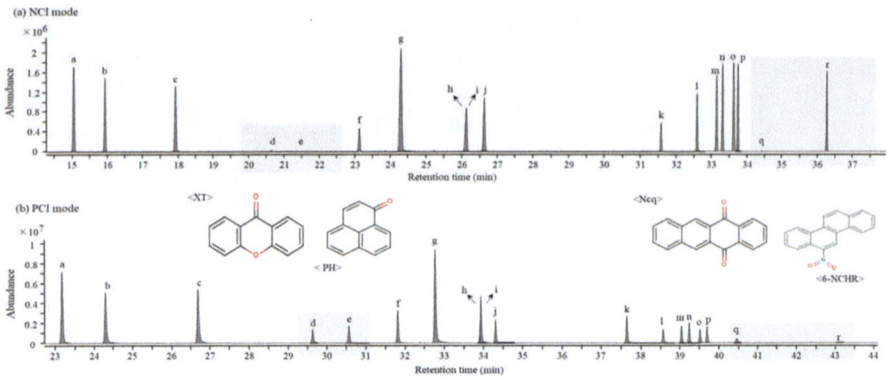

Fig. 12 Comparison between SIM and SRM modes of a standard solution of NPAHs and OPAHs (4 μg/mL). **a** SIM chromatogram under NCI condition on DB-5MS UI capillary column (30 m × 0.25 mm × 0.25 μm); **b** SRM chromatogram under PCI condition on DB-5MS UI capillary column (60 m × 0.25 mm × 0.25 μm). The gray shading indicates four congeners, with a significant difference in sensitivity between the two detection methods. The peak identities were as follows: **a**—1-nitronaphthalene (1-NNAP); **b**—2-nitronaphthalene (2-NNAP); **c**—9-fluorenone (9-Flu); **d**—xanthone (XT); **e**—phenalenone (PH); **f**—anthraquinone (Anq); **g**—1,8-naphthalic anhydride (1,8-NA); **h**—2-nitrofluorene (2-NFLUO); **i**—2-methylanthraquinone (2-Maq); **j**—9-nitroanthracene (9-NANT); **k**—benzo[b]fluoren-11-one (BbFLU); **l**—menzoanthrone (BZA); **m**—3-nitrofluoranthene (3-NFL); **n**—4-nitropyrene (4-NPYR); **o**—benz[a]anthracene-1,2-quinone (BAQ); **p**—1-nitropyrene (1-NPYR); **q**—5,12-naphthacenequinone (Ncq); **r**—6-nitrochrysene (6-NCHR) (Fig. 2 in Jo et al. 2023)

and pyrene, from human exposure was analyzed by HPLC-FLD (Chetiyanukornkul et al. 2002, 2006; Toriba et al. 2003) (Figs. 13 and 14).

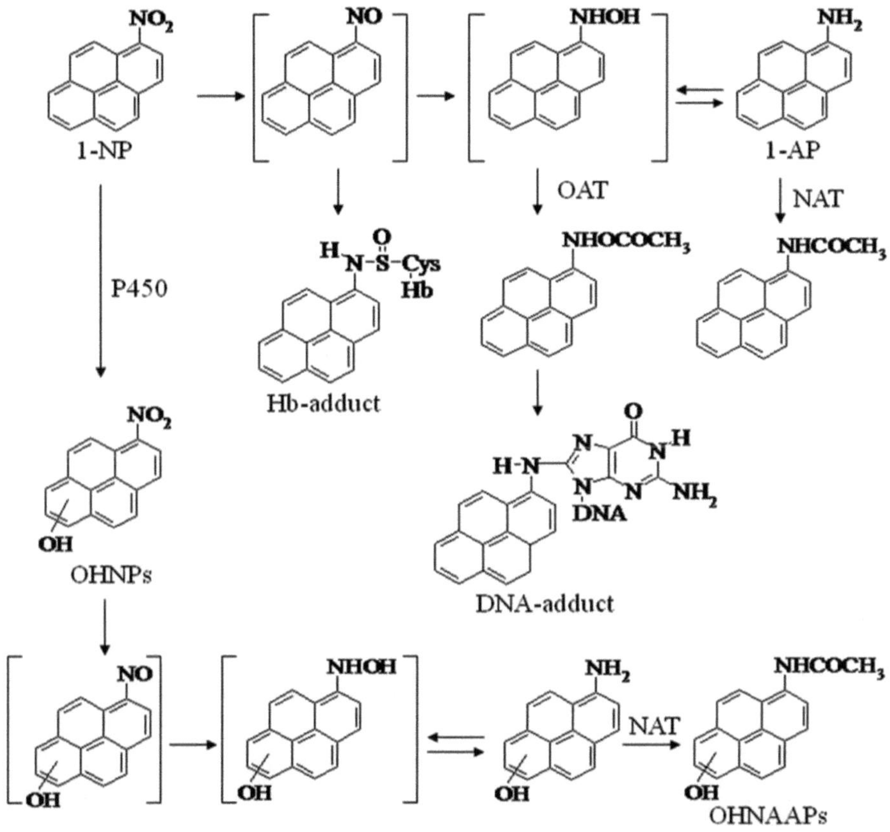

Fig. 13 Metabolism of 1-NP. P450, Cytochrome P450; NTA, *N*-Acetyltransferase; OTA, O-acetyltransferase (Fig. 10 in Hayakawa et al. 2007)

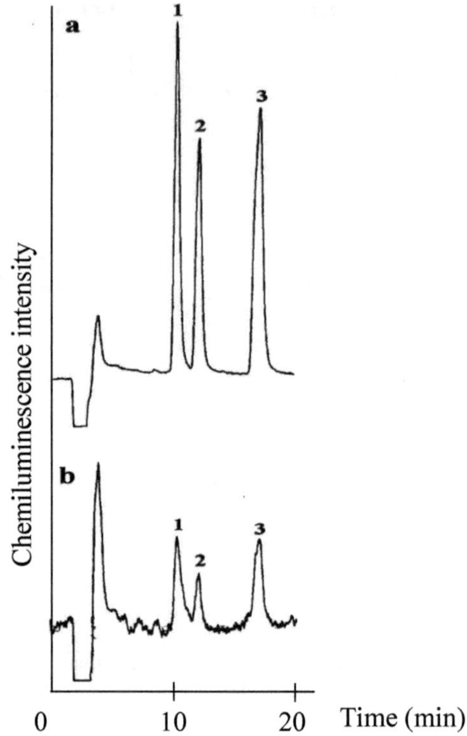

Fig. 14 HPLC-CLD chromatograms for **a** standard mixture of OHNPs and **b** rat urine spiked with 1-NP (3.3×10^{-3} M) after treatment with rat S9 mix (Fig. 11 in Hayakawa et al. 2007)

References

Abbas, I., Badran, G., Verdin, A., Ledoux, F., Roumié, M., Courcot, D., & Garçon, G. (2018). Polycyclic aromatic hydrocarbon derivatives in airborne particulate matter: Sources, analysis and toxicity. *Environmental Chemistry Letters*, *16*(2), 439–475. https://doi.org/10.1007/s10311-017-0697-0

Albinet, A., Nalin, F., Tomaz, S., Beaumont, J., & Lestremau, F. (2014). A simple QuEChERS-like extraction approach for molecular chemical characterization of organic aerosols: Application to nitrated and oxygenated PAH derivatives (NPAH and OPAH) quantified by GC–NICIMS. *Analytical and Bioanalytical Chemistry*, *406*(13), 3131–3148. https://doi.org/10.1007/s00216-014-7760-5

Atkinson, R., & Arey, J. (1994). Atmospheric chemistry of gas-phase polycyclic aromatic hydrocarbons: Formation of atmospheric mutagens. *Environmental Health Perspectives*, *102*(suppl 4), 117–126. https://doi.org/10.1289/ehp.94102s4117

Bandowe, B. A. M., & Meusel, H. (2017). Nitrated polycyclic aromatic hydrocarbons (nitro-PAHs) in the environment – A review. *Science of the Total Environment*, *581–582*, 237–257. https://doi.org/10.1016/j.scitotenv.2016.12.115

Bezabeh, D. Z., Bamford, H. A., Schantz, M. M., & Wise, S. A. (2003). Determination of nitrated polycyclic aromatic hydrocarbons in diesel particulate-related standard reference materials by using gas chromatography/mass spectrometry with negative ion chemical ionization. *Analytical and Bioanalytical Chemistry*, *375*(3), 381–388. https://doi.org/10.1007/s00216-002-1698-8

Boongla, Y., Orakij, W., Nagaoka, Y., Tang, N., Hayakawa, K., & Toriba, A. (2017). Simultaneous determination of polycyclic aromatic hydrocarbons and their nitro-derivatives in airborne particulates by using two-dimensional high-performance liquid chromatography with on-line reduction and fluorescence detection. *Asian Journal of Atmospheric Environment, 11*(4), 283–299. https://doi.org/10.5572/ajae.2017.11.4.283

Borges, B., Melo, A., Ferreira, I. M. P. L. V. O., & Mansilha, C. (2018). Dispersive liquid–liquid microextraction for the simultaneous determination of parent and nitrated polycyclic aromatic hydrocarbons in water samples. *Acta Chromatographica, 30*(2), 119–126. https://doi.org/10.1556/1326.2017.00126

Chetiyanukornkul, T., Toriba, A., Kizu, R., Makino, T., Nakazawa, H., Hayakawa, K. (2002). Determination of 1-hydroxypyrene in human urine by HPLC with fluorescence detection using a deuterated internal standard., *J. Chromatogr.* A, **961** (1)107–112 (2002). https://doi.org/10.1016/S0021-9673(02)00363-1

Chetiyanukornkul, T., Toriba, A., Kameda, T., Tang, N., Hayakawa, K. (2006). Simultaneous determination of urinary hydroxylated metabolites of naphthalene, fluorene, phenanthrene, fluoranthene and pyrene as multiple biomarkers of exposure to polycyclic aromatic hydrocarbons. *Anal. Bioanal. Chem.*, **386**, 712–718. https://doi.org/10.1007/s00216-006-0628-6

Daellenbach, K. R., Uzu, G., Jiang, J., Cassagnes, L.-E., Leni, Z., Vlachou, A., Stefenelli, G., Canonaco, F., Weber, S., Segers, A., Kuenen, J. J. P., Schaap, M., Favez, O., Albinet, A., Aksoyoglu, S., Dommen, J., Baltensperger, U., Geiser, M., El Haddad, I., Jaffrezo, J. L., Prévôt, A. S. H. (2020). Sources of particulate-matter air pollution and its oxidative potential in Europe. *Nature, 587*(7834), 414–419. https://doi.org/10.1038/s41586-020-2902-8

Deng, K., & Chan, W. (2017). Development of a QuEChERS-based method for determination of carcinogenic 2-nitroflfluorene and 1-nitropyrene in rice grains and vegetables: A comparative study with benzo[a]pyrene. Journal of Agricultural and Food Chemistry, *65*(9), 1992–1999. https://doi.org/10.1021/acs.jafc.7b00051

Delhomme, O., Herckes, P., & Millet, M. (2007). Determination of nitro-polycyclic aromatic hydrocarbons in atmospheric aerosols using HPLC fluorescence with a post-column derivatisation technique. *Analytical and Bioanalytical Chemistry, 389*(6), 1953–1959. https://doi.org/10.1007/s00216-007-1562-y

dos Santos, R. R., Vidotti Leal, L. D., De Lourdes Cardeal, Z., & Menezes, H. C. (2018). Determination of polycyclic aromatic hydrocarbons and their nitrated and oxygenated derivatives in coffee brews using an efficient cold fiber-solid phase microextraction and gas chromatography mass spectrometry method. *Journal of Chromatography A, 1584*, 64–71. https://doi.org/10.1016/j.chroma.2018.11.046

Gbeddy, G., Egodawatta, P., Goonetilleke, A., Ayoko, G., Jayarathne, A., Chen, L., & Russell, S. (2020). Optimized simultaneous pressurized fluid extraction and in-cell clean-up, and analysis of polycyclic aromatic hydrocarbons (PAHs), and nitro-, carbonyl-, hydroxy -PAHs in solid particles. *Analytica Chimica Acta, 1125*, 19–28. https://doi.org/10.1016/j.aca.2020.05.021

Gramblicka, T., Parizek, O., Stupak, M., Pulkrabova, J. (2023). Assessment of atmospheric pollution by oxygenated and nitrated derivatives ochf polycyclic aromatic hydrocarbons in two regions of the Czech Republic. *Atmos. Environ., 310*, 119981. https://doi.org/10.1016/j.atmosenv.2023.119981

Guiñez, M., Martinez, L. D., Fernandez, L., & Cerutti, S. (2017). Dispersive liquid–liquid microextraction based on solidification of floating organic drop and fluorescence detection for the determination of nitrated polycyclic aromatic hydrocarbons in aqueous samples. *Microchemical Journal, 131*, 1–8. https://doi.org/10.1016/j.microc.2016.10.020

Hayakawa, K., Terai, N., Suzuki, K., Dinning, P. G., Yamada, M., & Miyazaki, M. (1993). Chromatographic determination method for 1-nitropyrene and its metabolites in biological samples with fluorescence detection after on-line reduction. *Biomedical Chromatography, 7*(5), 262–266. https://doi.org/10.1002/bmc.1130070505

Hayakawa, K., Murahashi, T., Butoh, M., & Miyazaki, M. (1995). Determination of 1,3-, 1,6-, and 1,8-dinitropyrenes and 1-nitropyrene in urban air by high-performance liquid chromatography using chemiluminescence detection. *Environmental Science & Technology, 29*(4), 928–932. https://doi.org/10.1021/es00004a012

Hayakawa, K. (2000). Chromatographic methods for carcinogenic/mutagenic nitropolycyclic aromatic hydrocarbons – A review. *Biomedical Chromatography, 14*, 397–405. https://doi.org/10.1002/1099-0801(200010)14:6%3C397

Hayakawa, K., Noji, K., Tang, N., Toriba, A., Kizu, R., Sakai, S., & Matsumoto, Y. (2001). A high-performance liquid chromatographic system equipped with on-line reducer, clean-up and concentrator columns for determination of trace levels of nitropolycyclic aromatic hydrocarbons in airborne particulates. *Analytica Chimica Acta, 445*(2), 205–212. https://doi.org/10.1016/s0003-2670(01)01279-x

Hayakawa, K., Toriba, A., Kameda, T., Tang, N. (2007). A review: Development of analytical methods for hazardous nitropolycyclic aromatic hydrocarbons and studies on their environmental behavior. *Bunseki Kagaku, 56*(11), 905–920. https://doi.org/10.2116/bunsekikagaku.56.905

Hayakawa, K., Tang, N., Toriba, A., & Nagato, E. G. (2020). Calculating sources of combustion-derived particulates using 1-nitropyrene and pyrene as markers. *Environmental Pollution, 265*, 114730. https://doi.org/10.1016/j.envpol.2020.114730

Hayakawa, K., Tang, N., Matsuki, A., Inomata, Y., Toriba, A., & Nagato, E. G. (2021). Calculating source contributions to urban atmospheric polycyclic aromatic hydrocarbons and nitropolycyclic aromatic hydrocarbons using 1-nitropyrene and pyrene: An application to an Asian dust event. *Chemosphere, 280*, 130662. https://doi.org/10.1016/j.chemosphere.2021.130662

Hayakawa, K. (2016). Environmental behaviors and toxicities of polycyclic aromatic hydrocarbons and nitropolycyclic aromatic hydrocarbons – A review. *Chemical & Pharmaceutical Bulletin, 64*(2), 83–94. https://doi.org/10.1248/cpb.c15-00801

Hayakawa, K. (2022). Recent research progress on nitropolycyclic aromatic hydrocarbons in outdoor and indoor environments – A review. *Applied Sciences, 12*(21), 11259. https://doi.org/10.3390/app122111259

Hayakawa, K. (2022). Polycyclic aromatic hydrocarbons. In: H. Akimoto, & H. Tanimoto (Eds.), *Handbook of air quality and climate change*. Springer Nature Singapore Pte Ltd., pp. 1–17. https://doi.org/10.1007/978-981-15-2527-8_22-1

Hayakawa, K. (2023). Polycyclic aromatic hydrocarbons from vegetation burning and health effects, in Special issue "Air pollution from vegetation fire and its health effects in Asia". *Global Environmental Research, 27*, 13–20. https://airies.wikiplus.net/journal_27-1eng.html

Johnson Matthey. (2024). *Three-way catalysts (TWC)*. Retrieved August 23, 2024, from https://matthey.com/products-and-markets/transport/mobile-emission-control/gasoline-applications/three-way-catalysts

Jo, J., Lee, J.-Y., Jang, K.-S., Matsuki, A., Natsagdori, A. (2023). Development of quantitative chemical ionization using gas chromatography/mass spectrometry and gas chromatography/tandem mass spectrometry for ambient nitro- and oxy-PAHs and its applications. *Molecules*, **28**, 775. https://doi.org/10.3390/molecules28020775

Kameda, T., Azumi, E., Fukushima, A., Tang, N., Matsuki, A., Kamiya, Y., Toriba, A., & Hayakawa, K. (2016). Mineral dust aerosols promote the formation of toxic nitropolycyclic aromatic compounds. *Scientific Reports, 6*(1), 24427. https://doi.org/10.1038/srep24427

Kong, J., Han, M., Liu, Y., He, H., Gao, Z., Xian, Q., Yang, S., Sun, C., Li, S., & Zhang, L. (2018). Analysis of trace-level nitrated polycyclic aromatic hydrocarbons in water samples by solid-phase microextraction with gas chromatography and mass spectrometry. *Journal of Separation Science, 41*(12), 2681–2687. https://doi.org/10.1002/jssc.201701271

Kong, J., Zhu, F., Huang, W., He, H., Hu, J., Sun, C., Xian, Q., & Yang, S. (2019). Sol–gel based metal-organic framework zeolite imidazolate framework-8 fibers for solid-phase microextraction of nitro polycyclic aromatic hydrocarbons and polycyclic aromatic hydrocarbons in water samples. *Journal of Chromatography A, 1603*, 92–101. https://doi.org/10.1016/j.chroma.2019.06.063

Krzyszczak, A., & Czech, B. (2021). Occurrence and toxicity of polycyclic aromatic hydrocarbons derivatives in environmental matrices – A review. *Science of the Total Environment*, *788*, 147738. https://doi.org/10.1016/j.scitotenv.2021.147738

Li, J., Liu, Y., Su, H., Elaine Wong, Y.-L., Chen, X., Dominic Chan, T.-W., & Chen, Q. (2017). In situ hydrothermal growth of a zirconium-based porphyrinic metal-organic framework on stainless steel fibers for solid-phase microextraction of nitrated polycyclic aromatic hydrocarbons. *Microchimica Acta*, *184*(10), 3809–3815. https://doi.org/10.1007/s00604-017-2403-0

Li, L., Ho, S. S. H., Han, Y. M., Chow, J. C., Watson, J. G., Wang, L., Qu, L., & Cao, J. (2018). Quantification of nitrated-polycyclic aromatic hydrocarbons in atmospheric aerosol samples with in-injection port thermal desorption-gas chromatography/ negative chemical ionization mass spectrometry method. *Atmospheric Environment*, *192*, 84–93. https://doi.org/10.1016/j.atmosenv.2018.08.049

Liu, T.-Y., & Robbat, A. Jr. (1991). High-performance liquid chromatography retention index and detection of nitrated polycyclic aromatic hydrocarbons. *Journal of Chromatography A*, *539*(1), 1–14. https://doi.org/10.1016/s0021-9673(01)95355-5

Ministry of Environment, Government of Japan (2014). *Multicomponent measurement method for atmospheric polycyclic aromatic hydrocarbons in Hazardous air pollutants measurement methods manual*. Retrieved November 18, 2024, from https://www.env.go.jp/content/900402510.pdf

Nowakowski, M., Rykowska, I., Wolski, R., & Andrzejewski, P. (2022). Polycyclic aromatic hydrocarbons (PAHs) and their derivatives (O-PAHs, N-PAHs, OH-PAHs): Determination in suspended particulate matter (SPM) – a review. *Environmental Processes*, *9*(1), 2. https://doi.org/10.1007/s40710-021-00555-7

Ohno, T., Toriba, A., Kameda, T., Tang, N., & Hayakawa, K. (2009). Determination of 1-nitropyrene in low volume ambient air samples by high-performance liquid chromatography with fluorescence detection. *Journal of Chromatography A*, *1216*(21), 4625–4628. https://doi.org/10.1016/j.chroma.2009.03.070

Santos, L. O., dos Anjos, J. P., Ferreira, S. L. C., & de Andrade, J. B. (2017). Simultaneous determination of PAHs, nitro-PAHs and quinones in surface and groundwater samples using SDME/GC-MS. *Microchemical Journal*, *133*, 431–440. https://doi.org/10.1016/j.microc.2017.04.012

Santos, L. O., Santos, A. G., & de Andrade, J. B. (2018). Methodology to examine polycyclic aromatic hydrocarbons (PAHs) nitrated PAHs and oxygenated PAHs in sediments of the Paraguaçu River. *Marine Pollution Bulletin*, *136*, 248–256. https://doi.org/10.1016/j.marpolbul.2018.09

Scaramboni, C., Neris, J. B., do Nasciment, R. D. K. S. D., Chiaranda Da Rosa, N. L., Carvalho, J. S., Grosseli, G. M., Campos, M. L. A. M., Fadini, P. S., & Urban, R. C. (2021). Optimization of a low volume extraction method to determine polycyclic aromatic hydrocarbons in aerosol samples. *Frontiers in Environmental Science*, *9*, 742251. https://doi.org/10.3389/fenvs.2021.742251

Sigvardson, K. W., Kennish, J. M., & Birks, J. W. (1984). Peroxyoxalate chemiluminescence detection of polycyclic aromatic amines in liquid chromatography. *Analytical Chemistry*, *56*, 1096–1102. https://doi.org/10.1021/ac00271a001

Sun, C., Qu, L., Wu, L., Wu, X., Sun, R., & Li, Y. (2020). Advances in analysis of nitrated polycyclic aromatic hydrocarbons in various matrices. *TrAC Trends in Analytical Chemistry*, *127*, 115878. https://doi.org/10.1016/j.trac.2020.115878

Tang, Y., Yamamoto, S., Imasaka, T. (2019). Determination of nitrated polycyclic aromatic hydrocarbons in particulate matter 2.5 by laser ionization mass spectrometry using an on-line chemical-reduction system. *Analyst*, *144*(9), 2909–2913. https://doi.org/10.1039/c9an00308h

Tejada, S. B., Zweidinger, R. B., & Sigsby, J. E. (1986). Fluorescence detection and identification of nitro derivatives of polynuclear aromatic hydrocarbons by on-column catalytic reduction to aromatic amines. *Analytical Chemistry*, *58*(8), 1827–1834. https://doi.org/10.1021/ac00121a050

Toriba, A., Chetiyanukornkul, T., Kizu, R., Hayakawa, K. (2003). Quantification of 2-hydroxyfluorene in human urine by column-switching high performance liquid chromatography with fluorescence detection. *Analyst*, **128**, 605–610. https://doi.org/10.1039/B212738E

USEPA (1984). *Method 610: Polynuclear aromatic hydrocarbons*. Retrieved November 18, 2024, from https://www.epa.gov/sites/default/files/2015-10/documents/method_610_1984.pdf

Vu-Duc, N., Thi, L. A. P., Le-Minh , T., Nguyen, L.-A., Nguyen-Thi, H., Pham-Thi, L.-H., Doan-Thi, V.-A., Le-Quang, H., Nguyen-Xuan, H., Nguyen, T. T., Nguyen, P. T., Chu, B. T. (2021). Analysis of polycyclic aromatic hydrocarbon in airborne particulate matter samples by gas chromatography in combination with tandem Mass spectrometry (GC-MS/MS). J. Anal. Methods Chem., 6641326. https://doi.org/10.1155/2021/6641326

Wang, X., Sheng, W.-R., Jiao, X.-Y., Zhao, R.-S., Wang, M.-L., & Lin, J.-M. (2018). Zinc(II)-based metal–organic nanotubes coating for high sensitive solid phase microextraction of nitro-polycyclic aromatic hydrocarbons. *Talanta, 186*, 561–567. https://doi.org/10.1016/j.talanta.2018.02.062

Westblad, C., Levendis, Y. A., Richter, H., Howard, J. B., & Carlson, J. (2002). A study on toxic organic emissions from batch combustion of styrene. *Chemosphere, 49*(4), 395–412. https://doi.org/10.1016/s0045-6535(02)00311-9

Wise, S. A., Phinney, K. W., Sander, L. C., & Schantz, M. M. (2012). Role of chromatography in the development of standard reference materials for organic analysis – A review. *Journal of Chromatography A, 1261*, 3–22. https://doi.org/10.1016/j.chroma.2012.05.093

Yan, J., Wang, X., Gong, P., & Wang, C. (2021). Nitrated polycyclic aromatic compounds in the atmospheric environment: A review. *Critical Reviews in Environmental Science and Technology, 51*(11), 1159–1185. https://doi.org/10.1080/10643389.2020.1748486

Zajemska, M., Poskart, A., & Musial, D. (2015). The kinetics of nitrogen oxides formation in the flame gas. *Economics for Environmental Studies, 15*, 445–460. https://www.ees.uni.opole.pl

Biomonitoring of PAHs: Animals

Masato Honda

Abstract Biomonitoring studies of environmental polycyclic aromatic hydrocarbon contamination using animals are very useful and have been carried out worldwide. Many of these studies have used bivalves and fish in aquatic environments and land snails and crustaceans in terrestrial environments. A major advantage of polycyclic aromatic hydrocarbon biomonitoring is that, unlike simple analysis of environmental matrix samples, the use of biological samples makes it possible to investigate contamination spatially and temporally. At the same time, by analyzing the concentrations in living organisms, it is possible to evaluate the biological effects taking into account bioavailability data.

Keywords Polycyclic aromatic hydrocarbon · Biomonitoring · Mussel watch · Invertebrate · Vertebrate · Marine environment · Terrestrial environment · Sample pretreatment

1 Introduction

Most polycyclic aromatic hydrocarbons (PAHs) in the environment are generated by incomplete combustion of organic compounds such as fossil fuels and/or organic matter. Several anthropogenic sources, such as power plants, car exhaust, and controlled burning, can release PAHs into the environment (Liu et al. 2015; Mastral and Callen 2000). Additionally, PAHs can be accidentally released into the environment from fossil oils (Koyama et al. 2016; Uno et al. 2010). These are now common and ubiquitous pollutants in the environment (Honda et al. 2007; Ito et al. 2015; Magi et al. 2002; Miki et al. 2014; Nakata et al. 2014; Sarria-Villa et al. 2016; Slezakova et al. 2013). Consequently, environmental monitoring of PAHs is important worldwide.

M. Honda (✉)
Institute of Nature and Environmental Technology, Kanazawa University, Kakuma-Machi, Kanazawa, Japan
e-mail: mst-honda@se.kanazawa-u.ac.jp

K. Hayakawa and Y. Aoki (eds.), *A New Era of Polycyclic Aromatic Hydrocarbons*,
https://doi.org/10.1007/978-981-96-7695-8_12

For monitoring of PAHs pollution, it is necessary to determine the concentrations of these substances in environmental matrices, including soil, sediment, water, and air. Direct analysis of matrix samples can provide accurate data for set periods to understand the pollution situation. However, the concentrations in environmental matrices show large fluctuations over time, especially in water and air, which makes it difficult to understand the overall situation and make predictions about ongoing pollution.

Biological monitoring focusing on bioaccumulation of environmental pollutants has been developed to overcome the above issues. This type of monitoring has been investigated using various organisms in the field (Teixeira et al. 2024). In biomonitoring, organisms that usually live in and/or around the target area are used to monitor the concentrations and changes of pollutants in the environment. Pollutants are actively and/or passively taken in and concentrate in the body of the target organism through feeding, respiration, and dermal contact. Generally, typical environmental pollutants, such as persistent organic pollutants, have relatively high hydrophobicities and bioaccumulation potentials (Corsolini and Sarà 2017; Khairy et al. 2021); however, PAHs behave slightly different in terms of bioaccumulation. In many cases, increases in the concentrations of PAHs in animals at successive trophic levels in the food chain have not been observed (Baumard et al. 1998; Wan et al. 2007) generally and are not expected to occur because of the high metabolization/excretion potentials of most PAHs in various organisms. Generally, PAHs are biodegraded under aerobic conditions, with the rate of biodegradation decreasing dramatically with increases in the number of aromatic rings in the compound.

Many environmental pollutants, including PAHs, are relatively hydrophobic and therefore, bioaccumulation via water, air, and food. Biomonitoring has the following advantages compared with analysis in environmental media: (1) the concentration of a substance is generally higher in living organisms than in environmental media because of bioaccumulation and accumulation through food webs, which means that small sample volumes can be used for analysis; (2) for relatively/highly persistent substances, the intake amount is integrated over a long period and the time change is averaged, which means that a highly representative result can be obtained from a single measurement; and (3) because the results of biomonitoring depend on the accumulation potential of a pollutant in living organisms, it is possible to directly evaluate the accumulation and effects on organisms (Okamura and Aoyama 1998; Shibata and Takazawa 2014). Generally, animals are motile and if they prey on other organisms, the prey are motile as well. This means that analytical values in biomonitoring are regarded as average concentrations over a certain space (Shibata and Takazawa 2014).

When using living organisms for biomonitoring of environmental pollutants (i.e., PAHs), it is desirable that the organism to meet the following conditions: (1) the organism must accumulate the substance and the internal concentration of the substance must reach equilibrium relatively quickly; (2) knowledge must be available regarding the relationships among age, growth, feeding habits and other biological information; (3) the organism must have a clear, relatively narrow living area throughout its entire life cycle; and (4) the organism must be widely distributed throughout the world and easy to sample (Chandurvelan et al. 2015; Monirith et al. 2003).

Environmental biomonitoring of PAHs can be broadly divided into two types: (1) routine environmental surveys for general areas and those of concern and (2) surveys in response to accidents. The former are conducted by various research institutes and government agencies to monitor the environment on a daily basis (Ito et al. 2015; Miki et al. 2014; Tanaka and Onduka 2010), while the latter are carried out in response to sudden environmental pollution caused by oil spills, earthquakes, and fires (Bacosa and Inoue 2015; Koyama et al. 2004, 2016; Onozato et al. 2016; Pie et al. 2015; Pulster et al. 2020; Uno et al. 2010, 2017). PAH emission sources are broadly divided into petrogenic sources and pyrogenic sources, and their sources could be estimated by diagnostic ratio analysis using the abundance ratios of the PAHs (Tobiszewski and Namieśnik 2012). However, because PAHs in the body are affected by absorption, metabolism, and excretion pathways specific to each organism, the ratio of PAHs present is likely to vary greatly and diagnostic ratio analysis should be applied with caution.

Compared with environmental media samples (e.g., soil and water), organic samples are generally more difficult to analyze, except for by instrumental analysis. This means that organic samples must undergo complicated extraction and purification procedures because they usually contain more proteins, lipids, and other organic compounds.

In this chapter, examples of biomonitoring studies in the marine and terrestrial environments using animals are discussed and related analytical methods, including extraction/purification and instrumental analysis, are presented.

2 Biomonitoring Species

For use in biomonitoring surveys, it is desirable for biological species to satisfy all of the above conditions. Animals that have a high lipid content, such as carnivores and omnivores, are considered suitable for biomonitoring because they take in and can concentrate PAHs through food, water, and related environmental media.

It is simple to sample organisms living in a survey area and analyze them; however, factors, such as differences in species, sex, and age, and seasonal variation in biological activities such as uptake, metabolism, and excretion, can result in variability. Therefore, it is necessary to understand the biological and ecological characteristics of the target organisms when monitoring and then sample under conditions that control these factors as much as possible.

To reduce bias from factors such as differences in species, sex, age, and season, it is important to design a biological monitoring system that consists of the same species, or biologically closely related species, of the same sex and size, and with similar growth periods. From this perspective, small carnivorous/omnivorous organisms, including arthropods (crustaceans and insects), mollusks (bivalves and squid), fish, small mammals, and birds or their eggs are generally used as biomonitoring species

(Table 1). Organisms that have relatively short life spans are suitable for studying environmental changes. From the point of view of nature conservation, it is desirable to avoid endangered species and other protected species, and species where collection of even a small number of organisms may affect the local ecosystem. Appropriate species are those that are relatively small, live in large numbers in one area, and are easy to collect (Shibata and Takazawa 2014). Teixeira et al. (2024) listed the following organisms as useful biomonitoring indicator species for PAHs: benthos invertebrates, fish, amphibians, cephalopods, bivalves, terrestrial snails, birds, insects, and mammals (Al-Alam et al. 2019; Baroudi et al. 2020; Cunningham et al. 2022; Thuy et al. 2018; Wallace et al. 2020).

2.1 Aquatic Environment

Aquatic environments, and especially coastal areas close to human populations, tend to be highly polluted with PAHs from anthropogenic sources, for example, from domestic and industrial fossil fuel burning (Liu et al. 2015; Mastral and Callen 2000), oil spills (Koyama et al. 2004, 2016; Uno et al. 2010), artificial driftage (Ministry of the Environment 2012), tar balls (Zakaria et al. 2001), and plastic debris that has adsorbed PAHs (Rios et al. 2007). These pollution events may affect the ecosystem; therefore, biomonitoring of the marine environment is conducted worldwide.

Monitoring of PAHs in aquatic environments has focused on the marine environment and habitats. The most well-known and frequently using organisms are bivalves, which are used in the National Mussel Watch Program (NCCOS 2017). This program was started by the United States National Oceanic and Atmospheric Administration (NOAA) in 1986 for water contaminant monitoring (NCCOS 2017). The program monitors the status and spatial and temporal trends of environmental contaminants, including PAHs, in coastal water. A sentinel-based approach is used for monitoring with collection of sediment and bivalves (oysters and mussels) samples from a network of sites as surrogates for water pollution and bioaccumulation. Following on from the Mussel Watch Program, many associated biomonitoring surveys using bivalves have been performed worldwide. In these biomonitoring surveys, mussels (*Mytilus edulis*, *Mytilus californianus*, *Mytilus galloprovincialis*, and others), oysters (*Crassostrea Gigas*, *Crassostrea virginica*, and others), clams (*Ruditapes philippinarum*, and others), and other shellfish (*Turbo cornutus* and *Patella vulgata*) have been used as biomonitoring species (Bartolomé et al. 2011; Cacciatore et al. 2022; Fernández-Tajes et al. 2011; Gramlich et al. 2024; Kasiotis and Emmanouil 2015; Koyama et al. 2004; Onozato et al. 2016; Pérez et al. 2019; Tanaka and Onduka 2010). The physiological characteristics of bivalves closely match those of the indicator species mentioned above (Shibata and Takazawa 2014); moreover, they are also filter feeders and are highly susceptible to the surrounding environment because they are constantly exposed to large amounts of seawater (Shumway et al. 1985). These species are also frequently exposed to suspended solids in the bottom sediment, and

Table 1 Example of biomonitoring animals in previous studies

	Category	Common name	Scientific name	Study
Aquatic environment	Mollusk	Bay mussel	*Mytilus trossulus*	NCCOS (2017)
		Blue mussel	*Mytilus edulis*	Koyama et al. (2004), NCCOS (2017)
		California mussel	*Mytilus californianus*	NCCOS (2017)
		Mediterranean mussel	*Mytilus galloprovincialis*	Fernández-Tajes et al. (2011), NCCOS (2017), Onozato et al. (2016)
		Purplish bifurcate mussel	*Septifer virgatus*	Tanaka and Onduka (2010)
		Green mussel	*Perna viridis*	Dsikowitzky et al. (2016)
		Pacific oyster	*Crassostrea Gigas*	NCCOS (2017), Onozato et al. (2016)
		Eastern oyster	*Crassostrea virginica*	NCCOS (2017)
		Manila clam	*Ruditapes philippinarum*	Cacciatore et al. (2022), Onozato et al. (2016)
		Venus clam	*Venerupis pullastra*	Fernández-Tajes et al. (2011)
		Common limpet	*Patella vulgata*	Bartolomé et al. (2011)
		True limpet	*Cellana toreuma*	Koyama et al. (2004)
		Japanese grata limpet	*Cellana grata*	Koyama et al. (2004)
		Limpet	*Patella* spp.	Pérez et al. (2019)
		Common cockle	*Cerastoderma edule*	Fernández-Tajes et al. (2011)
		Spiny top shell	*Turbo cornutus*	Koyama et al. (2004)
	Fish	White sucker	*Catostomus commersonii*	Arens et al. (2017)
		Whitefish	*Coregonus clupeaformis*	Ohiozebau et al. (2016)
		Northern pike	*Esox luscius*	Ohiozebau et al. (2016)
		Walleye	*Sander vitreus*	Ohiozebau et al. (2016)
		Yellowfin Tuna	*Thunnus albacares*	Pulster et al (2020)
		Golden Tilefish	*Lopholatilus chamaeleonticeps*	Pulster et al (2020)
		Red Drum	*Sciaenops ocellatus*	Pulster et al (2020)

(continued)

Table 1 (continued)

	Category	Common name	Scientific name	Study
		Thumbprint emperor	*Lethrinus harak*	Uno et al. (2010)
		Mahogany snapper	*Lutjanus mahogoni*	Dsikowitzky et al. (2016)
		Mackerel	*Rastrelliger kanagurta*	Dsikowitzky et al. (2016)
		Sea catfish	*Netuma thalassina*	Dsikowitzky et al. (2016)
	Cephalopod	European flying squid	*Todarodes sagittatus*	Perugini et al. (2007)
		Common octopus	*Octopus vulgaris*	Semedo et al. (2014)
	Crustacean	Goose barnacle	*Capitulum mitella*	Koyama et al. (2004)
		Gazami crab	*Portunus trituberculatus*	Wan et al. (2007)
		Burrowing shrimp	*Upogebia* sp.	Wan et al. (2007)
		Norway lobster	*Nephrops norvegicus*	Perugini et al. (2007)
		Aquatic insects		Wayland et al. (2008)
	Polychaetas	Blow lugworm	*Arenicola marina*	Kaag et al. (1998)
	Amphibians	Wood frog	*Lithobates sylvaticus*	Mundy et al. (2019)
	Bird	Herring gull	*Larus argentatus*	Wan et al. (2007)
		Common tern	*Sterna hirundo*	Hebert et al. (2011)
		Ring-billed gull	*Larus delawarensis*	Hebert et al. (2011)
	Mammal	Sea otter	*Enhydra lutris*	Harris et al. (2011)
Terrestrial environment	Mollusk	Brown garden snail	*Cornu aspersum Müller*	Al-Alam et al. (2024)
		Garden snail	*Helix aspersa*	De Vaufleury et al. (2006), Reviewed in Baroudi et al. (2020)
		Grove snail	*Cepaea nemoralis*	Reviewed in Baroudi et al. (2020)
	Insect	Western honey bee	*Apis mellifera* L.	Di Fiore et al. (2023), Lambert et al. (2012)
	Crustacean	Wharf roach	*Ligia* spp.	Honda et al. (2018, 2021)

are therefore likely to be influenced by the bottom sediment (Menon and Menon 1999).

Fish are common in the marine environment and a major food source for humans. Consequently, their pollution status has attracted attention. Fish, including demersal fish, are frequently used for biomonitoring of PAHs in marine and freshwater environments (Table 1) (Arens et al. 2017; Kuhn et al. 2024; Ohiozebau et al. 2016; Pulster et al 2020; Uno et al. 2010). Because fish are very mobile, the contaminated area that they represent is not always clear. Several studies have reported relatively low accumulation of PAHs in fish (Dsikowitzky et al. 2016; Poliakova et al. 2000; Wan et al. 2007) compared with invertebrates. The metabolism and excretion rates of PAHs in fish are relatively fast, and because the concentrations of PAHs normally accumulated in fish are low, it is considered that fish can rapidly reflect changes in environmental concentrations after oil spills or other contamination incidents. To monitor the coastal environment, cephalopods (e.g., squid and octopus), crustaceans (e.g., crab and mysids), and benthos invertebrates (e.g., aquatic insects and polychaetas) have been used as biomonitoring species (Gauthier et al. 2014; Gramlich et al. 2024; Kaag et al. 1998; Mearns et al. 2019; Perugini et al. 2007; Semedo et al. 2014; Wallace et al. 2020). Aquatic mammals and birds have also been used as biomonitoring species in some studies (Harris et al. 2011; Hebert et al. 2011; Wallace et al. 2020).

2.2 *Terrestrial Environment*

In the terrestrial environment, anthropogenic PAHs originate from industrial activities (emissions, waste, and sewage), vehicular emissions, agricultural waste, open fires, domestic waste, oil spills, pharmaceutical waste, and forest fires (Teixeira et al. 2024). Thus, the terrestrial environment is contaminated by PAHs both directly and indirectly. Plants are generally used for biomonitoring of terrestrial environments, but some studies have used animals in particular environments. This chapter describes cases of terrestrial PAHs' biomonitoring using animals.

As in the aquatic environment, invertebrates are key species for biomonitoring PAHs in terrestrial environments. Land snails (e.g., *Cornu aspersum* Müller, *Helix aspersa*, and *Cepaea nemoralis*) are frequently used for monitoring of both general and polluted areas (Al-Alam et al. 2024; Baroudi et al. 2020). Snails inhabit a wide range of soils and plant surfaces and are exposed to the air through breathing. Therefore, they can absorb various pollutants through multiple pathways and accumulate them in their fatty tissues (De Vaufleury et al. 2006; Radwan et al. 2020). Land snails are considered suitable biomonitoring species because of their adaptability to the above conditions and their ability to accumulate contaminants from the air, soil, and flora. Terrestrial insects, especially honeybees, are also used as biomonitoring species (Di Fiore et al. 2023; Lambert et al. 2012). Honeybees are highly sensitive organisms and play an important role in the ecosystem. Because honeybees move between different plants and flowers to collect and transport pollen, they can become contaminated by environmental pollutants, including PAHs, by contact with contaminated air, soil, and water, or secondarily contaminated plants (Al-Alam et al. 2019). Honeybees are believed to reflect the widespread environmental pollution within a few kilometers of the hive and bring contaminants, such as PAHs, back to the hive, where they accumulate in the larvae and various bees by-products, such as honey, pollen, wax, and propolis (Al-Alam et al. 2019; Badiou-Bénéteau et al. 2013). Therefore, biomonitoring of contamination using honeybees allows for spatial and temporal monitoring of the area around the hive (Cochard et al. 2021; Kargar et al. 2017; Lambert et al. 2012).

The intertidal zone is the boundary between terrestrial and marine environments and often suffers from severe pollution by anthropogenic substances. Coastal areas, including the intertidal zone, are also an extremely important area for reproduction of organisms in the marine ecosystem. Consequently, biomonitoring of the intertidal zone has been carried out worldwide, including in biomonitoring surveys using bivalve mollusks. Bivalves are commonly used as an indicator species for monitoring and a huge amount of data has been accumulated (NCCOS 2017). However, these species mainly live in the subtidal zone, which is below the intertidal zone and underwater. Biomonitoring in the supratidal zone, where contamination is as widespread as in the subtidal zone (Naidoo et al. 2015), is lacking. The use of wharf roaches (*Ligia* sp.) was conceived to fill this data gap (Honda et al. 2018, 2021). Compared with bivalves, wharf roaches have a wider range that stretches from the intertidal zone to the supratidal zone and covers temperate and tropical areas of the world (Hurtado et al. 2010; Tsai and Dai 2001). Additionally, they are omnivorous, do not move from site to site, are easy to sample, tolerate a range of salinities, and can accumulate chemicals from the diet, water, and soil. Considering their geographical/geomorphological distributions and biological characteristics, wharf roaches are suitable for monitoring of environmental pollutants from the marine environment to the terrestrial environment. In initial study, as part of research into the Great East Japan Earthquake and the resulting tsunami damage, we investigated PAHs pollution of the intertidal and supratidal zones using wharf roaches collected from nine sites on the central and northern coasts of Japan in November 2011 (Honda et al. 2018). Ubiquitous PAH pollution was observed at the nine sampling sites (Fig. 1). Seven sampling sites had relatively low total concentrations of PAHs (57.7–130 ng/g dry weight (dw)), including sites in Miyagi Prefecture that were seriously damaged by the Great East Japan Earthquake and the following tsunami on March 11, 2011. Two sampling sites had higher concentrations of total PAHs (sites 2: 1025 ng/g dw and site 3: 591 ng/g dw). A diagnostic ratio analysis showed that sites 2 and 3 were contaminated with PAHs of pyrolytic origin that possibly came from vehicle exhaust, industrial fuel combustion, or both sources. The concentrations of PAHs in the wharf roaches differed among the sampling sites, and this was taken to reflect the differing contamination situation in water or soil environments at the sampling sites. The PAHs concentrations detected at this time in wharf roaches from the seven low-concentration sampling sites (57.7–130 ng/g dw) were similar to or lower than those found in bivalves in other studies, for example, 16 PAHs (15.2–527 ng/g dw in blue mussels and 52–293 ng/g dw in green mussels) in coastal areas around Japan (Tanaka and Onduka 2010). These results indicate that wharf roaches likely reflect environmental pollution from a wide vertical/horizontal range in coastal areas, including the intertidal and supratidal zones. In another study, we conducted a general environmental survey for PAHs using wharf roaches/mussels and estimated the exposure pathways in wharf roaches (Honda et al. 2021). Wharf roaches, mussels, and environmental media (water, soil/sand, and drifting seaweed) were collected from 12 sites in Japan along coastal areas of the Sea of Japan. PAH concentrations in wharf roaches were higher than those in mussels (median total of 15 PAHs: 48.5 ng/g dw in wharf roaches and 39.9 ng/g dw in mussels), except for in samples from Ishikawa (wharf roaches: 47.9 ng/g dw and mussels: 132 ng/g dw) (Fig. 2). Diagnostic ratio analysis showed that nearly all PAHs in soil and sand were of petrogenic origin. Correlation analysis of PAH concentrations between wharf roaches and the environmental media may indicate that wharf roach exposure to three- and four-ring PAHs was likely from food (drifting seaweed) and soil/sand, whereas exposure to four- and five-ring PAHs was from multiple environmental media including unknowns. These findings indicate that wharf roaches can reflect PAHs pollution over a wide range in the intertidal zone. Consequently, it is possible to use wharf roaches to investigate PAH pollution in the intertidal zone from a different perspective compared with bivalves.

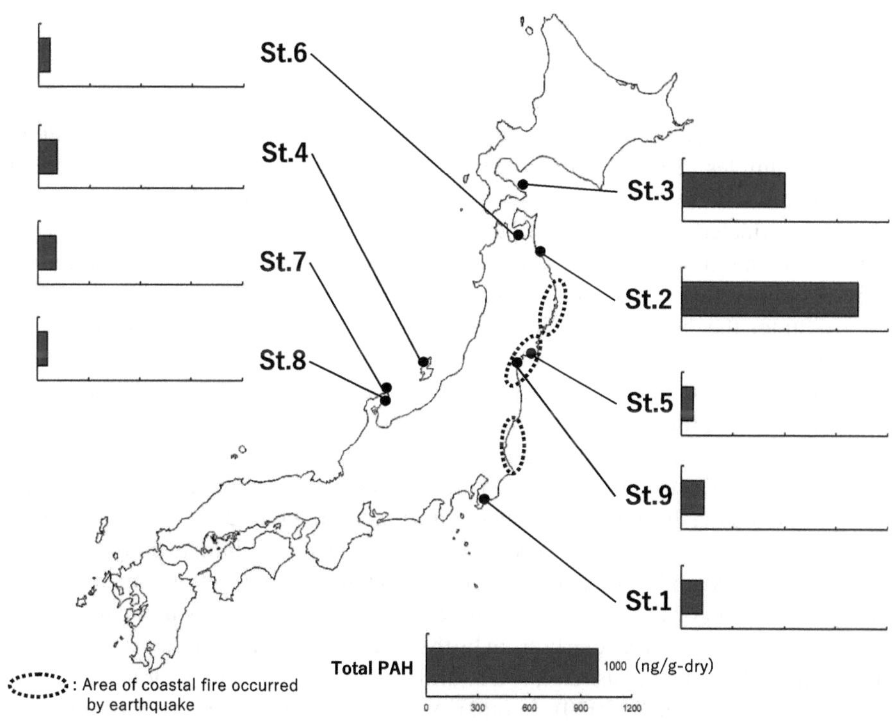

Fig. 1 Total PAH concentrations in wharf roaches from the nine sampling sites (modified from Fig. 1. in Honda et al. 2018)

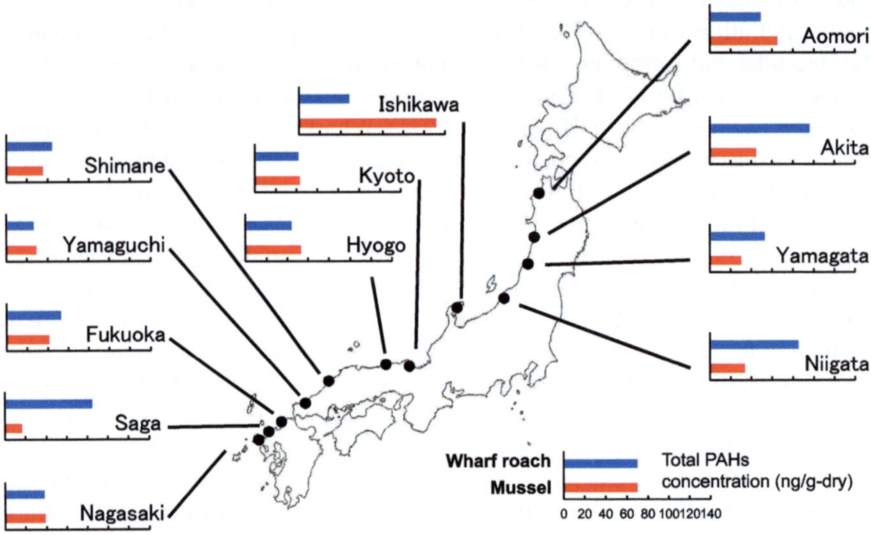

Fig. 2 Total PAH concentrations in wharf roaches and mussels from the twelve sampling sites (data modified from Table 2 in Honda et al. 2021)

Although it may be controversial to include humans in the animal section, biomonitoring of PAHs is commonly conducted in humans, and so such studies are presented here as examples. Several human activities generate PAHs and result in human contamination both directly and indirectly. Normally, human biomonitoring for PAHs is conducted using urine samples, which are analyzed for mono-hydroxylated PAHs. Human biomonitoring studies are commonly performed in Western countries (Angerer et al. 2007; Jacob and Seidel 2002; Louro et al. 2022; Souza et al. 2022). For example, the National Health and Nutrition Examination Survey conducted by the Centers for Disease Control and Prevention continuously surveyed environmental pollutants, including PAHs, in human specimens since 1999–2000 until now in the general population in the United States (CDC 2017). In Germany, PAHs in the human population have been monitored by the German Environmental Survey and the German Environmental Specimen Bank (Kolossa-Gehring et al. 2012). Few surveys and studies have been conducted in Asia (Guo et al. 2013). In Japan, some government projects to survey environmental pollutants in the human body have been conducted, for example, Japan Environment and Children's Study was conducted to clarify the effects of the environment on children's health. This study checked the health status of children regularly from birth and investigated the influence of various factors, including chemicals (e.g., PAHs), lifestyle habits, genetic factors, and social factors (Matsumura et al. 2020). In other Asian countries, surveys of differences in PAHs exposure with occupation and geography have been conducted (Chetiyanukornkul et al. 2006; Thai et al. 2015). Chetiyanukornkul et al. (2006) reported that rural villagers had higher contamination levels of PAHs than urban dwellers, and PAHs contamination likely arose from open burning, biomass burning for cooking and heating, as well as from internal combustion engines.

3 Analytical Methods

3.1 *Target PAHs*

Generally, the 16 priority PAHs selected by the United States Environmental Protection Agency are the target for monitoring of environmental PAHs pollution because of their toxicities and distributions. These PAHS are naphthalene, acenaphthylene, acenaphthene, fluorene, phenanthrene, anthracene, fluoranthene, pyrene, benz[*a*]anthracene, chrysene, benzo[*b*]fluoranthene, benzo[*k*]fluoranthene, benzo[*a*]pyrene, benzo[*ghi*]perylene, dibenz[*a,h*]anthracene, and indeno[1,2,3-*cd*]pyrene. Several other PAHs are also monitored because of their toxicities and abundances (Table 1).

3.2 *Sampling*

It is important that the sample size and period are uniform throughout the sampling. To compare concentration data, biological samples should be taken at the same time as environmental media samples. Clean sample collectors and containers should be used that are not contaminated with detectable levels of the target PAHs or interfering substances. Frequently, the sample collection sites for biological organisms are the muscles for fish and soft body for crustaceans and shellfish. Because bottom-dwelling shellfish, such as clams and mussels, ingest bottom sediment along with their food, they need to be soaked overnight in 3% salt water to expel the bottom sediment from their digestive tracts before analysis. If the amount of sample required for analysis cannot be obtained from a single organism, samples from multiple organisms collected at the same site are mixed to obtain the required amount.

Fat, muscle, and liver tissues, which are easy to obtain in large quantities, have been used for analysis of PAHs in large organisms. For small organisms, whole body analysis is often performed. In some species, where PAHs are easily metabolized and are not concentrated in the body, research is conducted using metabolites in bile and urine (Guo et al. 2013; Vuorinen et al. 2006).

Information about the organism body length (e.g., shell height, length, and width or carapace length and width), body weight, species, sex, developmental stage (age, sexually maturity/immaturity), lipid content, and intestinal contents, is collected while sampling.

3.3 Extraction and Purification

To analyze PAHs, it is necessary to extract and purify them before instrumental analysis. PAHs accumulate mainly in the fatty tissues of organisms and thorough sample cleanup is required to prevent interference from the sample matrix in the analytical results.

Purification is frequently performed by column chromatography using silica or alumina or size-exclusion high-performance liquid chromatography (HPLC) and gel permeation chromatography (Yusa et al. 2005). Solid-phase extraction with various sorbents (silica, alumina, magnesium silicate, Florisil, cyanopropyl) is also widely used, but in most cases the sample needs to be pre-cleaned before these purification methods. The traditional method for extracting PAHs from biological sample containing abundant lipophilic compounds is saponification with a strongly basic alcohol solution (e.g., with NaOH or KOH) and/or organic solvent extraction using acetone, *n*-hexane, ethanol, or dichloromethane (Honda et al. 2018; Yusa et al. 2005). The extraction methods include Soxhlet extraction (Janska et al. 2006; Nakata et al. 2014), ultrasound-assisted extraction (Honda et al. 2018, 2021), microwave-assisted liquid extraction (Cortazar et al. 2008; Zuloaga et al. 2009), and accelerated solvent extraction or pressurized liquid extraction (Carabias-Martínez et al. 2005; Yusa et al. 2005). Additionally, matrix solid-phase dispersion (Tu and Chen 2018) and QuEChERS extraction (Forsberg et al. 2011; João Ramalhosa et al. 2009) have recently been used for sample pretreatment for PAH analysis. In some cases, the uncertainty of single measurements for individual samples can be alleviated by sample pooling.

Instrumental analysis of PAHs is mainly conducted using gas chromatography mass spectrometry (GC/MS) or high-performance liquid chromatography coupled with fluorescence detection (HPLC-FLD). These methods are described in detail in Part IV. Chap. 11.

4 Conclusion

Biomonitoring studies of environmental PAH contamination using animals are very useful and have been carried out worldwide. Bivalves and fish in aquatic environments and land snails and crustaceans in terrestrial environments have been used in many of these studies. These environmental indicator species have been selected because of their specificities, locations, tissue types, metabolic capacities, accumulation rates, occurrences, species richness, climatic and seasonal conditions, biodiversity, and contamination status. Biomonitoring studies allow for detailed analysis of bioavailability and actual accumulation characteristics that cannot be obtained by analysis of environmental matrix samples alone. It is important to note that the absorption, metabolism, and excretion parameters differ for each species, which means that the concentrations in organisms do not simply reflect the environmental concentrations.

Additionally, considering the characteristics of anthropogenic PAH emissions into the environment, it is important to accumulate basic data through routine and regular biomonitoring and to conduct investigations into sudden events such as oil spills. Environmental assessments can then be conducted by comparing the basic data with those from sudden events.

Acknowledgements I would like to thank Gabrielle David, from Edanz Group (https://enauthor-services.edanz.com/) for editing a draft of this manuscript. Preparing this manuscript was supported by the Kurita Water and Environment Foundation, Japan (24B063).

References

Al-Alam, J., Chbani, A., Faljoun, Z., & Millet, M. (2019). The use of vegetation, bees, and snails as important tools for the biomonitoring of atmospheric pollution—a review. *Environmental Science and Pollution Research, 26*, 9391–9408.

Al-Alam, J., Millet, M., Khoury, D., Rodrigues, A., Akoury, E., Tokajian, S., & Wazne, M. (2024). Biomonitoring of PAHs and PCBs in industrial, suburban, and rural areas using snails as sentinel organisms. *Environmental Science and Pollution Research, 31*(3), 4970–4984.

Angerer, J., Ewers, U., & Wilhelm, M. (2007). Human biomonitoring: state of the art. *International journal of hygiene and environmental health, 210*(3-4), 201–228.

Arens, C. J., Arens, J. C., Hogan, N. S., Kavanagh, R. J., Berrue, F., Van Der Kraak, G. J., & van den Heuvel, M. R. (2017). Population impacts in white sucker (*Catostomus commersonii*) exposed to oil sands–derived contaminants in the Athabasca River. *Environmental toxicology and chemistry, 36*(8), 2058–2067.

Bacosa, H. P., & Inoue, C. (2015). Polycyclic aromatic hydrocarbons (PAHs) biodegradation potential and diversity of microbial consortia enriched from tsunami sediments in Miyagi, Japan. *Journal of hazardous materials, 283*, 689–697.

Badiou-Bénéteau, A., Benneveau, A., Géret, F., Delatte, H., Becker, N., Brunet, J. L., ... & Belzunces, L. P. (2013). Honeybee biomarkers as promising tools to monitor environmental quality. *Environment international, 60*, 31–41.

Baroudi, F., Al Alam, J., Fajloun, Z., & Millet, M. (2020). Snail as sentinel organism for monitoring the environmental pollution; a review. *Ecological indicators, 113*, 106240.

Bartolomé, L., Bustamante, M., Navarro, P., Tajadura, J., Gorostiaga, J. M., Díez, I., ... & Etxebarria, N. (2011). The use of limpets as monitor of PAHs pollution in the Cantabrian coast. *Continental Shelf Research, 31*(17), 1818–1826.

Baumard, P., Budzinski, H., Garrigues, P., Sorbe, J. C., Burgeot, T., & Bellocq, J. (1998). Concentrations of PAHs (polycyclic aromatic hydrocarbons) in various marine organisms in relation to those in sediments and to trophic level. Marine Pollution Bulletin, 36(12), 951–960.

Cacciatore, F., Moltedo, G., Bernarello, V., Formalewicz, M., Catalano, B., Martuccio, G., ... & Virno Lamberti, C. (2022). Before-during-after biomonitoring assessment for a pipeline construction in a coastal lagoon in the Northern Adriatic Sea (Italy). *Environments, 9*(7), 81.

Carabias-Martínez, R., Rodríguez-Gonzalo, E., Revilla-Ruiz, P., & Hernández-Méndez, J. (2005). Pressurized liquid extraction in the analysis of food and biological samples. *Journal of Chromatography A, 1089*(1–2), 1–17.

Centers for Disease Control and Prevention (CDC), (2017). In: Fourth National Report on Human Exposure to Environmentl Chemicals Updated Tables. January 2017, Volume One.

Chandurvelan, R., Marsden, I. D., Glover, C. N., & Gaw, S. (2015). Assessment of a mussel as a metal bioindicator of coastal contamination: relationships between metal bioaccumulation and multiple biomarker responses. *Science of the Total Environment, 511*, 663–675.

Chetiyanukornkul, T., Toriba, A., Kameda, T., Tang, N., & Hayakawa, K. (2006). Simultaneous determination of urinary hydroxylated metabolites of naphthalene, fluorene, phenanthrene, fluoranthene and pyrene as multiple biomarkers of exposure to polycyclic aromatic hydrocarbons. *Analytical and Bioanalytical Chemistry, 386*, 712–718.

Cochard, P., Laurie, M., Veyrand, B., Le Bizec, B., Poirot, B., & Marchand, P. (2021). PAH7 concentration reflects anthropization: A study using environmental biomonitoring with honeybees. *Science of the Total Environment, 751*, 141831.

Corsolini, S., & Sarà, G. (2017). The trophic transfer of persistent pollutants (HCB, DDTs, PCBs) within polar marine food webs. *Chemosphere, 177*, 189–199.

Cortazar, E., Bartolomé, L., Arrasate, S., Usobiaga, A., Raposo, J. C., Zuloaga, O., & Etxebarria, N. (2008). Distribution and bioaccumulation of PAHs in the UNESCO protected natural reserve of Urdaibai, Bay of Biscay. *Chemosphere, 72*(10), 1467–1474.

Cunningham, M. M., Tran, L., McKee, C. G., Polo, R. O., Newman, T., Lansing, L., ... & Guarna, M. M. (2022). Honey bees as biomonitors of environmental contaminants, pathogens, and climate change. *Ecological Indicators, 134*, 108457.

De Vaufleury, A., Coeurdassier, M., Pandard, P., Scheifler, R., Lovy, C., Crini, N., & Badot, P. M. (2006). How terrestrial snails can be used in risk assessment of soils. *Environmental Toxicology and Chemistry: An International Journal, 25*(3), 797–806.

Di Fiore, C., De Cristofaro, A., Nuzzo, A., Notardonato, I., Ganassi, S., Iafigliola, L., ... & Avino, P. (2023). Biomonitoring of polycyclic aromatic hydrocarbons, heavy metals, and plasticizers residues: Role of bees and honey as bioindicators of environmental contamination. *Environmental Science and Pollution Research, 30*(15), 44234–44250.

Dsikowitzky, L., Nordhaus, I., Andarwulan, N., Irianto, H. E., Lioe, H. N., Ariyani, F., ... & Schwarzbauer, J. (2016). Accumulation patterns of lipophilic organic contaminants in surface sediments and in economic important mussel and fish species from Jakarta Bay, Indonesia. *Marine pollution bulletin, 110*(2), 767–777.

Fernández-Tajes, J., Flórez, F., Pereira, S., Rábade, T., Laffon, B., & Méndez, J. (2011). Use of three bivalve species for biomonitoring a polluted estuarine environment. *Environmental Monitoring and Assessment, 177*, 289–300.

Forsberg, N. D., Wilson, G. R., & Anderson, K. A. (2011). Determination of parent and substituted polycyclic aromatic hydrocarbons in high-fat salmon using a modified QuEChERS extraction, dispersive SPE and GC–MS. *Journal of agricultural and food chemistry, 59*(15), 8108–8116.

Gauthier, P. T., Norwood, W. P., Prepas, E. E., & Pyle, G. G. (2014). Metal–PAH mixtures in the aquatic environment: a review of co-toxic mechanisms leading to more-than-additive outcomes. *Aquatic toxicology, 154*, 253–269.

Gramlich, K. C., Monteiro, F. C., & da Silva Carreira, R. (2024). Polycyclic aromatic hydrocarbons on the Atlantic coast of South America and the Caribbean: A systematic literature review on biomonitoring coastal regions employing marine invertebrates. *Regional Studies in Marine Science, 78*, 103792.

Guo, Y., Senthilkumar, K., Alomirah, H., Moon, H. B., Minh, T. B., Mohd, M. A., ... & Kannan, K. (2013). Concentrations and profiles of urinary polycyclic aromatic hydrocarbon metabolites (OH-PAHs) in several Asian countries. *Environmental science & technology, 47*(6), 2932–2938.

Harris, K. A., Nichol, L. M., & Ross, P. S. (2011). Hydrocarbon concentrations and patterns in free-ranging sea otters (*Enhydra lutris*) from British Columbia, Canada. *Environmental toxicology and chemistry, 30*(10), 2184–2193.

Hebert, C. E., Weseloh, D. C., MacMillan, S., Campbell, D., & Nordstrom, W. (2011). Metals and polycyclic aromatic hydrocarbons in colonial waterbird eggs from Lake Athabasca and the Peace–Athabasca Delta, Canada. *Environmental Toxicology and Chemistry, 30*(5), 1178–1183.

Honda, K., Mizukami, M., Ueda, Y., Hamada, N., & Seike, N. (2007). Residue level of polycyclic aromatic hydrocarbons in Japanese paddy soils from 1959 to 2002. *Chemosphere, 68*(9), 1763–1771.

Honda, M., Mukai, K., Nagato, E., Uno, S., & Oshima, Y. (2021). Correlation between polycyclic aromatic hydrocarbons in wharf roach (*Ligia* spp.) and environmental components of the intertidal and supralittoral zone along the Japanese Coast. *International journal of environmental research and public health, 18*(2), 630.

Honda, M., Qiu, X., Koyama, J., Uno, S., Undap, S. L., Shimasaki, Y., & Oshima, Y. (2018). The wharf roach, *Ligia* sp., A novel indicator of polycyclic aromatic hydrocarbon contamination in coastal areas. *International Journal of Environmental Research, 12*, 1–11.

Hurtado, L. A., Mateos, M., & Santamaria, C. A. (2010). Phylogeography of supralittoral rocky intertidal Ligia isopods in the Pacific region from central California to central Mexico. *PLoS One, 5*(7), e11633.

Ito, K., Uno, S., Tanaka, H., Miki, S., Kokushi, E., Yamamoto, M., & Koyama, J. (2015). Distribution of parent and alkylated PAHs in bivalves collected from Osaka Bay, Japan. *Japanese Journal of Environmental Toxicology, 18*(2), 11–24.

Jacob, J., & Seidel, A. (2002). Biomonitoring of polycyclic aromatic hydrocarbons in human urine. *Journal of Chromatography B, 778*(1–2), 31–47.

Janska, M., Tomaniova, M., Hajšlová, J., & Kocourek, V. (2006). Optimization of the procedure for the determination of polycyclic aromatic hydrocarbons and their derivatives in fish tissue: Estimation of measurements uncertainty. *Food additives and contaminants, 23*(3), 309–325.

João Ramalhosa, M., Paíga, P., Morais, S., Delerue-Matos, C., & Prior Pinto Oliveira, M. B. (2009). Analysis of polycyclic aromatic hydrocarbons in fish: evaluation of a quick, easy, cheap, effective, rugged, and safe extraction method. *Journal of separation science, 32*(20), 3529–3538.

Kaag, N. H. B. M., Scholten, M. T., & Van Straalen, N. M. (1998). Factors affecting PAH residues in the lugworm Arenicola marina, a sediment feeding polychaete. *Journal of sea research, 40*(3–4), 251–261.

Kargar, N., Matin, G., Matin, A. A., & Buyukisik, H. B. (2017). Biomonitoring, status and source risk assessment of polycyclic aromatic hydrocarbons (PAHs) using honeybees, pine tree leaves, and propolis. *Chemosphere, 186*, 140–150.

Kasiotis, K. M., & Emmanouil, C. (2015). Advanced PAH pollution monitoring by bivalves. *Environmental Chemistry Letters, 13*, 395–411.

Khairy, M., Brault, E., Dickhut, R., Harding, K. C., Harkonen, T., Karlsson, O., ... & Lohmann, R. (2021). Bioaccumulation of PCBs, OCPs and PBDEs in marine mammals from West Antarctica. *Frontiers in Marine Science, 8*, 768715.

Kolossa-Gehring, M., Becker, K., Conrad, A., Schröter-Kermani, C., Schulz, C., & Seiwert, M. (2012). Environmental surveys, specimen bank and health related environmental monitoring in Germany. *International Journal of Hygiene and Environmental Health, 215*(2), 120–126.

Koyama, J., Uno, S., Nagai, Y., & Anukorn, B. (2016). Early monitoring of spilled oil contamination in Rayong, Thailand. *Japanese Journal of Environmental Toxicology, 19*(1), 25–33.

Koyama, J., Uno, S., & Kohno, K. (2004). Polycyclic aromatic hydrocarbon contamination and recovery characteristics in some organisms after the Nakhodka oil spill. *Marine pollution bulletin, 49*(11–12), 1054–1061.

Kuhn, A. V., Dal Pont, G., Cozer, N., & Sadauskas-Henrique, H. (2024). The concentrations of polycyclic aromatic hydrocarbons in fish: A systematic review. *Marine Pollution Bulletin, 198*, 115778.

Lambert, O., Veyrand, B., Durand, S., Marchand, P., Le Bizec, B., Piroux, M., ... & Pouliquen, H. (2012). Polycyclic aromatic hydrocarbons: bees, honey and pollen as sentinels for environmental chemical contaminants. *Chemosphere, 86*(1), 98–104.

Liu, Y., Wang, S., Lohmann, R., Yu, N., Zhang, C., Gao, Y., ... & Ma, L. (2015). Source apportionment of gaseous and particulate PAHs from traffic emission using tunnel measurements in Shanghai, China. *Atmospheric Environment, 107*, 129–136.

Louro, H., Gomes, B. C., Saber, A. T., Iamiceli, A. L., Göen, T., Jones, K., ... & Silva, M. J. (2022). The use of human biomonitoring to assess occupational exposure to pahs in Europe: A comprehensive review. *Toxics, 10*(8), 480.

Magi, E., Bianco, R., Ianni, C., & Di Carro, M. (2002). Distribution of polycyclic aromatic hydrocarbons in the sediments of the Adriatic Sea. *Environmental pollution, 119*(1), 91–98.

Mastral, A. M., & Callen, M. S. (2000). A review on polycyclic aromatic hydrocarbon (PAH) emissions from energy generation. *Environmental Science & Technology, 34*(15), 3051–3057.

Matsumura, K., Hamazaki, K., Tsuchida, A., Inadera, H., & Japan Environment and Children's Study (JECS) Group. (2020). Prospective Association of Air-Purifier Usage during Pregnancy with Infant Neurodevelopment: A Nationwide Longitudinal Study—Japan Environment and Children's Study (JECS). *Journal of Clinical Medicine, 9*(6), 1924.

Mearns, A. J., Bissell, M., Morrison, A. M., Rempel-Hester, M. A., Arthur, C., & Rutherford, N. (2019). Effects of pollution on marine organisms. *Water Environment Research, 91*(10), 1229–1252.

Menon, N. N., & Menon, N. R. (1999). Uptake of polycyclic aromatic hydrocarbons from suspended oil borne sediments by the marine bivalve *Sunetta scripta. Aquatic toxicology, 45*(1), 63–69.

Miki, S., Uno, S., Ito, K., Koyama, J., & Tanaka, H. (2014). Distributions of polycyclic aromatic hydrocarbons and alkylated polycyclic aromatic hydrocarbons in Osaka Bay, Japan. *Marine pollution bulletin, 85*(2), 558–565.

Ministry of the Environment, Government of Japan (2012) Annual report of coastal driftage, 2012. Available via webpage of Ministry of the Environment. https://www.env.go.jp/water/marine_litter/mat14_24nenkan.pdf (in Japanese). Accessed 1 Jan 2025.

Monirith, I., Ueno, D., Takahashi, S., Nakata, H., Sudaryanto, A., Subramanian, A., ... & Tanabe, S. (2003). Asia-Pacific mussel watch: monitoring contamination of persistent organochlorine compounds in coastal waters of Asian countries. *Marine Pollution Bulletin, 46*(3), 281–300.

Mundy, L. J., Bilodeau, J. C., Schock, D. M., Thomas, P. J., Blais, J. M., & Pauli, B. D. (2019). Using wood frog (Lithobates sylvaticus) tadpoles and semipermeable membrane devices to monitor polycyclic aromatic compounds in boreal wetlands in the oil sands region of northern Alberta, Canada. *Chemosphere, 214*, 148–157.

Naidoo, T., Glassom, D., & Smit, A. J. (2015). Plastic pollution in five urban estuaries of KwaZulu-Natal, South Africa. *Marine pollution bulletin, 101*(1), 473–480.

Nakata, H., Uehara, K., Goto, Y., Fukumura, M., Shimasaki, H., Takikawa, K., & Miyawaki, T. (2014). Polycyclic aromatic hydrocarbons in oysters and sediments from the Yatsushiro Sea, Japan: comparison of potential risks among PAHs, dioxins and dioxin-like compounds in benthic organisms. *Ecotoxicology and environmental safety, 99*, 61–68.

National Centers for Coastal Ocean Science (NCCOS) (2017) website: https://coastalscience.noaa.gov/science-areas/pollution/mussel-watch/. Accessed 1 Jan 2025.

Ohiozebau, E., Tendler, B., Hill, A., Codling, G., Kelly, E., Giesy, J. P., & Jones, P. D. (2016). Products of biotransformation of polycyclic aromatic hydrocarbons in fishes of the Athabasca/Slave river system, Canada. *Environmental geochemistry and health, 38*, 577–591.

Okamura, H., & Aoyama, I. (1998). Toxicity Assessment of Water Environment using Biomonitoring. *Journal of Pesticide Science, 23*(2), 166–173.

Onozato, M., Nishigaki, A., & Okoshi, K. (2016). Polycyclic aromatic hydrocarbons in sediments and bivalves on the Pacific Coast of Japan: influence of tsunami and fire. *PloS one, 11*(5), e0156447.

Perugini, M., Visciano, P., Giammarino, A., Manera, M., Di Nardo, W., & Amorena, M. (2007). Polycyclic aromatic hydrocarbons in marine organisms from the Adriatic Sea, Italy. *Chemosphere, 66*(10), 1904–1910.

Pie, H. V., Heyes, A., & Mitchelmore, C. L. (2015). Investigating the use of oil platform marine fouling invertebrates as monitors of oil exposure in the Northern Gulf of Mexico. *Science of the Total Environment, 508*, 553–565.

Poliakova, O. V., Lebedev, A. T., Petrosyan, V. S., Hanninen, O., Renzoni, A., Sawa, D., & Walker, C. (2000). Accumulation of persistent organic pollutants in the food chain of Lake Baikal. *Toxicological & Environmental Chemistry, 75*(3–4), 235–243.

Pulster, E. L., Gracia, A., Armenteros, M., Toro-Farmer, G., Snyder, S. M., Carr, B. E., ... & Murawski, S. A. (2020). A first comprehensive baseline of hydrocarbon pollution in Gulf of Mexico fishes. *Scientific Reports, 10*(1), 6437.

Pérez, S., Sánchez-Marín, P., Bellas, J., Viñas, L., Besada, V., & Fernández, N. (2019). Limpets (*Patella* spp. Mollusca, Gastropoda) as model organisms for biomonitoring environmental quality. *Ecological indicators, 101*, 150–162.

Radwan, M. A., El-Gendy, K. S., & Gad, A. F. (2020). Biomarker responses in terrestrial gastropods exposed to pollutants: A comprehensive review. *Chemosphere, 257*, 127218.

Rios, L. M., Moore, C., & Jones, P. R. (2007). Persistent organic pollutants carried by synthetic polymers in the ocean environment. *Marine pollution bulletin, 54*(8), 1230–1237.

Sarria-Villa, R., Ocampo-Duque, W., Páez, M., & Schuhmacher, M. (2016). Presence of PAHs in water and sediments of the Colombian Cauca River during heavy rain episodes, and implications for risk assessment. *Science of the Total Environment, 540*, 455–465.

Semedo, M., Oliveira, M., Gomes, F., Reis-Henriques, M. A., Delerue-Matos, C., Morais, S., & Ferreira, M. (2014). Seasonal patterns of polycyclic aromatic hydrocarbons in digestive gland and arm of octopus (*Octopus vulgaris*) from the Northwest Atlantic. *Science of the Total Environment, 481*, 488–497.

Shibata, Y., & Takazawa, Y. (2014). Environmental monitoring of chemical pollution by short-lived wild-organisms. *Global Environmental Research, 19*(2), 155–163.

Shumway, S. E., Cucci, T. L., Newell, R. C., & Yentsch, C. M. (1985). Particle selection, ingestion, and absorption in filter-feeding bivalves. *Journal of experimental marine biology and ecology, 91*(1–2), 77–92.

Slezakova, K., Pires, J. C. M., Castro, D., Alvim-Ferraz, M. D. C. M., Delerue-Matos, C., Morais, S., & Pereira, M. D. C. (2013). PAH air pollution at a Portuguese urban area: carcinogenic risks and sources identification. *Environmental Science and Pollution Research, 20*, 3932–3945.

Souza, M. C. O., Rocha, B. A., Ximenez, J. P. B., Devóz, P. P., Santana, A., Campíglia, A. D., & Barbosa Jr, F. (2022). Urinary levels of monohydroxylated polycyclic aromatic hydrocarbons in Brazilian children and health risk assessment: a human biomonitoring-based study. *Environmental Science and Pollution Research, 29*(31), 47298–47309.

Tanaka, H., & Onduka, T. (2010). Background levels of PAHs in the coastal waters of Japan based on residual concentrations of bivalves. *Journal of Environmental Chemistry*, 20(2), 137–148.

Teixeira, J., Delerue-Matos, C., Morais, S., & Oliveira, M. (2024). Environmental contamination with polycyclic aromatic hydrocarbons and contribution from biomonitoring studies to the surveillance of global health. *Environmental Science and Pollution Research, 31*(42), 54339–54362.

Thai, P. K., Li, Z., Sjödin, A., Fox, A., Diep, N. B., Binh, T. T., & Mueller, J. F. (2015). Biomonitoring of polycyclic aromatic hydrocarbons exposure in small groups of residents in Brisbane, Australia and Hanoi, Vietnam, and those travelling between the two cities. *Chemosphere, 139*, 358–364.

Thuy, H. T. T., Loan, T. T. C., & Phuong, T. H. (2018). The potential accumulation of polycyclic aromatic hydrocarbons in phytoplankton and bivalves in Can Gio coastal wetland, Vietnam. *Environmental Science and Pollution Research, 25*(18), 17240–17249.

Tobiszewski, M., & Namieśnik, J. (2012). PAH diagnostic ratios for the identification of pollution emission sources. *Environmental pollution, 162*, 110–119.

Tsai, M. L., & Dai, C. F. (2001). Life history plasticity and reproductive strategy enabling the invasion of *Ligia exotica* (Crustacea: Isopoda) from the littoral zone to an inland creek. *Marine Ecology Progress Series, 210*, 175–184.

Tu, X., & Chen, W. (2018). A review on the recent progress in matrix solid phase dispersion. *Molecules, 23*(11), 2767.

Uno, S., Koyama, J., Kokushi, E., Monteclaro, H., Santander, S., Cheikyula, J. O., ... & Matsuoka, T. (2010). Monitoring of PAHs and alkylated PAHs in aquatic organisms after 1 month from the Solar I oil spill off the coast of Guimaras Island, Philippines. *Environmental monitoring and assessment, 165*, 501–515.

Uno, S., Kokushi, E., Añasco, N. C., Iwai, T., Ito, K., & Koyama, J. (2017). Oil spill off the coast of Guimaras Island, Philippines: Distributions and changes of polycyclic aromatic hydrocarbons in shellfish. *Marine pollution bulletin, 124*(2), 962–973.

Vuorinen, P. J., Keinänen, M., Vuontisjärvi, H., Baršienė, J., Broeg, K., Förlin, L., ... & Schiedek, D. (2006). Use of biliary PAH metabolites as a biomarker of pollution in fish from the Baltic Sea. *Marine Pollution Bulletin, 53*(8–9), 479–487.

Wallace, S. J., De Solla, S. R., Head, J. A., Hodson, P. V., Parrott, J. L., Thomas, P. J., ... & Langlois, V. S. (2020). Polycyclic aromatic compounds (PACs) in the Canadian environment: Exposure and effects on wildlife. *Environmental Pollution, 265*, 114863.

Wan, Y., Jin, X., Hu, J., & Jin, F. (2007). Trophic dilution of polycyclic aromatic hydrocarbons (PAHs) in a marine food web from Bohai Bay, North China. *Environmental science & technology, 41*(9), 3109–3114.

Wayland, M., Headley, J. V., Peru, K. M., Crosley, R., & Brownlee, B. G. (2008). Levels of polycyclic aromatic hydrocarbons and dibenzothiophenes in wetland sediments and aquatic insects in the oil sands area of Northeastern Alberta, Canada. *Environmental Monitoring and Assessment, 136*, 167–182.

Yusa, V., Pardo, O., Martí, P., & Pastor, A. (2005). Application of accelerated solvent extraction followed by gel performance chromatography and high-performance liquid chromatography for the determination of polycyclic aromatic hydrocarbons in mussel tissue. *Food additives and contaminants, 22*(5), 482–489.

Zakaria, M. P., Okuda, T., & Takada, H. (2001). Polycyclic aromatic hydrocarbon (PAHs) and hopanes in stranded tar-balls on the coasts of Peninsular Malaysia: applications of biomarkers for identifying sources of oil pollution. *Marine Pollution Bulletin, 42*(12), 1357–1366.

Zuloaga, O., Prieto, A., Usobiaga, A., Sarkar, S. K., Chatterjee, M., Bhattacharya, B. D., ... & Satpathy, K. K. (2009). Polycyclic aromatic hydrocarbons in intertidal marine bivalves of Sunderban mangrove wetland, India: an approach to bioindicator species. *Water, air, and soil pollution, 201*, 305–318.

The NP Method: A Source Analysis Technique for Combustion-Derived Particulates, Polycyclic Aromatic Hydrocarbons, and Nitro-Derivatives

Kazuichi Hayakawa

Abstract Airborne particulate matter (PM) contains hazardous polycyclic aromatic hydrocarbons (PAHs) and nitropolycyclic aromatic hydrocarbons (NPAHs) emitted from various combustion sources. The lack of effective methods for identifying these sources has impeded research in atmospheric pollution and public health. This chapter introduces a novel source analysis method, the NP method, for identifying combustion-derived atmospheric particulates (P_c), PAHs, and NPAHs. The method bases on the temperature dependence of nitrogen oxide (NO_x) formation during organic material combustion and the subsequent conversion of PAHs to NPAHs. Using 1-nitropyrene (1-NP) and pyrene as key markers, the NP method classifies combustion sources into high-temperature (vehicles) and low-temperature (coal/biomass) categories. Formulas are derived for the fraction of P_c in PM [y ($0 < y < 1$)], with P_c further divided into high-temperature (P_h) and low-temperature (P_l) combustion fractions [P_c values of x ($0 < x < 1$) and ($1 - x$), respectively]. Additionally, the fractions of high-temperature combustion-derived 1-NP and pyrene in atmospheric 1-NP and pyrene are calculated as a ($0 < a < 1$) and b ($0 < b < 1$), respectively. This chapter details the theory, performance, and applications of the NP method, which surpasses conventional source appointment techniques.

Keywords NP method · Source appointment · Vehicles · Coal combustion · Polycyclic aromatic hydrocarbon · Nitropolycyclic aromatic hydrocarbon · Combustion-derived particulate matter

---

K. Hayakawa (✉)
Institute of Nature and Environmental Technology, Kanazawa University, Kakuma-Machi, Kanazawa, Japan
e-mail: hayakawa@p.kanazawa-u.ac.jp

K. Hayakawa and Y. Aoki (eds.), *A New Era of Polycyclic Aromatic Hydrocarbons*,
https://doi.org/10.1007/978-981-96-7695-8_13

1 Introduction

Polycyclic aromatic hydrocarbons (PAHs) and their derivatives, such as nitropolycyclic aromatic hydrocarbons (NPAHs), are produced via the incomplete combustion of organic substances. In the air, many of these compounds exist in fine particles with a diameter of $\leq$ 2.5 μm [particulate matter $(PM)_{2.5}$]. These particles are inhaled into the body, leading to adverse health effects, including carcinogenicity and mutagenicity. The International Agency for Research on Cancer (IARC) has reviewed the carcinogenicity of over 60 PAHs and NPAHs, classifying benzo[*a*]pyrene (BaP) as Group 1 (carcinogenic to humans). Four PAHs (cyclopenta[*cd*]pyrene, dibenz[*a,h*] anthracene, dibenzo[*a,l*]pyrene, and dibenz[*a,j*]acridine) and one NPAH, 1-nitropyrene (1-NP), are classified as Group 2A (probably carcinogenic to humans), whereas fifteen PAHs and their hetero-derivatives (naphthalene, benz[*j*]aceanthrylene, benz[*a*]anthracene, benzo[*b*]fluoranthene, benzo[*j*]fluoranthene, benzo[*k*]fluoranthene, benzo[*c*]phenanthrene, chrysene, 5-methylchrysene, dibenzo[*a,h*]pyrene, dibenzo[*a,i*]pyrene, indeno[1,2,3-*cd*] pyrene, benzofuran, dibenz[*c,h*]acridine, and dibenz[*a,h*]acridine) and five NPAHs, 5-nitroacenaphthene, 2-nitrofluorene, 4-NP, 1,6-dinitropyrene (DNP), and 1,8-DNP, are classified as Group 2B (possibly carcinogenic to humans) (IARC).

Primary sources of these pollutants include vehicles, ships, airplanes, power plants, factories, incinerators, forest fires, slash-and-burn agriculture, smoking, and cooking, among others (Hayakawa 2022). The World Health Organization (WHO) reported that ambient and household air pollution contribute to 7 million premature deaths annually (WHO 2024). To develop effective pollution control measures, it is crucial to identify the contribution of each source at contaminated sites.

Several methods have been developed for source appointment of atmospheric PAHs and NPAHs, including chemical mass balance (CMB) methods, such as positive matrix factor analysis (PMF) and principal component analysis (PCA) (Brown et al. 2015), as well as the diagnostic ratio (DR) method (Masclet et al. 1987; Schauer et al. 1996; Yunker et al. 2002; Manoli et al. 2004; Davis et al. 2019). The PMF method calculates the contribution of anthropogenic sources to atmospheric organic carbon; however, the application of CMB methods to PAHs and NPAHs is limited owing to the extensive combustion source datasets required. The PCA method can identify vehicle contributions to PAHs and NPAHs in urban atmospheres (Kalisa et al. 2018; Zhuo et al. 2017; Zhao et al. 2020b). The DR method, based on PAH concentration ratios, is also used to approximate major PAH sources. Ratios such as fluoranthene to fluoranthene + pyrene ([FR]/[FR + Pyr]), benz[*a*]anthracene to benz[*a*]anthracene + chrysene ([BaA]/[BaA + Chr]), benz[*a*]pyrene to benzo[*ghi*]perylene ([BaP]/[BgPe]), and indeno[1,2,3-*cd*]pyrene to indeno[1,2,3-*cd*]pyrene + benzo[*ghi*]perylene ([IDP]/[IDP + BgPe]) are commonly used (Masclet et al. 1987; Rogge et al. 1993; Schauer et al. 1996; Yunker et al. 2002; Tobiszewski and Namieśnik 2012; Davis et al. 2019). However, different sources are

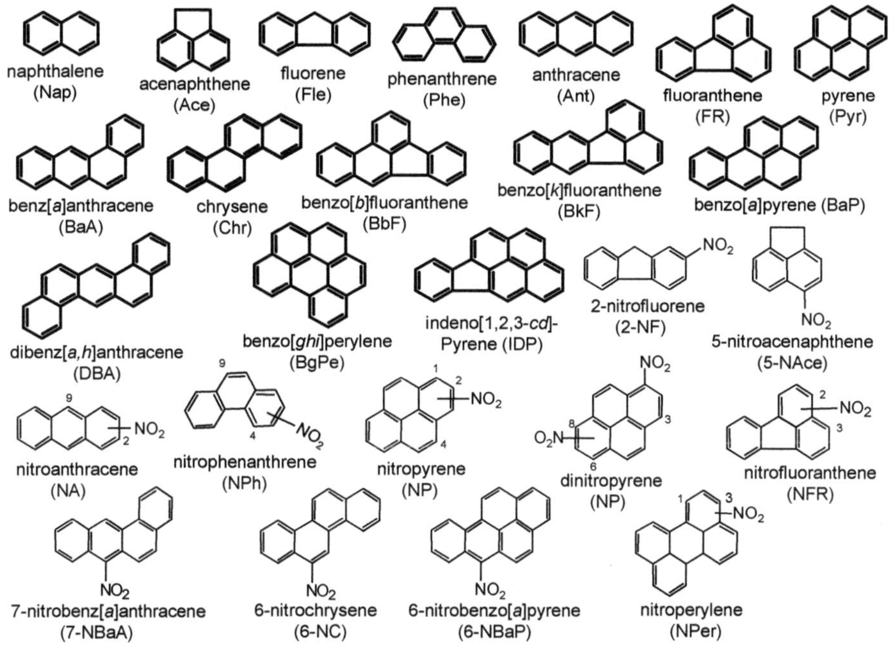

Fig. 1 Chemical compositions and abbreviations of PAHs and NPAHs

often identified based on the PAHs selected as markers, and the percentage contribution of each source to the mixture remains unknown. Additionally, this method cannot be applied for NPAHs.

We have developed the NPAH–PAH combination (NP) method, a novel approach to calculate the contribution of major sources to atmospheric combustion-derived PM (P_c), PAHs, and NPAHs, using 1-NP and pyrene (Pyr) as representative markers (Hayakawa et al. 2020a, 2021b). This method is based on the temperature-dependent formation of NPAHs from PAHs. This chapter details the theory, performance, and applications of the NP method, which surpasses conventional source appointment techniques. To aid understanding, the chemical structures and abbreviations of PAHs and NPAHs are illustrated in Fig. 1.

2 PAHs for Source Appointment

In Kanazawa, total PAH concentration decreased from 4.65 pg m^{-3} in winter 1997 to 0.64 pg m^{-3} in winter 2021, less than one-seventh of the original level. During this period, the fraction (%) of six-ring PAHs, BghiP and IDP, in total PAHs decreased from 25 to 21% (Fig. 2, partly modified from Hayakawa et al. 2024). The amount and composition of PAHs generated through combustion depend heavily on combustion

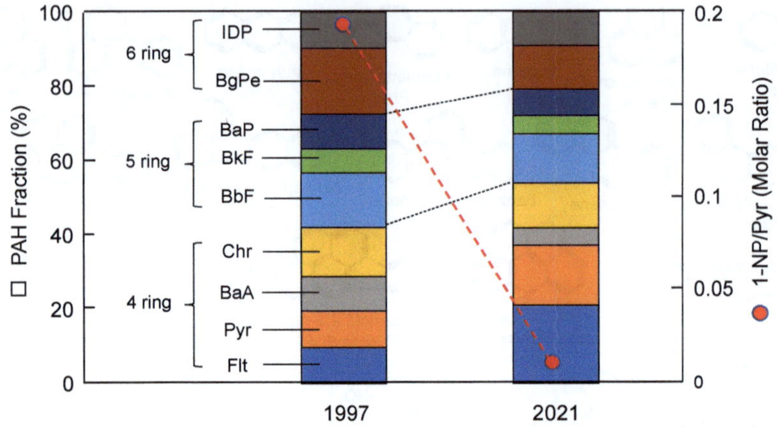

Fig. 2 Change in winter atmospheric PAH composition and concentration ratio of 1-NP to Pyr in Kanazawa between 1999 and 2021 (Hayakawa et al. 2024, with several modifications)

conditions. Higher combustion temperatures increase the yield of high-molecular-weight PAHs (Westblad et al. 2002; McGrath et al. 2003). Emission rates of six- and seven-ring PAHs are higher from vehicles than from coal or wood burning (Ravindra et al. 2008). To investigate the cause of these changes, the DR method was applied to PAH concentrations. Common PAH pairs include [FR]/[FR + Pyr], [BaA]/[BaA + Chr], [IDP]/[IDP + BgPe] and [BaP]/[BgP]. A [FR]/[FR + Pyr] ratio below 0.4 indicates petroleum sources, whereas 0.4–0.5 suggests vehicles. Ratios above 0.5 indicate coal combustion. A [BaA]/[BaA + Chr] ratio between 0.2 and 0.35 indicates vehicles, whereas values exceeding 0.35 suggest coal and biomass burning (Tobiszewski et al. 2012). In Kanazawa in winter 1997, the [FR]/[FR + Pyr] and [BaA]/[BaA + Chr] ratios were 0.52 and 0.39, respectively, whereas in winter 2021, they were 0.44 and 0.28, respectively. These values indicate coal/biomass combustion as the major source in 1997 and vehicles in 2021 (Fig. 3a).

On the other hand, a [IDP]/[IDP + BgPe] ratio below 0.2 indicates petroleum sources, whereas 0.2–0.5 suggests petroleum combustion/vehicles. Ratios above 0.5 indicate coal/biomass combustion. [BaP]/[BgP] ratios below 0.6 and over 0.6 indicate vehicles and coal combustion, respectively (Yunker et al. 2002). In Kanazawa, the [IDP]/[IDP + BgPe] and [BaP]/[BgP] ratios in winters 1997 and 2021 were 0.43 and 0.67, and 0.41 and 0.61, respectively. The values were close in the two years, but indicating different sources, petroleum combustion/vehicles or coal combustion (Fig. 3b). Moreover, there is a discrepancy between (a) and (b). Despite its simplicity, the DR method often yields different sources depending on the PAH markers used, and it cannot identify NPAH sources or the contribution (%) of each source.

Nitrogen oxide (NO_x) formation in flame gases and subsequent NPAH formation from parent PAHs are temperature-dependent (Sicre et al. 1987), increasing NPAH concentrations in PM. Additionally, NPAHs such as 2-nireofluoranthene (2-NFR) and 2-NP are formed through secondary atmospheric processes (Krzyszczak and

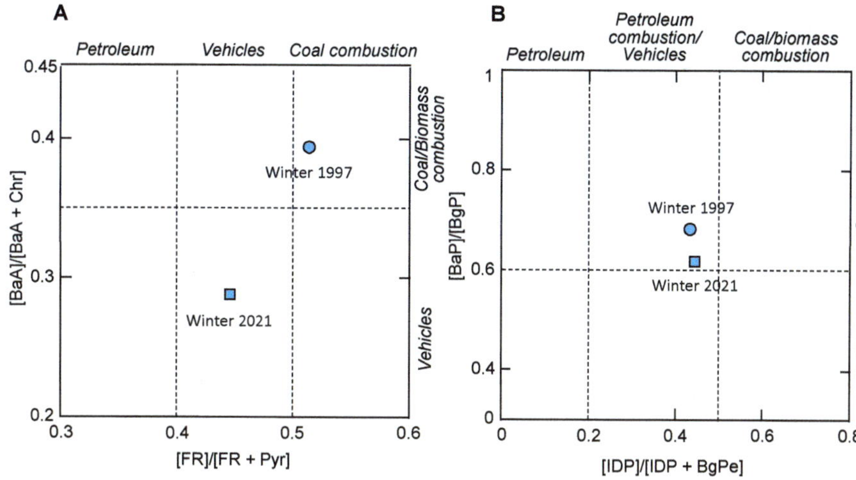

Fig. 3 Source appointment of atmospheric PAHs in Kanazawa in 1997 and 2021 by diagnostic ratios method. **A** [FR]/[FR + Pyr] versus [BaA]/[BaA + Chr]; **b** [IDP]/[IDP + BgPe] versus [BaP]/[BgP]

Czech 2021), as explained in the following sections. To estimate the contributions of these formation and emission processes, both PAH composition and the ratio of NPAHs to PAHs are useful indicators.

3 Combination of PAHs and NPAHs for Primary Source Analysis

Figure 4 compares concentrations of PAHs and NPAHs in PM emitted from wood stoves, coal stoves, and diesel vehicles. PAH concentrations decreased in the following order: wood stoves > coal stoves > diesel vehicles (Hayakawa et al. 2024). However, as noted earlier, NPAH-to-PAH concentration ratios are temperature-dependent, resulting in the reverse order for NPAHs: wood stoves < coal stoves < diesel vehicles (Hayakawa et al. 2020a). Figure 5 presents the concentration ratios of [1-NP]/[Pyr], 6-nitrochrysene (6-NC) to chrysene (Chr) ([6-NC]/[Chr]), 7-nitrobenz[*a*]anthracene (7-NBaA) to BaA ([7-NBaA]/[BaA]), and 6-nitrobenzo[*a*]pyrene (6-NBaP) to BaP ([6-NBaP]/[BaP]) in PM emitted from diesel and gasoline-engine vehicles (combustion temperature: 2700–3000 °C), coal stoves (1100–1200 °C), and wood stoves (500 °C–600 °C). All [NPAH]/[PAH] ratios significantly increase with combustion temperature (Hayakawa 2016).

Among these NPAHs, 1-NP had the highest concentration, making it the best marker for identifying traffic emissions (Miller-Schulze et al. 2016; Srivastava et al.

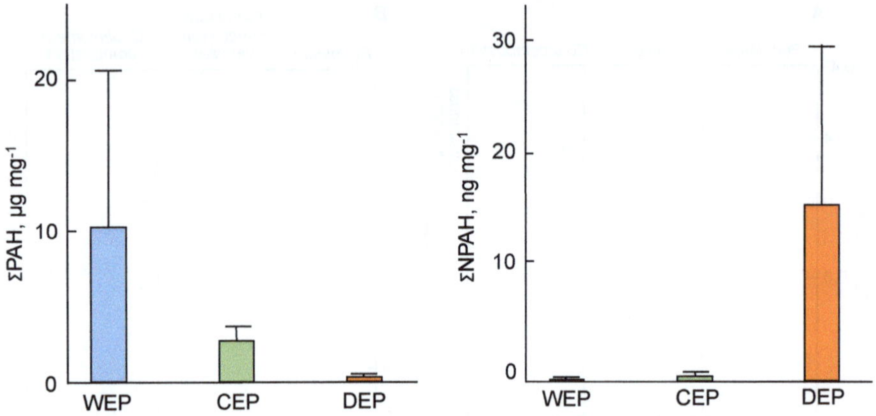

Fig. 4 Concentrations of PAHs and NPAHs in particulates emitted from different combustions. WEP, wood burning; CEP, coal burning; DEP; diesel-engine vehicle. ΣPAH = FR + Pyr + BaA + Chr + BbF + BkF + BaP + BgPe + IDP. ΣNPAH = 1,3-DNP + 1,6-DNP + 1,8-DNP + 1-NP (Hayakawa 2024a, with several modifications)

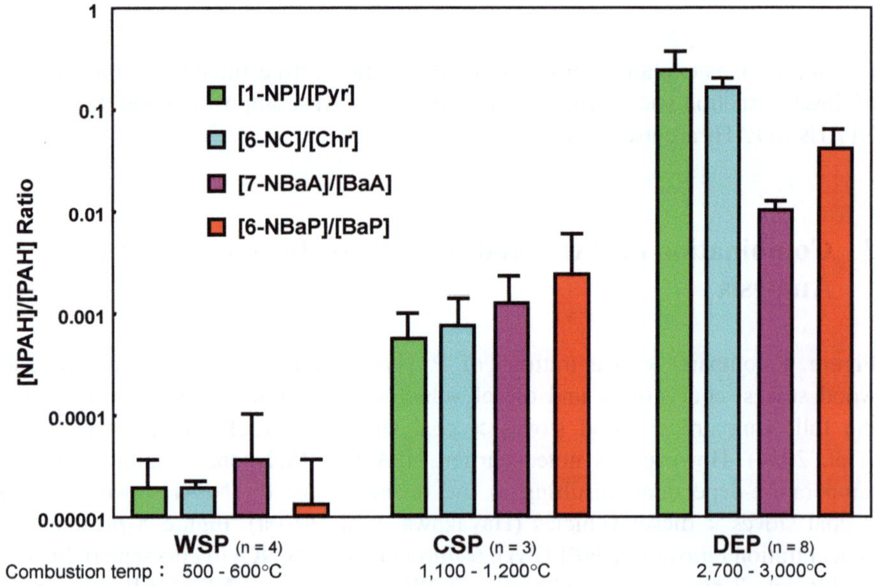

Fig. 5 [NPAH]/[PAH] ratios in diesel vehicle exhaust particulates (DEP), coal stove exhaust particulates (CEP) and wood stove particulates (WSP) (partly modified from Hayakawa 2016)

2018a, b). Pyr is also a highly concentrated PAH in PM. Given the stable detection of Pyr and 1-NP through high-performance liquid chromatography or gas chromatography–mass spectrometry, the [1-NP]/[Pyr] ratio (k) serves as an effective.

proxy to differentiate combustion sources by temperatures. Figure 5 shows the substantial difference in k values between diesel vehicles and coal stoves (Hattori et al. 2007). Since 1975, the author and collaborators have collected PM samples from more than 20 cities worldwide. Although k values varied by more than two orders of magnitude, they always fell between coal/wood combustion and vehicle emissions. Consequently, 1-NP and Pyr were selected as representative markers for developing the NP method, detailed in Sect. 5.

4 NPAHs for Secondary Formation Analysis

Several NPAHs, such as 2-NP and 2-NFR, are formed secondarily in the atmosphere, as mentioned earlier. These NPAHs are not detected in vehicle exhausts but are generated in the atmosphere over time (Hayakawa et al. 2000). The concentration ratios of 2-NFR to 1-NP ([2-NFR]/[1-NP]) and 2-NFR to 2-NP ([2-NFR]/[2-NP]) are often used to distinguish between two secondary formation processes in the atmosphere. A ratio close to 10 suggests OH^- radical–initiated formation, whereas a ratio above 100 suggests NO_3^--initiated formation (Bamford and Baker 2003). Many studies using these thresholds have concluded that the OH^- radical-initiated formation process is the primary pathway (Hayakawa et al. 2022).

5 NP Method

5.1 Theory

As described earlier, the NP method was developed using 1-NP and Pyr based on the temperature dependence of the k value (Hayakawa et al. 2020a, b). Figures 6 and 7 illustrate this method, where PM in the air is divided into combustion-derived PM (P_c) and noncombustion-derived PM (P_n). P_c is further divided into particulates emitted from high-temperature combustion (P_h) and low-temperature combustion (P_l).

When the proportion of P_h in P_c is x ($0 < x < 1$) and the proportion of P_c in total PM is y ($0 < y < 1$) in the air, Equations (1)–(3) can be written as follows:

$$[1-\mathrm{NP}] = [1-\mathrm{NP_h}][P_c]x + [1-\mathrm{NP_l}][P_c](1-x) \quad (1)$$

$$[\mathrm{Pyr}] = [\mathrm{Pyr_h}][\mathrm{P_c}]x + [\mathrm{Pyr_l}][\mathrm{P_c}](1-x) \quad (2)$$

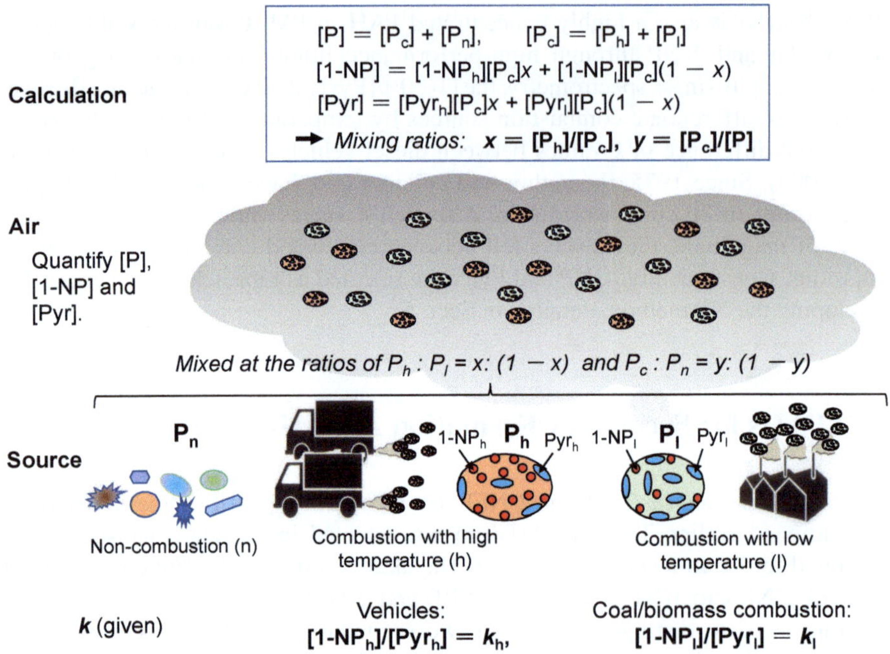

Fig. 6 Principles of the NP method: calculation of atmospheric combustion-derived particulate (P_c) and its emission sources

$$y = [P_c]/([P_c] + [P_n]) \tag{3}$$

The concentration ratios of 1-NP and Pyr in particulates emitted the two major combustion contributors were determined experimentally: $k_h = [1\text{-NP}_h]/[\text{Pyr}_h] = 0.425$ (mole ratio) or 0.520 (g ratio) for vehicles and $k_l = [1\text{-NP}_l]/[\text{Pyr}_l] = 0.0013$ (mole ratio) or 0.00159 (g ratio) for coal/biomass (Hayakawa et al. 2020a). According to a recent report on PAHs and NPAHs in emissions from gasoline vehicles (Zhao et al. 2020a), the ratio was calculated to be 0.3 for $[1\text{-NP}_h]/[\text{Pyr}_h]$, which was slightly smaller than the above. This is probably due to technological development and energy change for vehicles. Therefore, the ratios have been changed for PM samples collected after 2015. Using these values in Equations (1)–(3), x and y can be calculated, facilitating the estimation of atmospheric concentrations for P_h, P_l, and P_c (Fig. 6).

On the other hand, the contributions of these two major emission sources to atmospheric PAHs and NPAHs can be calculated without weighing atmospheric PM. Given that 1-NP and Pyr are also common markers for NPAH and PAH, respectively, when the proportion of 1-NP_h in total 1-NP ($1\text{-NP}_h + 1\text{-NP}_l$) in the air is a ($0 < a < 1$), and the proportion of Pyr_h in total Pyr ($\text{Pyr}_h + \text{Pyr}_l$) in the air is b ($0 < b < 1$), the following equations hold:

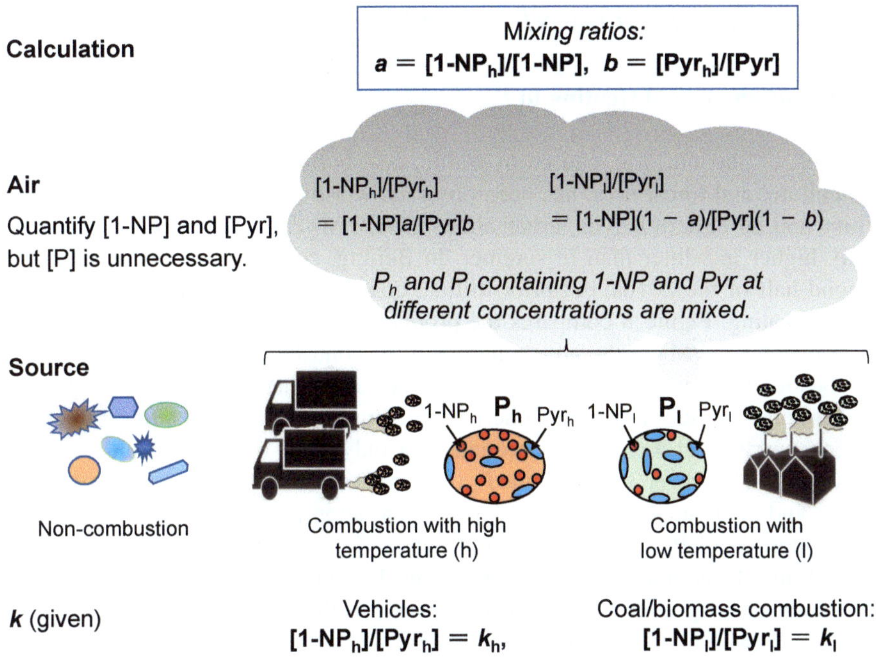

Fig. 7 Principles of the NP method: calculation of emission sources for atmospheric 1-NP and Pyr

$$[1-\mathrm{NP_h}] = [1-\mathrm{NP}]a \text{ and } [1-\mathrm{NP_l}] = [1-\mathrm{NP}](1-a) \tag{4}$$

$$[\mathrm{Pyr_h}] = [\mathrm{Pyr}]b \text{ and } [\mathrm{Pyr_l}] = [\mathrm{Pyr}](1-b) \tag{5}$$

Dividing Eq. (4) by Eq. (5) gives the following equations for the concentration ratio of 1-NP to Pyr for each source:

$$[1-\mathrm{NP_h}]/[\mathrm{Pyr_h}] = [1-\mathrm{NP}]a/[\mathrm{Pyr}]b \tag{6}$$

$$[1-\mathrm{NP_l}]/[\mathrm{Pyr_l}] = [1-\mathrm{NP}](1-a)/[\mathrm{Pyr}](1-b) \tag{7}$$

The values of [1-$\mathrm{NP_h}$]/[$\mathrm{Pyr_h}$] and [1-$\mathrm{NP_l}$]/[$\mathrm{Pyr_l}$] can be determined by quantifying 1-NP and Pyr in PM samples collected from high- and low-temperature combustion. By applying these values to Eqs. (6) and (7), a and b can be calculated. Subsequently, the atmospheric concentrations of 1-$\mathrm{NP_h}$, 1-$\mathrm{NP_l}$, Pyr, and $\mathrm{Pyr_l}$ can be determined (Fig. 7). This method is highly effective for estimating the contributions of the two major sources to PAHs, NPAHs, and P_c, as well as for determining the fraction of P_c in atmospheric PM.

5.2 Applications of the NP Method

Contribution of Coal Heating to P_c

The "haze" phenomenon, caused by traffic, coal-burning facilities, slash-and-burn agriculture, and forest fires, has been observed worldwide. In central and northern China and far-eastern Russia, urban atmospheric $PM_{2.5}$ concentrations are substantially higher in winter than in summer. In Beijing, coal use was restricted in the second half of the 2010s, but prior to that, coal boilers were commonly used for winter heating. Figure 8 compares the proportion of P_c in total suspended particulate matter (TSPM) in the atmosphere and the contributions of vehicles and coal combustion to P_c in Beijing during the summer and winter of 2010, where the size of the circles represent the atmospheric concentration of pollutants. In summer, the P_c concentration (1.4 $\mu g\ m^{-3}$) accounted for only 1.5% of the TSP concentration (96 $\mu g\ m^{-3}$). However, in winter, the P_c concentration (88 $\mu g\ m^{-3}$) was substantially higher and made up 42% of the TSPM concentration (207 $\mu g\ m^{-3}$). Furthermore, the proportion of P_l (83 $\mu g\ m^{-3}$) to P_c was 94%, indicating that most of the P_c was emitted from coal combustion in winter. In winter, biomass burning, such as rice-straw burning after harvest, is uncommon in the Beijing area. This seasonal difference was primarily attributed to the large amounts of coal used for winter coal heating (Hayakawa et al. 2020b).

Vehicles and Coal Combustion Make Different Contributions in Western Pacific

Since the 1990s, my international research group has been collecting seasonal atmospheric PM (TSPM or $PM_{2.5}$) samples in 10 cities across the Western Pacific region. We found that Chinese and far-eastern Russian cities exhibit much higher concentrations of PM, PAHs, and NPAHs compared with Japanese cities. Moreover, these Chinese and Russian cities display stronger seasonal variation (winter > summer). The ten cities were divided into three groups: Group 1, showing lower $[P_c]$ values but larger $[P_h]/[P_c]$ ratios, included Sapporo, Kanazawa, Tokyo, Sagamihara (Japan), Shanghai (China), and Busan (Korea), predominantly commercial cities where vehicles are the major contributors to P_c. Group 2, exhibiting much higher winter $[P_c]$ values with larger $[P_l]/[P_c]$ ratios, included Beijing, Shenyang (China), and Vladivostok (Russia), where coal heating is the main P_c source. Group 3, showing lower $[P_c]$ values with larger $[P_l]/[P_c]$ ratios, included Kitakyushu, where coke oven plants, used in steel manufacturing, are the main P_c source. Atmospheric pollutant concentrations in Japanese cities were found to be consistently lower than those in Chinese cities.

In this section, four cities, namely Kanazawa and Kitakyushu in Japan and Shenyang and Shanghai in China, are selected from the three groups to compare the source contributions to PAHs and NPAHs. The characteristics of [PM] and $[P_c]$

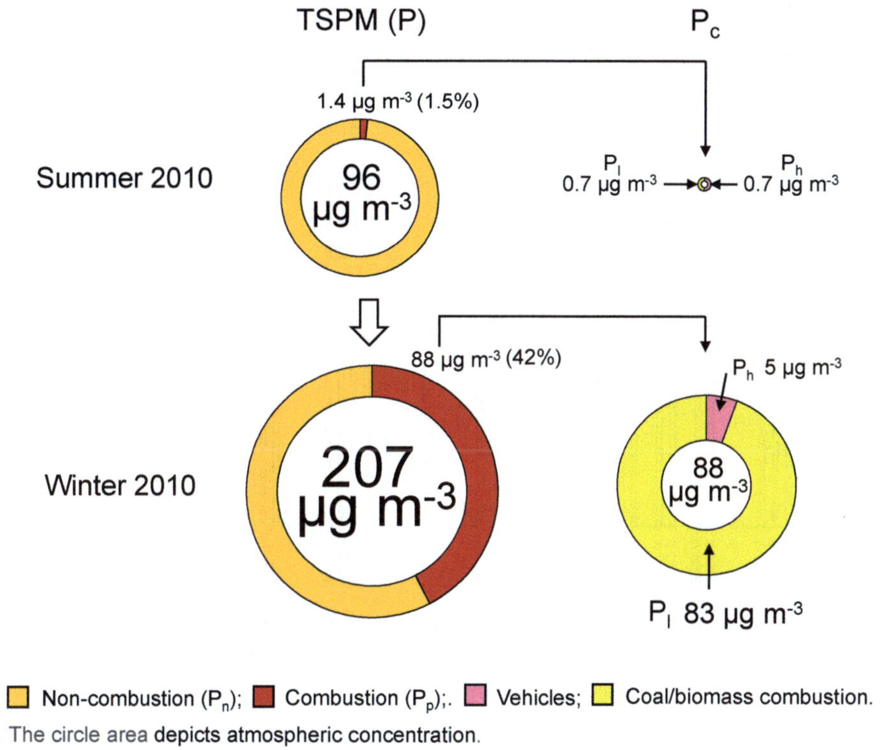

Fig. 8 Contribution of coal winter heating to combustion-derived particulate (P_c) in Beijing in 2010 (Hayakawa et al. 2020b, with several modifications)

variations in these cities, along with their respective source contributions (a1, a2, b1, and b2 in Figs. 9, 10, 11, and 12). In Kanazawa, [P_c] was 11.0 μg m^{-3} in winter 1999, representing 20% of [PM]. However, starting in the 2000s, [P_c] levels declined, dropping below 1 μg m^{-3}, with seasonal change observed (summer < winter). By summer 2007, the fraction of [P_h] in [P_c] exceeded 80%, but after winter 2018, it had decreased to less than two-third (Fig. 9). A similar decline in the [P_h]/[P_c] ratio was observed in other Japanese cities, such as Sapporo, Tokyo, and Sagamihara, primarily attributed to the reduced contribution from vehicles (Hayakawa et al. 2022). In Kitakyushu, the [PM] levels were similar to those in Kanazawa, but the fraction of [P_c] in [PM] remained below 5% throughout the monitoring period, except for the winter of 2014 (13%). The average fraction of [P_h] in [P_c] (34%) was substantially lower than in the other Japanese cities. Muroran, another steel manufacturing city in Japan, also showed a small [P_h] fraction in [P_c] (11%) (Hayakawa et al. 2016). These findings suggest that coke oven plants, which consume large quantities of coal, were the primary contributors to P_c (Fig. 10).

In China, Shenyang and Beijing showed much higher winter [P_c] levels (exceeding 40 μg m^{-3}) and stronger seasonal variations (winter > summer) compared with

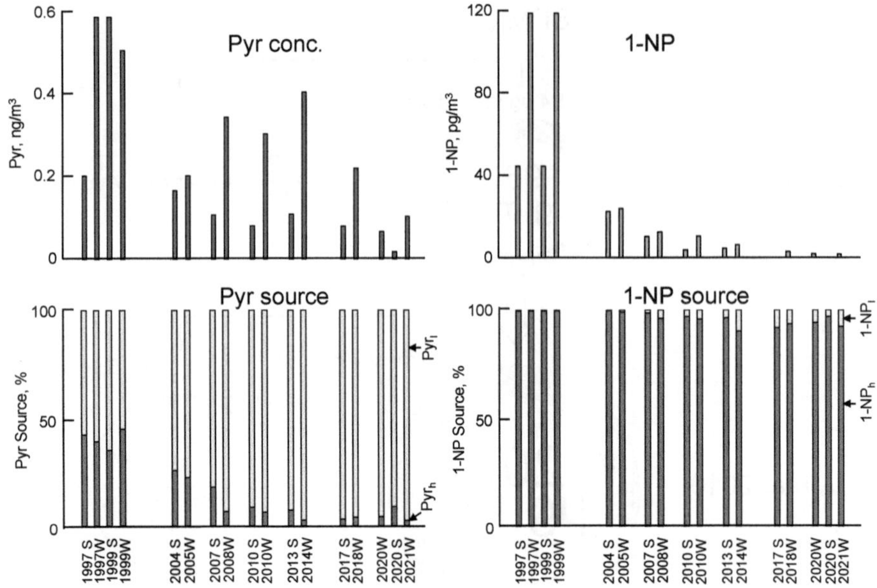

Fig. 9 Long-term change in atmospheric concentrations of PM, P_c, Pyr, and 1-NP and their sources in Kanazawa. Abbreviations of seasons: S = summer, W = winter (Hayakawa 2024b, with several modifications)

Shanghai (3.37 μg m^{-3}). The seasonal difference in the former cities was due to higher [P_c] fractions in [PM] and larger [P_l] fractions in [P_c] during winter. Although the [P_l] fraction in [P_c] in Shenyang did not differ markedly from that in Kitakyushu, its seasonal difference (winter > summer) was more pronounced. These results are mainly attributed to extensive coal burning for winter heating (Fig. 11). In Shanghai, where heating is not used in winter, the [P_c] levels were as low as those in Japanese cities, and the seasonal variation was not as strong as that observed in Shenyang. However, Shanghai exhibited a larger [P_h] fraction in [P_c] compared to Shenyang, with a decreasing trend observed, similar to that in Kanazawa (Fig. 12).

In summary, the NP method effectively highlighted differences in the contributions of vehicles and coal combustion to P_c across the Western Pacific region. In Kanazawa and Shanghai, the primary contributors were a mixture of vehicle emissions and coal combustion. In contrast, coal heating was the dominant source of high [P_c] levels in Shenyang during winter. Coal combustion also played a major role in Kitakyushu, where coke oven plants are used for steel manufacturing, although the [P_c] levels in this city were markedly lower than those in Shenyang.

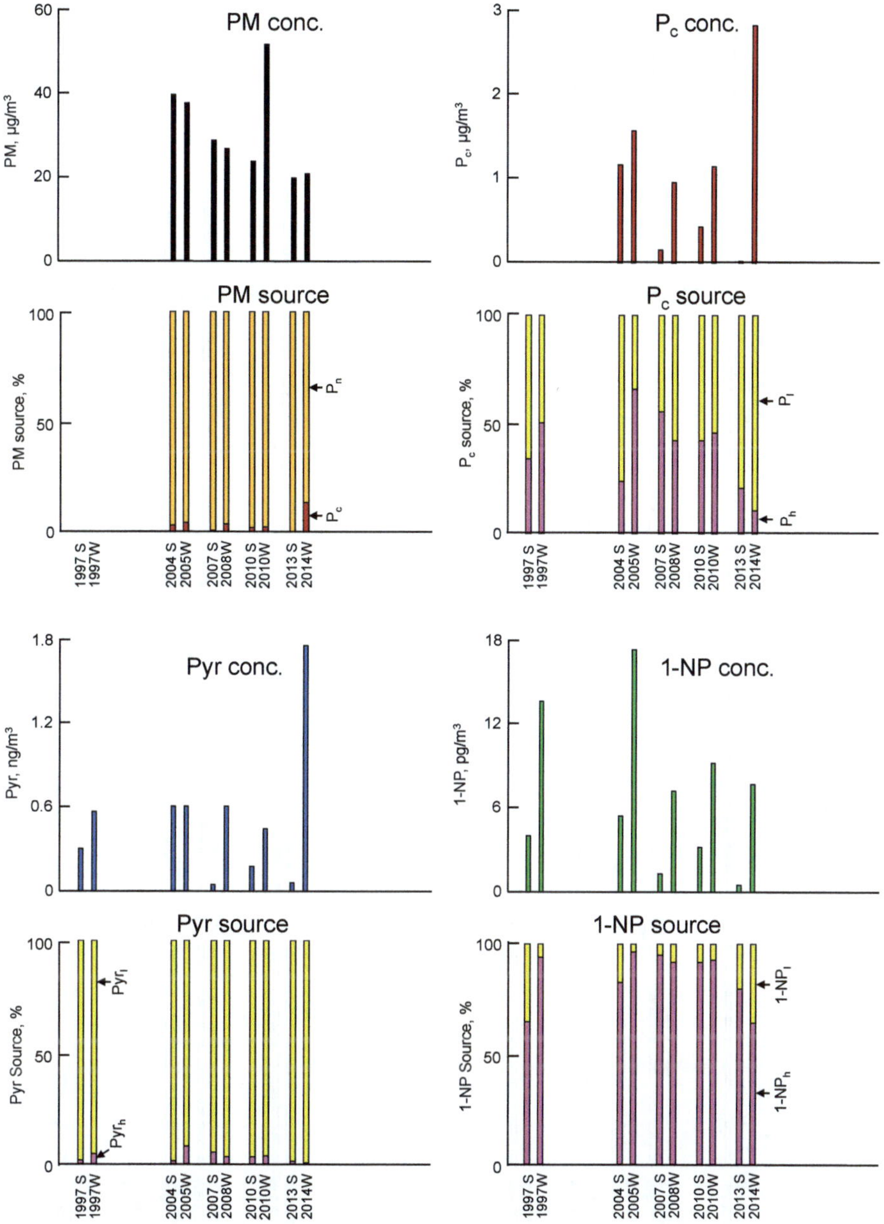

Fig. 10 Long-term change in atmospheric concentrations of PM, P_c, Pyr, and 1-NP and their sources in Kitakyushu. Abbreviations of seasons are the same as in Fig. 9 (Hayakawa 2024b, with several modifications)

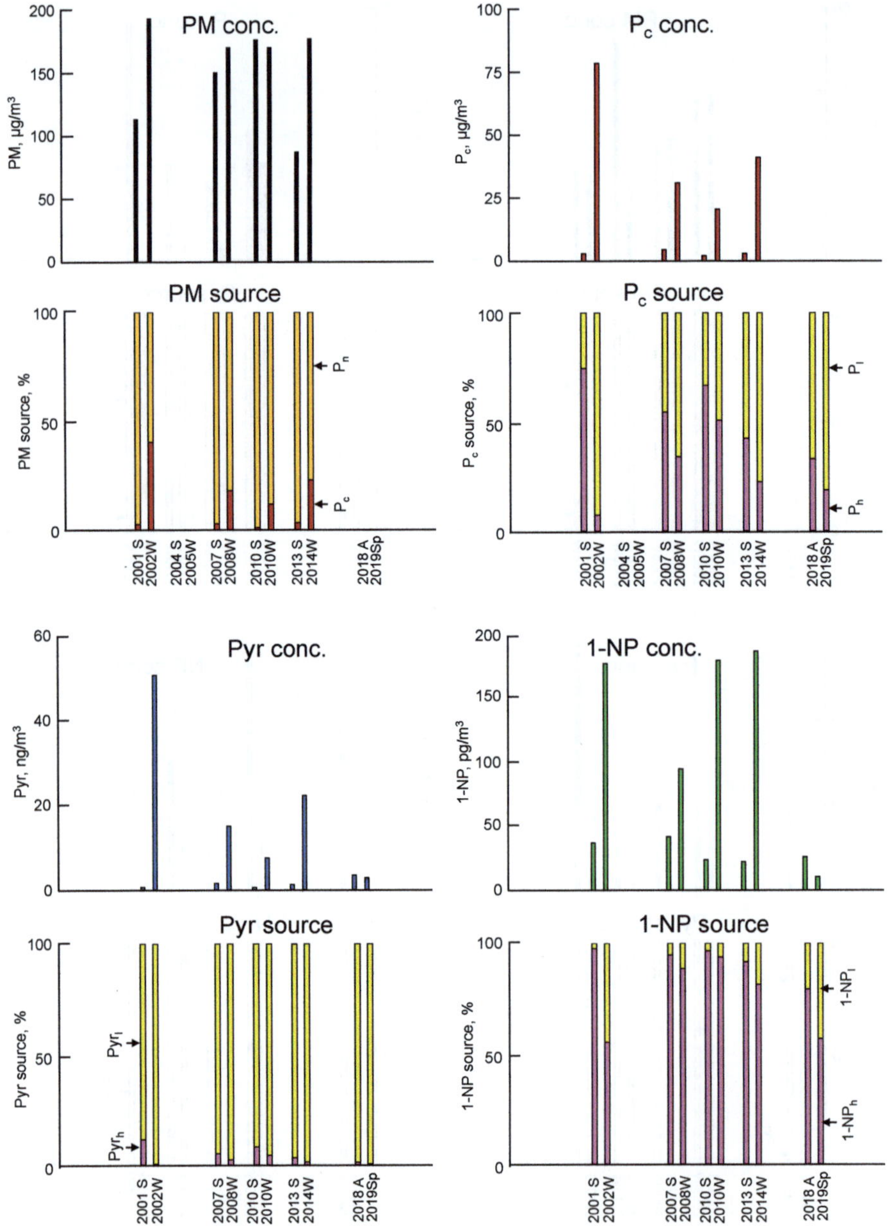

Fig. 11 Long-term change in atmospheric concentrations of PM, P_c, Pyr, and 1-NP and their sources in Shenyang. Abbreviations of seasons: S = spring, W = winter, A = autumn, Sp = spring (Hayakawa 2024b, with several modifications)

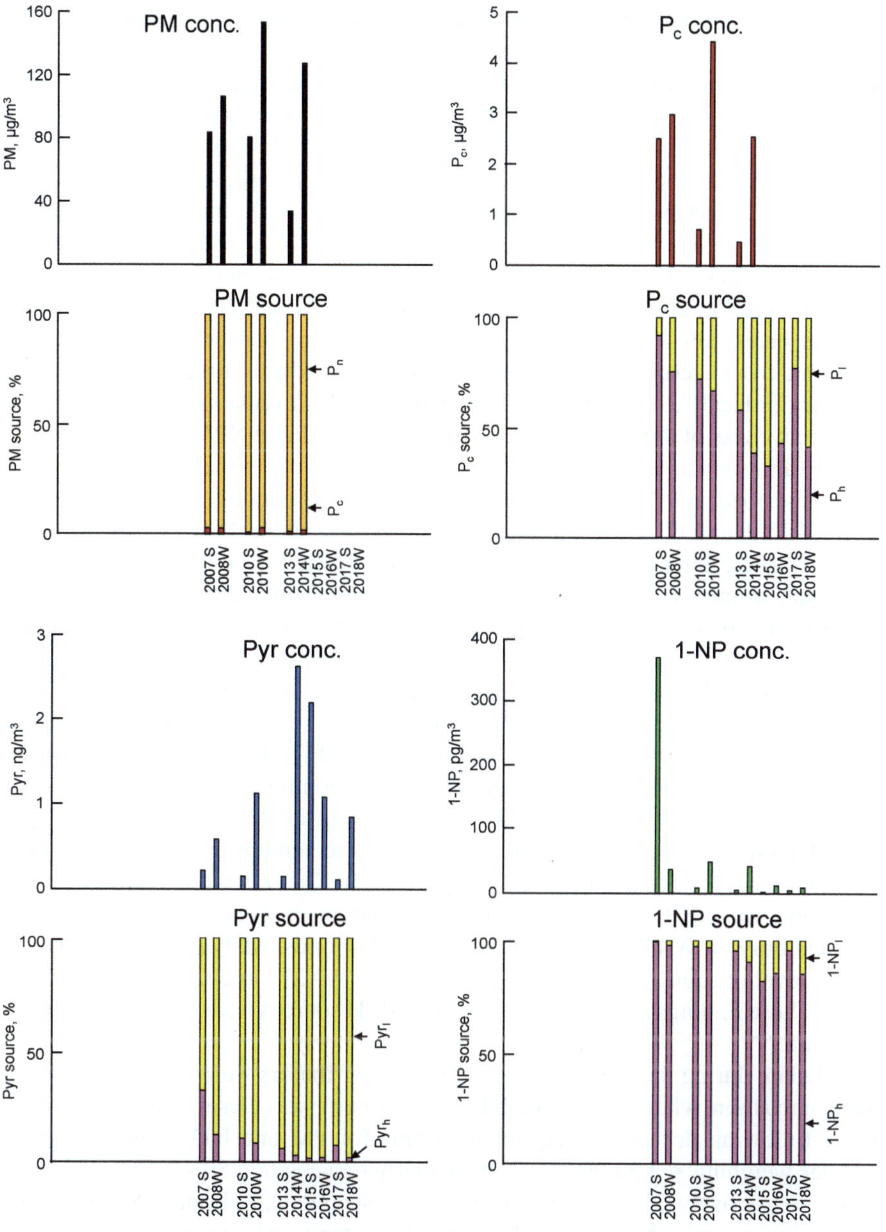

Fig. 12 Long-term change in atmospheric concentrations of PM, P_c, Pyr, and 1-NP and their sources in Shanghai. Abbreviations of seasons are the same as in Fig. 9 (Hayakawa 2024b, with several modifications)

Vehicle Emission Control Has Markedly Altered Source Contributions

To reduce carbon monoxide (CO), hydrocarbons (HCs), and NO_x in vehicle emissions, a three-way catalyst system was developed. The introduction of catalysts reduced PAH emissions from gasoline-engine vehicles to less than 10% (Rogge et al. 1993). In Japan, this system began being implemented in new gasoline engines in the 1970s, with governmental regulations on CO, HC, and NO_x emissions for gasoline vehicles starting in 1973. Although these regulations were gradually strengthened, their effectiveness remained limited. However, with subsequent technical advancements in diesel engines and fuels, the emission of PM and NO_x from diesel vehicles has been substantially reduced. Governmental regulations on PM and NO_x emissions from diesel vehicles have been tightened several times since the 1990s (Ministry of Health, Labor and Welfare, Japan 2022). It remains crucial to estimate the effectiveness of these regulations on reducing PAH and NPAH concentrations.

The variations in atmospheric [Pyr] and [1-NP] levels in the same four cities as above described were compared with their source contributions (c1, c2, d1, and d2 in Figs. 9, 10, 11, and 12). In Kanazawa, [Pyr] levels were above 0.6 ng m^{-3} in the 1990s, gradually decreasing to less than 0.1 ng m^{-3} by the 2020s, with seasonal variation observed (summer < winter). Notably, [1-NP] levels, which exceeded 40 pg m^{-3} in the 1990s, decreased even more rapidly. Interestingly, the source contribution trends to Pyr and 1-NP differed markedly. The fraction of [Pyr_h] in [Pyr] was close to 50% in the 1990s, falling to less than 10% in the 2020s. Conversely, the fraction of [1-NP_h] in [1-NP] remained above 99% throughout the monitoring period, indicating that vehicles were the main contributors to NPAHs (Fig. 9).

Since 2004, we have conducted atmospheric monitoring at the Wajima atmospheric monitoring site (WAMS), a remote site on the Noto Peninsula in Japan, located about 100 km north of Kanazawa along the same route of westerly winds from the Asian continent. In 2005, the [Pyr] level in Kanazawa was more than threefold higher than that at WAMS, but by 2021, the levels in Kanazawa had dropped to match those at WAMS and the difference had diminished. On the other hand, the [1-NP] level in Kanazawa was about twice that at WAMS in 2021 (Hayakawa et al. 2024). These findings suggest that PAH and NPAH sources have seen major changes over time.

In Japan, during the 1990s, the major PAH contributors were diesel vehicles and coal combustion, whereas the main NPAH contributor was diesel vehicles. However, in the following decade, the regulatory limits for PM and NO_x from new diesel vehicles weighing $\leq$ 3.5 tons were substantially reduced to 2.8%–3.8% and 13.7%–29.4%, respectively (Hayakawa et al. 2014). Consequently, BaP and 6-NBaP concentrations in the atmosphere declined to around one-tenth or less of the previous levels within 12 years (Hama et al. 2012), indicating a major reduction in the contribution of diesel vehicles to atmospheric PAHs and NPAHs. As a result, the main source of PAHs has shifted from vehicles to coal combustion, largely due to long-range transport from the Asian continent. The recent atmospheric Py concentration became much lower than before, but the seasonal difference (winter < summer) became much clearer due to Pyr long-range transported (Hayakawa et al. 2021a). A similar declining trend in [1-NP], as well as in the fractions of [P_h] in [P_c] and [Pyr_h] in [Pyr], was observed in Shanghai (Fig. 12), likely reflecting the effects of delayed vehicle emission controls, similar to those in Kanazawa.

As described above, the fraction of [P_l] in [P_c] was higher in Kitakyushu and Shenyang than in Kanazawa and Shanghai, although both [P_c] and [Pyr] levels were markedly higher in these cities. Additionally, the fraction of [1-NP_l] in [1-NP] was larger in Kitakyushu and Shenyang than in Kanazawa and Shanghai, reflecting the greater contribution of coal combustion in these cities. Thus, the NP method has revealed the historical changes in the source contributions of urban atmospheric pollutants related to PAHs and NPAHs in cities across the Western Pacific.

6 Comparison of the NP Method and Previously Established Methods

Here, the NP method is compared with the DR method. Figure 13 shows the [FR]/[FR + Pyr] and [BaA]/[BaA + Chr] plots for Kanazawa, Japan, from April 24 to 30, 2017. The [FR]/[FR + Pyr] ratio indicates that vehicles were the major contributors from April 24 to 27, whereas coal combustion was the primary contributor from April 28 to 30. Although the DR method identifies vehicles and coal/biomass combustion as contributors, the primary source varies based on the specific PAH diagnostic pair used. Additionally, the DR method does not quantify the fraction (%) contributed by each source (Hayakawa et al. 2021b).

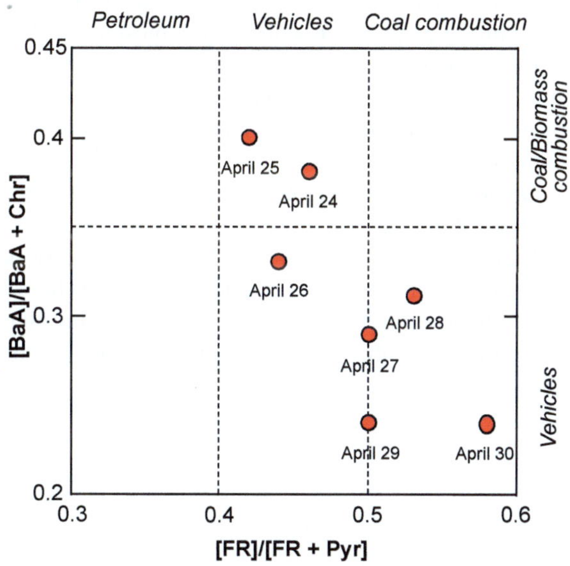

Fig. 13 [FR]/[FR + Pyr] versus [BaA]/[BaA + Chr] plots for atmospheric $PM_{2.5}$ samples collected in Kanazawa from April 24 to 30, 2017 (partly modified from Hayakawa et al. 2021b)

Daily change in source contribution to not only PAHs but also NPAHs and PM can be clarified by the NP method. Figure 14 shows (a) daily variations in atmospheric $PM_{2.5}$, P_c, Pyr, and 1-NP concentrations and (b) the contributions of the two sources during the same period shown in Fig. 13, as calculated via the NP method. Compared with the first four days, $PM_{2.5}$ concentrations more than doubled in the last three days due to westerly winds. Although the fraction of $[P_c]$ in $[PM_{2.5}]$ remained low ($\leq 2.8\%$) throughout the week, the daily variations in $[P_c]$ mirrored those in $[PM_{2.5}]$. This indicates that the increase in $[P_c]$ during the latter three days was mainly driven by a rise in $[P_l]$. Given that there was no major local source of P_l around Kanazawa, these results suggest that P_l was mainly transported over long distances in coal/biomass combustion emissions from the Asian continent, as describe earlier. Figure 14c shows that the fraction of $[Pyr_h]$ in [Pyr] remained low ($\leq$ 2.9%) throughout the week. In contrast, $[Pyr_l]$ accounted for nearly all of the [Pyr], and its variation closely followed that of $[P_c]$. This indicates that Pyr_l was largely emitted from coal/biomass combustion in the Asian continent. However, Fig. 14d shows that [1-NP] did not increase to the same extent as [Pyr] in the latter three days, suggesting that long-range transport had a smaller impact on [1-NP] in Kanazawa. The NP method also revealed that the proportion of $[Pyr_l]$ in Kanazawa increased in winter due to the long-range transportation of substantial PAH levels generated by coal heating in northern and central China (Hayakawa et al. 2021b).

7 Conclusion

The NP method was developed to estimate the contributions of high- and low-temperature combustion sources to atmospheric P_c, PAHs, and NPAHs, using Pyr and 1-NP as markers. This method is based on the premise that the concentration ratio of NPAHs to corresponding PAHs is dependent on combustion temperature. By employing the method, it is possible to determine the contributions of two major sources, vehicles and coal/biomass combustion, to these pollutants. The NP method is simple, fast, and offers numerous advantages over traditional source appointment methods. Although the global contribution of vehicles to human exposure to PAHs and their derivatives, such as NPAHs, is steadily decreasing, the contribution of biomass burning, including the increase in forest fires, has not diminished (Hayakawa 2022). To develop effective strategies for reducing air pollution, it is increasingly vital to accurately identify the sources of hazardous chemicals.

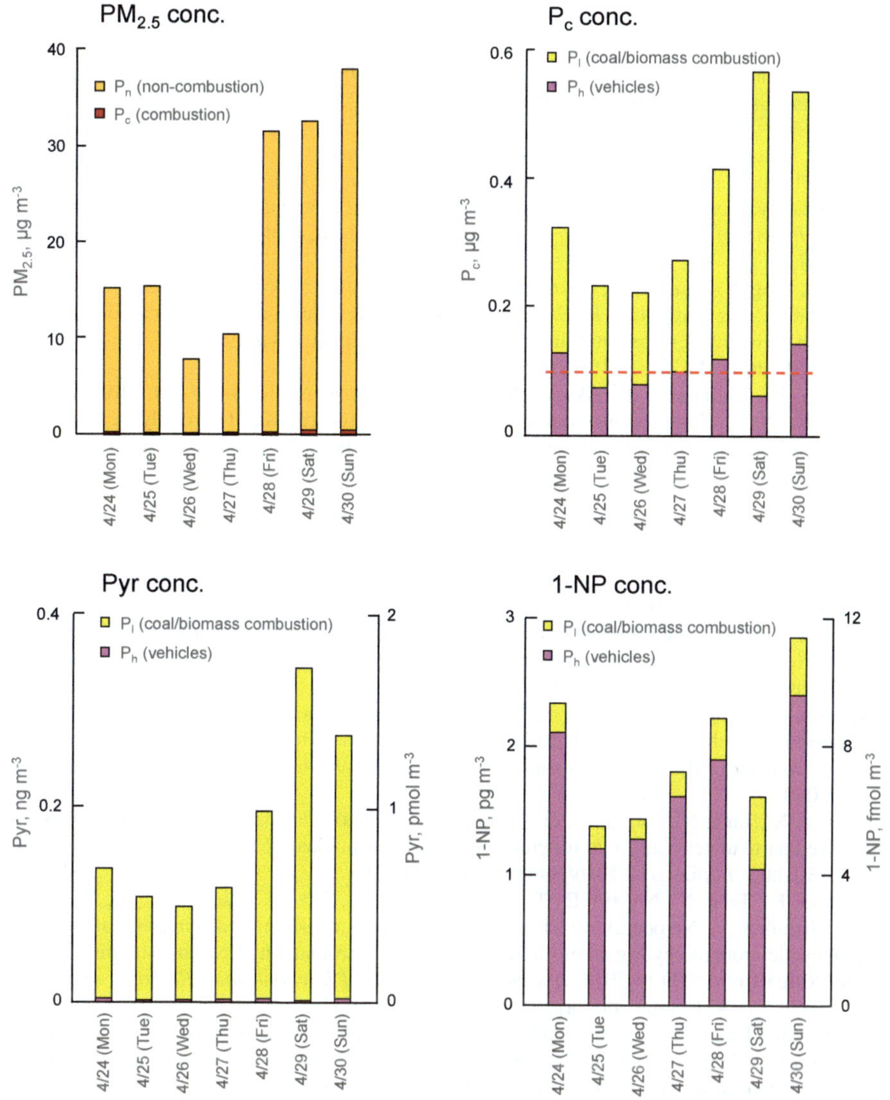

Fig. 14 Daily atmospheric concentrations of $PM_{2.5}$, P_c, Pyr, and 1-NP with source contributions in Kanazawa from April 24 to 30, 2017 (partly modified from Hayakawa et al. 2021b)

References

Bamford, H. A., Baker, J. E. (2003). Nitro-polycyclic aromatic hydrocarbon concentrations and sources in urban and suburban atmospheres of the Mid-Atlantic region. *Atmospheric Environment*, *37*(15), 2077–2091. https://doi.org/10.1016/s1352-2310(03)00102-x

Brown, S. G., Eberly, S., Paatero, P., & Norris, G. A. (2015). Methods for estimating uncertainty in PMF solutions: Examples with ambient air and water quality data and guidance on reporting

PMF results. *Science of the Total Environment*, *518*, 626–635. https://doi.org/10.1016/j.scitotenv.2015.01.022

Davis, E., Walker, T. R., Adams, M., & Willis, R. (2019). Estimating PAH sources in harbor sediments using diagnostic ratios. *Remediation Journal*, *29*(3), 51–62. https://doi.org/10.1002/rem.21600

Hama, H., Tokuda, T., Izaki, A., Ohno, T., Watanabe, Y., Kanda, T., Tang, N., Kameda, T., Toriba, A., & Hayakawa, K. (2012). Variation in polycyclic aromatic hydrocarbons and nitropolycyclic aromatic hydrocarbons in airborne particulates collected in urban Kanazawa, Japan, in last 12 years. *Journal of Japan Society for Atmospheric Environment*, *47*, 1–8. https://doi.org/10.11298/taiki.47.1

Hattori, T., Tang, N., Tamura, K., Hokoda, A., Yang, X., Igarashi, K., Ohno, M., Okada, Y., Kameda, T., Toriba, A., & Hayakawa, K. (2007). Particulate polycyclic aromatic hydrocarbons and their nitrated derivatives in three cities in Liaoning Province, China. *Environmental Forensics*, *8*(1–2), 165–172. https://doi.org/10.1080/15275920601180701

Hayakawa, K., Murahashi, T., Akutsu, K., Kanda, T., Tang, N., Kakimoto, H., Toriba, A., & Kizu, R. (2000). Comparison of polycyclic aromatic hydrocarbons and nitropolycyclic aromatic hydrocarbons in airborne and automobile exhaust particulates. *Polycyclic Aromatic Compounds*, *20*(1–4), 179–190. https://doi.org/10.1080/10406630008034784

Hayakawa, K., Tang, N., Nagato, E. G., Toriba, A., Sakai, S., Kano, F., Goto, S., Endo, O., Arashidani, K.-I., & Kakimoto, H. (2014). Long term trends in atmospheric concentrations of polycyclic aromatic hydrocarbons and nitropolycyclic aromatic hydrocarbons: A study of Japanese cities from 1997 to 2014. *Environmental Pollution*, *233*, 474–482. https://doi.org/10.1016/j.envpol.2017.10.038

Hayakawa, K. (2016). Environmental behaviors and toxicities of polycyclic aromatic hydrocarbons and nitropolycyclic aromatic hydrocarbons. *Chemical & Pharmaceutical Bulletin*, *64*(2), 83–94. https://doi.org/10.1248/cpb.c15-00801

Hayakawa, K., Tang, N., Morisaki, H., Toriba, A., Akutagawa, T., & Sakai, S. (2016). Atmospheric polycyclic and nitropolycyclic aromatic hydrocarbons in an iron-manufacturing city. *Asian Journal of Atmospheric Environment*, *10*(2), 90–98. https://doi.org/10.5572/ajae.2016.10.2.090

Hayakawa, K., Tang, N., Toriba, A., & Nagato, E. G. (2020a). Calculating sources of combustion-derived particulates using 1-nitropyrene and pyrene as markers. *Environmental Pollution*, *265*, 114730. https://doi.org/10.1016/j.envpol.2020.114730

Hayakawa, K., Tang, N., Nagato, E., Toriba, A., Lin, J.-M., Zhao, L., Zhou, Z., Qing, W., Yang, X., Mishukov, V., Neroda, A., & Chung, H.-Y. (2020b). Long-term trends in urban atmospheric polycyclic aromatic hydrocarbons and nitropolycyclic aromatic hydrocarbons: China, Russia, and Korea from 1999 to 2014. *International Journal of Environmental Research and Public Health*, *17*(2), 431. https://doi.org/10.3390/ijerph17020431

Hayakawa, K., Tang, N., Xing, W., Oanh, P. K., Hara, A., & Nakamura, H. (2021a). Concentrations and sources of atmospheric PM, polycyclic aromatic hydrocarbons and nitropolycyclic aromatic hydrocarbons in Kanazawa, Japan. *Atmosphere*, *12*(2), 256. https://doi.org/10.3390/atmos12020256

Hayakawa, K., Tang, N., Matsuki, A., Inomata, Y., Toriba, A., & Nagato, E. G. (2021b). Calculating source contributions to urban atmospheric polycyclic aromatic hydrocarbons and nitropolycyclic aromatic hydrocarbons using 1-nitropyrene and pyrene: An application to an Asian dust event. *Chemosphere*, *280*, 130662. https://doi.org/10.1016/j.chemosphere.2021.130662

Hayakawa, K., Tang, N., & Toriba, A. (2022). Long-term and seasonal changes in sources of urban atmospheric particulates in the Western Pacific. *Applied Sciences*, *12*(4), 2149. https://doi.org/10.3390/app12042149

Hayakawa, K. (2022). Recent research progress on nitropolycyclic aromatic hydrocarbons in outdoor and indoor environments. *Applied Sciences*, *12*(21), 11259. https://doi.org/10.3390/app122111259

Hayakawa, K. (2024a). Polycyclic aromatic hydrocarbons from vegetation burning and health effects. *Global Environmental Research*, 27(1), 13–20. https://doi.org/10.57466/ger.27.1_13

Hayakawa, K. (2024b). Long-term and seasonal changes in emission sources of atmospheric particulate-bound pyrene and 1-nitropyrene in four selected cities in the Western Pacific. *Atmosphere*, **15**, 634. https://doi.org/10.3390/atmos15060634

Hayakawa, K., Kim-Oanh, P., Takami, A., Odajima, H., Nagato, E. G., Hara, A., & Nakamura, H. (2024). Contributions of long-range transport from the Asian continent and local emissions on atmospheric $PM_{2.5}$, polycyclic aromatic hydrocarbons, and their nitro-derivatives in Kanazawa and Noto Peninsula, Japan. *Heliyon, 10*(5), e27320. https://doi.org/10.1016/j.heliyon.2024.e27320

IARC Monograph on the Identification of Carcinogenic Hazards to Humans. https://monographs.iarc.who.int

Kalisa, E., Nagato, E. G., Bizuru, E., Lee, K. C., Tang, N., Pointing, S. B., Hayakawa, K., Archer, S. D. J., & Lacap-Bugler, D. C. (2018). Characterization and risk assessment of atmospheric $PM_{2.5}$ and PM_{10} particulate-bound PAHs and NPAHs in Rwanda, Central-east Africa. *Environmental Science & Technology*, *52*(21), 12179–12187. https://doi.org/10.1021/acs.est.8b03219

Krzyszczak, A., & Czech, B. (2021). Occurrence and toxicity of polycyclic aromatic hydrocarbons derivatives in environmental matrices. *Science of the Total Environment*, *788*, 147738. https://doi.org/10.1016/j.scitotenv.2021.147738

Manoli, E., Kouras, A., & Samara, C. (2004). Profile analysis of ambient and source emitted particle-bound polycyclic aromatic hydrocarbons from three sites in northern Greece. *Chemosphere*, *56*(9), 867–878. https://doi.org/10.1016/j.chemosphere.2004.03.013

Masclet, P., Bresson, M. A., & Mouvier, G. (1987). Polycyclic aromatic hydrocarbons emitted by power stations, and influence of combustion conditions. *Fuel*, *66*(4), 556–562. https://doi.org/10.1016/0016-2361(87)90163-3

McGrath, T. E., Chan, W. G., & Hajaligol, M. R. (2003). Low temperature mechanism for the formation of polycyclic aromatic hydrocarbons from the pyrolysis of cellulose. *Journal of Analytical and Applied Pyrolysis*, *66*(1–2), 51–70. https://doi.org/10.1016/s0165-2370(02)00105-5

Miller-Schulze, J. P., Paulsen, M., Kameda, T., Toriba, A., Hayakawa, K., Cassidy, B., Naeher, L., Villalobos, M. A., & Simpson, C. D. (2016). Nitro-PAH exposures of occupationally-exposed traffic workers and associated urinary 1-nitropyrene metabolite concentrations. *Journal of Environmental Sciences*, *49*, 213–221. https://doi.org/10.1016/j.jes.2016.06.007

Ministry of Health, Labour and Welfare, Japan. (2022). *Demographic statistics Japanese*. https://www.mhlw.go.jp/toukei/saikin/hw/jinkou/geppo/nengai03/index.html

Ravindra, K., Sokhi, R., & Vangrieken, R. (2008). Atmospheric polycyclic aromatic hydrocarbons: Source attribution, emission factors and regulation. *Atmospheric Environment*, *42*(13), 2895–2921. https://doi.org/10.1016/j.atmosenv.2007.12.010

Rogge, W. F., Hildemann, L. M., Mazurek, M. A., Cass, G. R., & Simoneit, B. R. T. (1993). Sources of fine organic aerosol. 2. Noncatalyst and catalyst-equipped automobiles and heavy-duty diesel trucks. *Environmental Science & Technology*, *27*(4), 636–651. https://doi.org/10.1021/es00041a007

Schauer, J. J., Rogge, W. F., Hildemann, L. M., Mazurek, M. A., Cass, G. R., & Simoneit, B. R. T. (1996). Source apportionment of airborne particulate matter using organic compounds as tracers. *Atmospheric Environment*, *30*(22), 3837–3855. https://doi.org/10.1016/1352-2310(96)00085-4

Sicre, M. A., Marty, J. C., Saliot, A., Aparicio, X., Grimalt, J., & Albaiges, J. (1987). Aliphatic and aromatic hydrocarbons in different sized aerosols over the Mediterranean Sea: Occurrence and origin. *Atmospheric Environment (1967)*, *21*(10), 2247–2259. https://doi.org/10.1016/0004-6981(87)90356-8

Srivastava, D., Favez, O., Bonnaire, N., Lucarelli, F., Haeffelin, M., Perraudin, E., Gros, V., Villenave, E., & Albinet, A. (2018). Speciation of organic fractions does matter for aerosol source apportionment. Part 2: Intensive short-term campaign in the Paris area (France). *Science of the Total Environment*, *634*, 267–278. https://doi.org/10.1016/j.scitotenv.2018.03.296

Srivastava, D., Tomaz, S., Favez, O., Lanzafame, G. M., Golly, B., Besombes, J.-L., Alleman, L. Y., Jaffrezo, J.-L., Jacob, V., Perraudin, E., Villenave, E., & Albinet, A. (2018). Speciation of organic fraction does matter for source apportionment. Part 1: A one-year campaign in Grenoble (France). *Science of the Total Environment*, *624*, 1598–1611. https://doi.org/10.1016/j.scitotenv.2017.12.135

Tobiszewski, M., & Namieśnik, J. (2012). PAH diagnostic ratios for the identification of pollution emission sources. *Environmental Pollution*, *162*, 110–119. https://doi.org/10.1016/j.envpol.2011.10.025

Westblad, C., Levendis, Y. A., Richter, H., Howard, J. B., & Carlson, J. (2002). A study on toxic organic emissions from batch combustion of styrene. *Chemosphere*, *49*(4), 395–412. https://doi.org/10.1016/s0045-6535(02)00311-9

WHO. (2024). *Health consequences of air pollution on populations*. https://www.who.int/news/item/25-06-2024-what-are-health-consequences-of-air-pollution-on-populations

Yunker, M. B., Macdonald, R. W., Vingarzan, R., Mitchell, R. H., Goyette, D., & Sylvestre, S. (2002). PAHs in the Fraser River basin: A critical appraisal of PAH ratios as indicators of PAH source and composition. *Organic Geochemistry*, *33*(4), 489–515. https://doi.org/10.1016/s0146-6380(02)00002-5

Zhao, T., Yang, L., Huang, Q., Zhang, Y., Bie, S., Li, J., Zhang, W., Duan, S., Gao, H., & Wang, W. (2020a). $PM_{2.5}$-bound polycyclic aromatic hydrocarbons (PAHs) and their derivatives (nitrated-PAHs and oxygenated-PAHs) in a road tunnel located in Qingdao, China: Characteristics, sources and emission factors. *Science of the Total Environment*, *720*, 137521. https://doi.org/10.1016/j.scitotenv.2020.137521

Zhao, T., Yang, L., Huang, Q., Zhang, W., Duan, S., Gao, H., Wang, W. (2020b). $PM_{2.5}$-bound polycyclic aromatic hydrocarbons (PAHs) and nitrated-PAHs (NPAHs) emitted by gasoline vehicles: Characterization and health risk assessment. *Sci. Total Environ.* **727**, 138631. https.//doi.org/https://doi.org/10.1016/j.scitotenv.2020.138631

Zhuo, S., Du, W., Shen, G., Li, B., Liu, J., Cheng, H., Xing, B., & Tao, S. (2017). Estimating relative contributions of primary and secondary sources of ambient nitrated and oxygenated polycyclic aromatic hydrocarbons. *Atmospheric Environment*, *159*, 126–134. https://doi.org/10.1016/j.atmosenv.2017.04.003

MIX
Papier aus verantwortungsvollen Quellen
Paper from responsible sources
FSC® C105338

If you have any concerns about our products,
you can contact us on
ProductSafety@springernature.com

In case Publisher is established outside the EU,
the EU authorized representative is:
Springer Nature Customer Service Center GmbH
Europaplatz 3, 69115 Heidelberg, Germany

Printed by Libri Plureos GmbH
in Hamburg, Germany